Hugo Hens

Applied Building Physics

Hugo Hens

Applied Building Physics

Ambient Conditions, Building Performance and Material Properties

Second Edition

Professor Hugo S. L. C. Hens
University of Leuven (KULeuven)
Department of Civil Engineering
Building Physics
Kasteelpark Arenberg 40
3001 Leuven
Belgium

Cover: Existing dwelling, retrofitted and now a low energy building

Photo: Hugo Hens

This second edition is the result of a thorough revision of the first edition, published in 2010. Where appropriate, the text was corrected, reworked and extended with new information

Library of Congress Card No.: applied for

British Library Cataloguing-in-Publication Data
A catalogue record for this book is available from the British Library.

Bibliographic information published by the Deutsche Nationalbibliothek
The Deutsche Nationalbibliothek lists this publication in the Deutsche Nationalbibliografie; detailed bibliographic data are available on the Internet at <http://dnb.d-nb.de>.

Coverdesign: Sophie Bleifuß, Berlin, Germany
Typesetting: Thomson Digital, Noida, India
Printing and Binding: betz-druck GmbH, Darmstadt, Germany

Printed in the Federal Republic of Germany.
Printed on acid-free paper.

2. completely revised Edition

Print ISBN: 978-3-433-03147-6
oBook ISBN: 978-3-433-60711-4
ePDF ISBN: 978-3-433-60712-1
ePub ISBN: 978-3-433-60714-5
eMobi ISBN: 978-3-433-60723-7

To my wife, children and grandchildren

In remembrance of Professor A. de Grave who introduced Building Physics as a new discipline at the University of Leuven (KULeuven) Belgium in 1952

Contents

Preface

Until the first energy crisis of 1973, building physics was a rather dormant field within building engineering, with seemingly limited applicability. While soil mechanics, structural mechanics, building materials, building construction and HVAC were perceived as essential, designers sought advice on room acoustics, moisture tolerance, summer comfort or lighting only when really necesary or when problems arose. Energy was even not a concern, while thermal comfort and indoor environmental quality were presumed to be guaranteed thanks to infiltration, window operation and the heating and cooling system installed. The energy crises of the 1970s, persisting moisture problems, complaints about sick buildings, thermal, visual and olfactory discomfort and the move towards more sustainability changed it all. Societal pressure to diminish energy consumption in buildings without degrading usability activated the notion of performance based design and construction. As a result, today, building physics – and its potential to quantify related performance requirements – is at the forefront of building innovation.

As with all engineering sciences, building physics is orientated towards application, which is why, after the first volume on the fundamentals, this second volume examines performance metrics and requirements as the basis for sound building engineering. Choices have been made, among others to limit the text to the heat, air and moisture performances. Subjects treated are: the outdoor and indoor ambient conditions, the performance concept, performance at the building level, performance metrics at the building enclosure level and the heat-air-moisture material properties of building, insulation and finishing materials. The book reflects 38 years of teaching architectural, building and civil engineers, bolstered by close to 50 years' experience in research and consultancy. Where needed, information from international sources was used, which is why each chapter ends with an extended reading list.

The book uses SI units. Undergraduate and graduate students in architectural and building engineering should benefit, but also mechanical engineers studying HVAC and practising building engineers, who want to refresh their knowledge. The level of discussion presumes that the reader has a sound knowledge of the fundamentals treated in the first volume, along with a background in building materials and building construction.

Acknowledgements

The book reflects the work of many people, not just the author. Therefore, I would like to thank the thousands of students I have had during my 38 years of teaching. They have given me the opportunity to optimize the content. Also, were I not standing on the shoulders of those who precede me, this book would not be what it is. Although I started my career as a structural engineer, my predecessor, Professor Antoine de Grave, planted the seeds that fed my interest in building physics. The late Bob Vos of TNO, the Netherlands, and Helmut Künzel of the Fraunhofer Institut für Bauphysik, Germany, showed the importance of experimental work and field testing for understanding building performance, while Lars Erik Nevander of Lund University, Sweden, taught that solving problems does not always require complex modelling, mainly because reality in building construction is always much more complex than any model could simulate.

During my four decades at the Laboratory of Building Physics, several researchers and PhD students have been involved. I am very grateful to Gerrit Vermeir, Staf Roels, Dirk Saelens and Hans Janssen, colleagues at the university; also to Jan Carmeliet, professor at the ETH, Zürich; Piet Standaert, principal at Physibel Engineering; Jan Lecompte; Filip Descamps, principal at Daidalos Engineering and part-time professor at the Free University Brussels (VUB); Arnold Janssens, professor at the University of Ghent (UG); Rongjin Zheng, associate professor at Zhejiang University, China; Bert Blocken, full professor at the Technical University Eindhoven (TU/e); Griet Verbeeck, associate professor at the University of Hasselt; and Wout Parys, all of whom contributed through their work. The experiences gained as a structural engineer and building site supervisor at the start of my career, as building assessor over the years, as researcher and operating agent of four Annexes of the IEA, Executive Committee on Energy in Buildings and Communities forced me to rethink my engineering-based performance approach time and again. The many ideas I exchanged and received in Canada and the USA from Kumar Kumaran, the late Paul Fazio, Bill Brown, William B. Rose, Joe Lstiburek and Anton Ten Wolde were also of great help.

Finally, I thank my family, my wife Lieve, who manages to live with a busy engineering professor, my three children who had to live with that busy father and my many grandchildren who do not know that their grandfather is still busy.

Leuven, January 2016 *Hugo S.L.C. Hens*

0 Introduction

0.1 Subject of the book

This is the second volume in a series of four books:

- Building Physics: Heat, Air and Moisture
- **Applied Building Physics: Boundary Conditions, Building Performance and Material Properties**
- Performance Based Building Design: from below grade construction to cavity walls
- Performance Based Building Design: from timber-framed construction to partition walls.

Subjects discussed in this volume are: outdoor and indoor ambient conditions, performance concept, performance at the urban, building and building envelope level and the heat–air–moisture material properties. The book figures as a hinge between 'Building Physics: Heat, Air and Moisture' and the two volumes on 'Performance Based Building Design'. Although it does not deal with acoustics and lighting in detail, they form an integral part of the performance arrays and are mentioned as and when necessary.

The outdoor and indoor ambient conditions and related design approaches are highlighted in Chapter 1. Chapter 2 advances the performance concept with its hierarchical structure, going from the urban environment across the building and building assemblies down to the layer and material level. In Chapter 3, the main heat, air and moisture linked performances at the building level are discussed, while Chapter 4 analyses related metrics of importance for a well-performing building envelope and fabric. Chapter 5 deals with timber-framed walls as an exemplary case, and Chapter 6 lists the main heat, air and moisture material property values needed to predict the response of building assemblies.

A performance approach helps designers, consulting engineers and contractors to better ensure building quality. Of course, physical integrity is not the only factor adding value to buildings. Functionality, spatial quality and aesthetics – aspects belonging to the architect's responsibility – are of equal importance, though they should not cause us to neglect the importance of an overall outstanding building performance.

0.2 Building physics vs. applied building physics

Readers who would like to know more about the engineering field 'building physics', its importance and history, should consult the first volume, 'Building Physics: Heat, Air and Moisture'. It might seem that adding the term 'applied' to this second volume is unnecessary – building physics is, by definition, applied. Rather, the word stresses the focus of this book: entirely directed towards its use in building design and construction.

Applied Building Physics: Ambient Conditions, Building Performance and Material Properties,
Second Edition. Hugo Hens.

0.3 Units and symbols

The book uses the SI system (internationally mandated since 1977). Its base units are the metre (m), the kilogram (kg), the second (s), the kelvin (K), the ampere (A) and the candela. Derived units of importance when studying applied building physics are:

Force:	newton (N);	$1\,N = 1\,kg \cdot m \cdot s^{-2}$
Pressure:	pascal (Pa);	$1\,Pa = 1\,N/m^2 = 1\,kg \cdot m^{-1} \cdot s^{-2}$
Energy:	joule (J);	$1\,J = 1\,N \cdot m = 1\,kg \cdot m^2 \cdot s^{-2}$
Power:	watt (W);	$1\,W = 1\,J \cdot s^{-1} = 1\,kg \cdot m^2 \cdot s^{-3}$

For symbols, the ISO-standards (International Standardization Organization) are followed. If a quantity is not included, the CIB-W40 recommendations (International Council for Building Research, Studies and Documentation, Working Group 'Heat and Moisture Transfer in Buildings') and the list edited by Annex 24 of the IEA EBC (International Energy Agency, Executive Committee on Energy in Buildings and Communities) apply.

Table 0.1 List with symbols and quantities.

Symbol	Meaning	SI units
a	Acceleration	m/s^2
a	Thermal diffusivity	m^2/s
b	Thermal effusivity	$W/(m^2 \cdot K \cdot s^{0.5})$
c	Specific heat capacity	$J/(kg \cdot K)$
c	Concentration	kg/m^3, g/m^3
e	Emissivity	—
f	Specific free energy	J/kg
	Temperature ratio	—
g	Specific free enthalpy	J/kg
g	Acceleration by gravity	m/s^2
g	Mass flux	$kg/(m^2 \cdot s)$
h	Height	m
h	Specific enthalpy	J/kg
h	Surface film coefficient for heat transfer	$W/(m^2 \cdot K)$
k	Mass related permeability (mass could be moisture, air, salt . . .)	s
l	Length	m
l	Specific enthalpy of evaporation or melting	J/kg

Table 0.1 (*Continued*)

Symbol	Meaning	SI units
m	Mass	kg
n	Ventilation rate	s^{-1}, h^{-1}
p	Partial pressure	Pa
q	Heat flux	W/m^2
r	Radius	m
s	Specific entropy	$J/(kg \cdot K)$
t	Time	s
u	Specific latent energy	J/kg
v	Velocity	m/s
w	Moisture content	kg/m^3
x,y,z	Cartesian coordinates	m
A	Water sorption coefficient	$kg/(m^2 \cdot s^{0.5})$
A	Area	m^2
B	Water penetration coefficient	$m/s^{0.5}$
D	Diffusion coefficient	m^2/s
D	Moisture diffusivity	m^2/s
E	Irradiation	W/m^2
F	Free energy	J
G	Free enthalpy	J
G	Mass flow (mass = vapour, water, air, salt)	kg/s
H	Enthalpy	J
I	Radiation intensity	J/rad
K	Thermal moisture diffusion coefficient	$kg/(m \cdot s \cdot K)$
K	Mass permeance	s/m
K	Force	N
L	Luminosity	W/m^2
M	Emittance	W/m^2
P	Power	W
P	Thermal permeance	$W/(m^2 \cdot K)$
P	Total pressure	Pa
Q	Heat	J
R	Thermal resistance	$m^2 \cdot K/W$

Table 0.1 (*Continued*)

Symbol	Meaning	SI units
R	Gas constant	J/(kg · K)
S	Entropy	J/K
S	Saturation degree	—
T	Absolute temperature	K
T	Period (of a vibration or a wave)	s, days, . . .
U	Latent energy	J
U	Thermal transmittance	$W/(m^2 \cdot K)$
V	Volume	m^3
W	Air resistance	m/s
X	Moisture ratio	kg/kg
Z	Diffusion resistance	m/s
α	Thermal expansion coefficient	K^{-1}
α	Absorptivity	—
β	Surface film coefficient for diffusion	s/m
β	Volumetric thermal expansion coefficient	K^{-1}
η	Dynamic viscosity	$N \cdot s/m^2$
θ	Temperature	°C
λ	Thermal conductivity	W/(m · K)
μ	Vapour resistance factor	—
ν	Kinematic viscosity	m^2/s
ρ	Density	kg/m^3
ρ	Reflectivity	—
σ	Surface tension	N/m
τ	Transmissivity	—
ϕ	Relative humidity	—
α, ϕ, Θ	Angle	rad
ξ	Specific moisture capacity	kg/kg per unit of moisture potential
Ψ	Porosity	—
Ψ	Volumetric moisture ratio	m^3/m^3
Φ	Heat flow	W

Table 0.2 List with suffixes and notations.

Symbol	Meaning	Symbol	Meaning
Indices			
A	Air	m	Moisture, maximal
c	Capillary, convection	o	Operative
e	Outside, outdoors	r	Radiant, radiation
h	Hygroscopic	sat	Saturation
I	Inside, indoors	s	Surface, area, suction
cr	Critical	v	Water vapour
CO_2, SO_2	Chemical symbol for gasses	w	Water
		ϕ	Relative humidity

Notation	*Meaning*
[], bold,	Matrix, array, value of a complex number
dash (ex..: $\bar{a}$)	Vector

Further reading

CIB-W40 (1975) *Quantities, symbols and units for the description of heat and moisture transfer in buildings: Conversion factors*, IBBC-TNP, report BI-75-59/03.8.12, Rijswijk.

De Freitas, V.P. and Barreira, E. (2012) *Heat, Air and Moisture Transfer Terminology*, Parameters and Concepts, report CIB W040, 52 pp.

ISO-BIN (1985) Standards series X02-101 to X023-113.

Kumaran, K. (1996) *Task 3: Material Properties*, Final Report IEA EXCO ECBCS Annex 24, ACCO, Leuven, 135 pp.

1 Outdoor and indoor ambient conditions

1.1 Overview

In building physics, the outdoor and indoor ambient conditions play a role comparable to the loads in structural engineering, which is why the term 'ambient loads' is often used. Their knowledge is essential to make correct design decisions. The components that shape them are:

Outdoors		Indoors	
Air temperature (also called dry bulb temperature)	θ_e	Air temperature (also called dry bulb temperature)	θ_i
		Radiant temperature	θ_R
Relative humidity	ϕ_e	Relative humidity	ϕ_i
(Partial water) vapour pressure	p_e	(Partial water) vapour pressure	p_i
Solar radiation	E_S		
Under-cooling	q_{rL}		
Wind	v_w	Air speed	v
Rain and snow	g_r		
Air pressure	$P_{a,e}$	Air pressure	$P_{a,i}$

In the paragraphs that follow, these components are discussed separately. Bear in mind though that the greater the decoupling between the outdoor and indoor temperatures, and sometimes the relative humidities, the stricter will be the envelope and HVAC performance requirements. Otherwise, much more energy will be needed to maintain those differences.

Predicting future outdoor conditions is hardly possible. Not only are most components measured in only a few locations but the future is never the same as the past. Unfortunately, climate does not obey the paradigm 'the longer the data chain available, the better the forecast'. Moreover, global warming is affecting everything, see Figure 1.1.

A typical way of bypassing the problem is by using reference values and reference years for each performance check that needs climate data, such as the heating and cooling load, end energy consumption, overheating, moisture tolerance and other durability issues.

Much of the data illustrating the facts and trends discussed in the following paragraphs comes from the weather station at Uccle, Belgium (50° 51′ north, 4° 21′ east). This is

Applied Building Physics: Ambient Conditions, Building Performance and Material Properties, Second Edition. Hugo Hens.

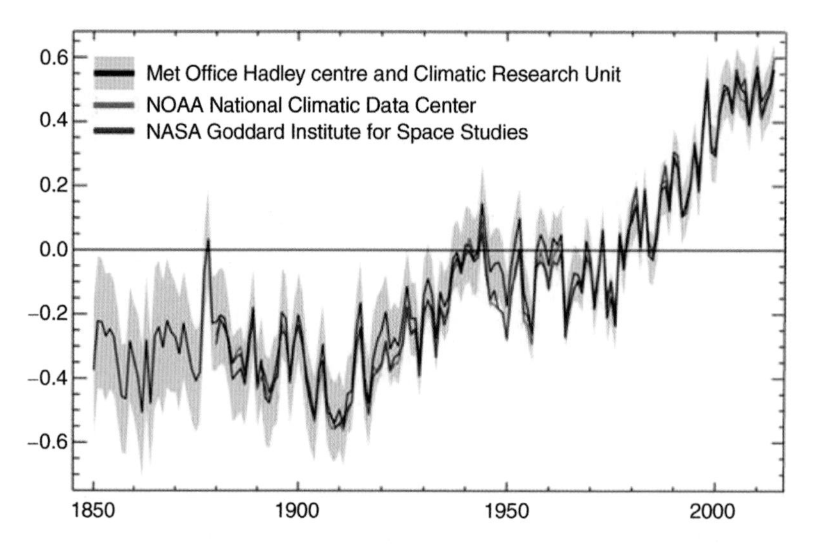

Figure 1.1 Increase in the world's average annual temperature between 1850 and 2014.

because of the large number of observations available, which allowed the synthesis of the weather there over the past century.

1.2 Outdoors

Geographical location defines, to a large extent, the outdoor climate: northern or southern latitude, proximity of the sea, presence of a warm or cold sea current and height above sea level. Of course, microclimatic factors also intervene. The urban heat island effect means that air temperature in city centres is higher, relative humidity lower and solar radiation less intense than in the countryside. Table 1.1 lists the monthly mean dry bulb temperature measured in a thermometer hut for the period 1901–1930 at Uccle and Sint Joost. Both weather stations are situated in the Brussels region, though the Uccle one overlooks a green area, while the Sint Joost one is in the city centre.

The outdoor climate further shows periodic fluctuations, linked to the earth's inclination and its elliptic orbit around the sun, the year and its succession of winter, spring, summer and autumn, and the wet and dry seasons in the equatorial band. Then there is the sequence of high and low pressure fronts; in temperate and cold climates, high brings

Table 1.1 Monthly mean dry bulb temperature at Uccle and Sint Joost, Brussels region, for the period 1901–1930 (°C).

Location	Month											
	J	F	M	A	M	J	J	A	S	O	N	D
Uccle	2.7	3.1	5.5	8.2	12.8	14.9	16.8	16.4	14.0	10.0	5.2	3.7
Sint Joost	3.8	4.2	6.8	9.4	14.6	16.7	18.7	18.0	15.4	11.2	6.4	4.7

warmth in summer and cold in winter, low gives cool and wet in summer and fresh and wet in winter. Finally, there is the effect of day and night, a consequence of the earth's rotation around its axis.

Design focuses on annual cycles, daily cycles and daily averages, but the meteorological references are 30-year averages, for the 20–21st century: 1901–1930, 1931–1960, 1961–1990, 1991–2020, 2021–2050 . . . These vary due to long-term climate changes as induced by solar activity and global warming, but the data is also affected by relocation of weather stations, more accurate measuring devices and the way averages are calculated. Up to 1930, the 'daily mean temperature' was the average of that day's minimum and maximum, as logged by a minimum/maximum mercury thermometer. Today, the air temperature at many weather stations is logged at 10′ intervals, and the daily mean is calculated as the average of these 144 values.

1.2.1 Air temperature

Knowing the air temperature helps in estimating the heating and cooling load and the related annual end energy consumption with the loads fixing the size and cost of the HVAC system needed and the end energy used being part of the annual costs. From day to day the air temperature also participates in the heat, air and moisture stresses that enclosures endure, while high values increase the indoor overheating risk. Measurement takes place in a thermometer hut in open field, 1.5 m above grade. The accuracy imposed by the World Meteorological Organization is ±0.5 °C. Table 1.2 covers over 30 years of averaged monthly means for several weather stations across Europe and North America. All look well represented by an annual mean and single harmonic, although adding a second harmonic gives a better fit:

Single harmonic

$$\theta_e = \overline{\theta}_e + A_{1,1}\sin\left(\frac{2\pi t}{365.25}\right) + B_{1,1}\cos\left(\frac{2\pi t}{365.25}\right) \tag{1.1}$$

Two harmonics

$$\begin{aligned}\theta_e = \overline{\theta}_e &+ A_{2,1}\sin\left(\frac{2\pi t}{365.25}\right) + B_{2,1}\cos\left(\frac{2\pi t}{365.25}\right)\\ &+ A_{2,2}\sin\left(\frac{4\pi t}{365.25}\right) + B_{2,2}\cos\left(\frac{4\pi t}{365.25}\right)\end{aligned} \tag{1.2}$$

In both formulas, $\overline{\theta}_e$ is the annual mean and t time. For three of the locations listed, the two-harmonic equation gives (°C, see Figure 1.2):

	$\overline{\theta}_e$	$A_{2,1}$	$B_{2,1}$	$A_{2,2}$	$B_{2,2}$
Uccle	9.8	–2.4	–7.4	0.45	–0.1
Kiruna	–1.2	–4.2	–11.6	1.2	0.5
Catania	17.2	–4.1	–6.6	0.8	0.2

Table 1.2 Monthly mean air temperatures for several locations (°C).

Location	Month											
	J	F	M	A	M	J	J	A	S	O	N	D
Uccle (B)	2.7	3.1	5.5	8.2	12.8	14.9	16.8	16.4	14.0	10.0	5.2	3.7
Den Bilt (NL)	1.3	2.4	4.3	8.1	12.1	15.3	16.1	16.1	14.2	10.7	5.5	1.2
Aberdeen (UK)	2.5	2.7	4.5	6.8	9.0	12.1	13.7	13.3	11.9	9.3	5.3	3.7
Eskdalemuir (UK)	1.8	1.9	3.9	5.8	8.9	11.8	13.1	12.9	10.9	8.5	4.3	2.7
Kew (UK)	4.7	4.8	6.8	9.0	12.6	15.6	17.5	17.1	14.8	11.6	7.5	5.6
Kiruna (S)	−12.2	−12.4	−8.9	−3.5	2.7	9.2	12.9	10.5	5.1	−1.5	−6.8	−10.1
Malmö (S)	−0.5	−0.7	1.4	6.0	11.0	15.0	17.2	16.7	13.5	8.9	4.9	2.0
Västerås (S)	−4.1	−4.1	−1.4	4.1	10.1	14.6	17.2	15.8	11.3	6.3	1.9	−1.0
Lulea (S)	−11.4	−10.0	−5.6	−0.1	6.1	12.8	15.3	13.6	8.2	2.9	−4.0	−8.9
Oslo (N)	−4.2	−4.1	−0.2	4.6	10.8	15.0	16.5	15.2	10.8	6.1	0.8	−2.6
München (D)	−1.5	−0.4	3.4	8.1	11.9	15.6	17.5	16.7	13.9	8.8	3.6	−0.2
Potsdam (D)	−0.7	−0.3	3.5	8.0	13.1	16.6	18.1	17.5	13.8	9.2	4.1	0.9
Roma (I)	7.6	9.0	11.3	13.9	18.0	22.3	25.2	24.7	21.5	16.8	12.1	8.9
Catania (I)	10.0	10.4	12.0	14.0	18.0	22.0	25.2	25.6	23.2	18.4	15.2	11.6
Torino (I)	1.6	3.5	7.6	10.8	15.4	19.0	22.3	21.6	17.9	12.3	6.2	2.4
Bratislava (Sk)	−2.0	0.0	4.3	9.6	14.2	17.8	19.3	18.9	15.3	10.0	4.2	0.1
Copenhagen (Dk)	−0.7	−0.8	1.8	5.7	11.1	15.1	16.2	16.0	12.7	9.0	4.7	1.1
Montreal	−9.9	−8.5	−2.4	5.7	13.1	18.4	21.1	19.5	14.6	8.5	1.8	−6.5
New York	0.6	2.2	6.1	11.7	17.2	22.2	25.0	24.4	20.0	13.9	8.9	3.3
Chicago	−5.6	−3.3	2.8	9.4	15.0	20.6	23.3	22.2	18.3	11.7	4.4	−2.8
Los Angeles	15.0	14.9	20.3	17.3	18.8	20.7	22.9	23.5	22.8	20.3	16.9	14.2

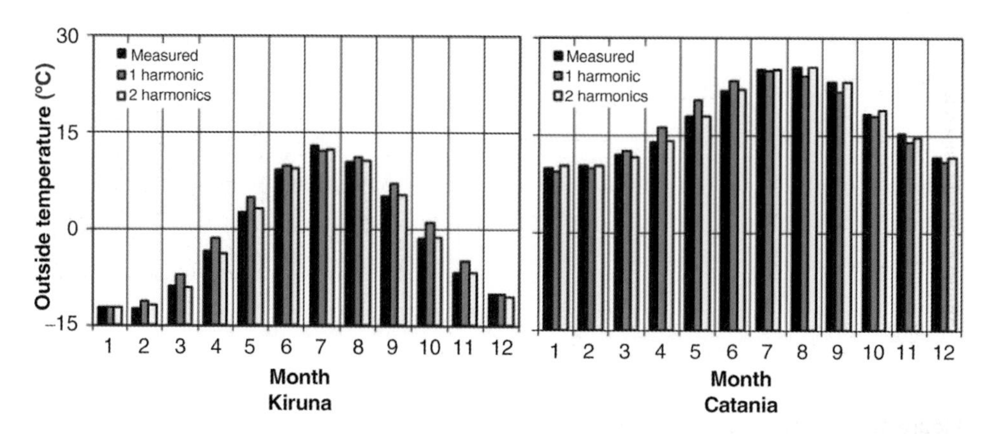

Figure 1.2 Air temperature: annual course, single and two harmonics.

For Uccle the average difference between the monthly mean daily minimums and maximums ($\theta_{e,max,day} - \theta_{e,min,day}$) during the period 1931 – 1960 looked as (°C):

J	F	M	A	M	J	J	A	S	O	N	D
5.6	6.6	7.9	9.3	10.7	10.8	10.6	10.1	9.8	8.0	6.2	5.2

A combination with the annual course could give (time in hours):

$$\begin{aligned}\theta_e &= \overline{\theta}_e + \hat{\theta}_e \cos\left[\frac{2\pi(t-h_1)}{8766}\right] \\ &+ \frac{1}{2}\left\{\Delta\overline{\theta}_{e,dag} + \Delta\hat{\theta}_{e,dag} \cos\left[\frac{2\pi(t-h_2)}{8766}\right]\right\}\sin\left[\frac{2\pi(t-h_3)}{24}\right]\end{aligned} \tag{1.3}$$

with:

$\Delta\overline{\theta}_{e,dag}$ (°C)	$\Delta\hat{\theta}_{e,dag}$ (°C)	h_1 (h)	h_2 (h)	h_3 (h)
8.4	2.8	456	–42	8

an equation, assuming that the daily values fluctuate harmonically. This is not the case. The gap between the daily minimum and maximum swings considerably, without even a hint of a harmonic course. To give an example, in Leuven, Belgium, that gap for January and July 1973 was purely random, with averages of 4.0 °C and 8.9 °C and standard deviation percentages 60 in January and 39 in July.

A question is whether the air temperature recorded during the past decades in any weather station reflects global warming? For that, the data recorded between 1997 and 2013 at the outskirts of Leuven were tabulated. Figure 1.3 shows the annual means and the monthly minima and maxima measured.

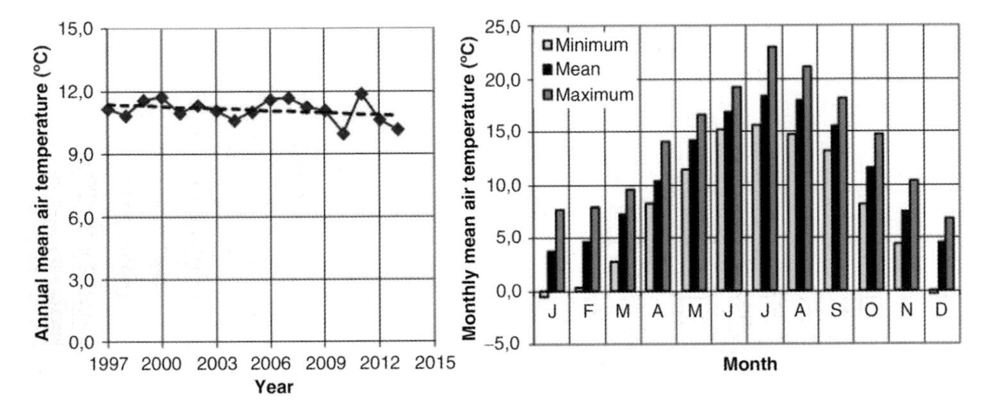

Figure 1.3 Leuven weather station (Belgium): air temperature between 1996 and 2013: annual mean (left) and average, minimum and maximum monthly mean (right).

The least square straight line through the annual means equals:

$$\overline{\theta}_{e,ann} = 11.1 - 0.034 \times \text{year}$$

With an of average 11.1 °C and a slightly negative slope, the expected increase seems absent. Not so at Uccle, 30 km west of Leuven. There the overall mean between 1901 and 1930 was 9.8 °C, that is 1.4 °C lower than measured at Leuven from 1997 to 2013. Between 1952 and 1971 that mean remained 9.8 °C but since then the moving 20-year mean has increased slowly, with the highest values noted between 1992 and 2011.

1.2.2 Solar radiation

Solar radiation means free heat gains. These lower the energy used for heating but may create cooling needs, as too many gains increase the overheating risk. The sun further lifts the outside surface temperature of irradiated envelope assemblies. Although this enhances drying, it also activates solar-driven vapour flow to the inside of moisture stored in rain buffering outer layers, whereas the associated drop in relative humidity aggravates hygrothermal stress in thin outer finishes.

The sun is a 5762 K hot black body, 150 000 000 km from the earth. Due to the large distance, the rays approach the earth in parallel. Above the atmosphere the solar spectrum follows the thin line in Figure 1.4, while the total irradiation is approximately:

$$E_{ST} = 5.67\left(\frac{T_S}{100}\right)^4\left(\frac{r_S}{D_{SE}}\right)^2 = 5.67 \times (57.64)^4 \times \left(\frac{0.695 \cdot 10^6}{1.496 \cdot 10^8}\right)^2 = 1332\,\text{W/m}^2 \quad (1.4)$$

with r_s the solar radius and D_{SE} the distance between sun and earth, both in km.

This 1332 W/m² represents the average solar constant (E_{STo}), that is the mean radiation per square metre that the earth would receive perpendicular to the beam if there were no atmosphere. The related energy flow is thinly spread. Burning 1 litre

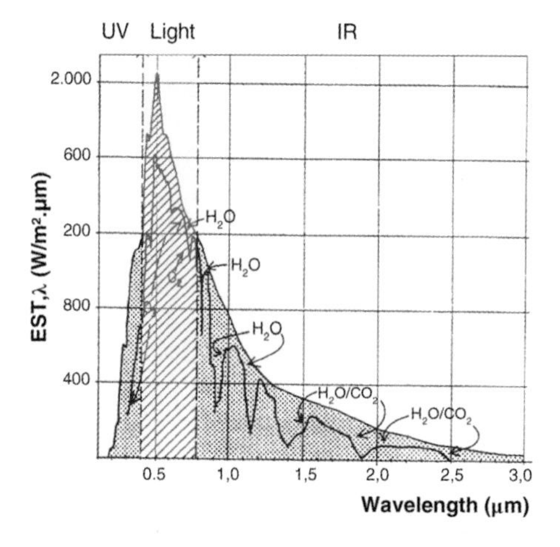

Figure 1.4 Solar spectrums before (thin) and after passing the atmosphere (thicker line).

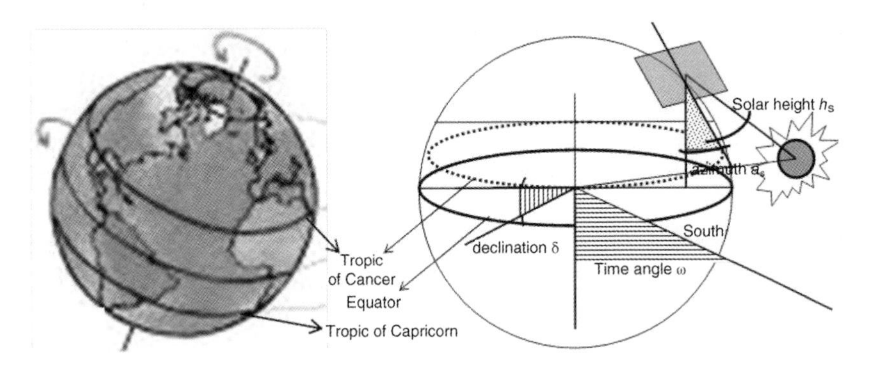

Figure 1.5 Solar angles.

of fuel gives 4.4×10^7 J, but collecting the same amount from the sun just above the atmosphere on a square metre perpendicular to the beam would take 9 hours of constant irradiation. This thinness explains why collecting solar energy for heat or electricity production requires such large surface areas. A more exact calculation of the solar constant accounts for the annual variation in distance between earth and sun and the annual cycle in solar activity (d the number of days from December 31/ January 1 at midnight):

$$E_{\text{STo}} = 1373\left\{1 + 0.03344\cos\left[\frac{2\pi}{365.25}(d - 2.75)\right]\right\}(\text{W/m}^2) \tag{1.5}$$

What fixes the solar position at the sky is either the azimuth (a_s) and solar height (h_s) or the time angle (ω) and declination (δ), that is the angle between the Tropics of Capricorn or Cancer, where the solar height reaches 90°, zenith position, and the equator plane, see Figure 1.5. The first two describe the sun's movement as seen locally, the second two relate it to the equator.

The time angle (ω) goes from 180° at 0 a.m. through 0° at noon to −180° at 12 p.m. One hour therefore corresponds to 15°. The declination (δ) in radians is given by:

$$\delta = arcsin\left[-\sin\left(\frac{\pi}{180}23.45\right)\cos\left[\frac{2\pi}{365.25}(d + 10)\right]\right] \tag{1.6}$$

where ±23.45 is the latitude of the Tropics in degrees. Solar height in radians at any moment equals:

$$h_s = \max\left[0, \pi/2 - \arccos(\cos\varphi\cos\delta\cos\omega + \sin\varphi\sin\delta)\right] \tag{1.7}$$

The maximum value ($h_{s,\max}$) in degrees or radians follows from:

$$\text{Degrees}: h_{s.\max} = 90 - \varphi(°) + \delta(°) \quad \text{Radians}: h_{s,\max} = \pi/2 - \varphi + \delta \tag{1.8}$$

with φ latitude, positive for the northern, negative for the southern hemisphere. The addition (°) means values in degrees celsius.

1.2.2.1 Beam radiation

During its passage through the atmosphere, selective absorption by ozone, oxygen, hydrogen, carbon dioxide and methane interferes with the solar beam and changes its spectrum, while scattering disperses a part of it. The longer the distance traversed through the atmosphere, the more the radiation is affected, as represented by the air factor m, which is the ratio between the real distance traversed to sea level through the atmosphere, with the sun at height h_s and the distance traversed to any location at and above sea level, assuming that the sun stands in zenith position there (Figure 1.6).

For a location z km above sea level the air factor can be written as (all formulas below with solar height in radians):

$$m = \frac{L}{L_o} = \frac{1 - 0.1\,z}{\sin(h_s) + 0.15(h_s + 3.885)^{-1.253}} \tag{1.9}$$

Beam radiation on a surface perpendicular to the solar rays then becomes:

$$E_{SD,n} = E_{STo}\exp(-md_R T_{Atm})$$

where T_{Atm} is atmospheric turbidity and d_R the optic factor, a measure for the scatter per unit of distance traversed:

$$\begin{aligned} d_R = {} & 1.4899 - 2.1099\cos(h_s) + 0.6322\cos(2h_s) + 0.0253\cos(3h_s) \\ & - 1.0022\sin(h_s) + 1.0077\sin(2h_s) - 0.2606\sin(3h_s) \end{aligned} \tag{1.10}$$

On a clear day with average air pollution, atmospheric turbidity is given by:

$$T_{Atm} = 3.372 + 3.037h_s - 0.296\cos(0.5236\,\text{mo})$$

With minimal air pollution, it becomes:

$$T_{Atm} = 2.730 + 1.549\,h_s - 0.198\cos(0.5236\,\text{mo})$$

In both formulas mo is the month, 1 for January, 12 for December.

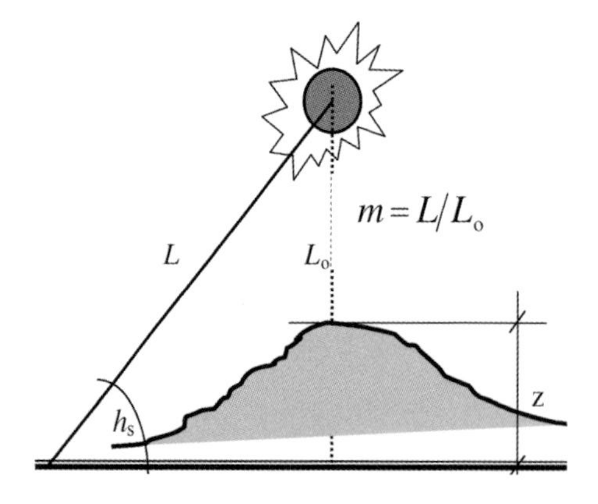

Figure 1.6 *L* the distance traversed through the atmosphere to sea level, *z* height of a location in km, L_o the distance traversed to *z* for the sun in zenith position.

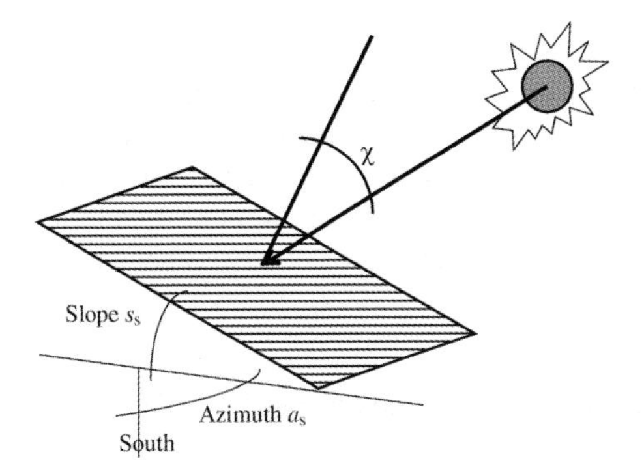

Figure 1.7 Direct radiation on a surface with slope s_S.

Beam radiation on a tilted surface, whose normal forms an angle χ with the solar rays, is calculated as (Figure 1.7):

$$E_{\mathrm{SD,s}} = \max\left(0, E_{\mathrm{SD,n}} \cos\chi\right) \tag{1.11}$$

with the zero to be applied before sunrise and after sunset and:

$$\begin{aligned}\cos\chi = {} & \sin\delta\sin\varphi\cos s_s - \sin\delta\cos\varphi\sin s_s\cos a_s \\ & + \cos\delta\cos\varphi\cos s_s\cos\omega + \cos\delta\sin\varphi\sin s_s\cos a_s\cos\omega \\ & + \cos\delta\sin s_s\sin a_s\sin\omega\end{aligned} \tag{1.12}$$

Here s_s is the surface's slope, 0° when horizontal, 90° ($\pi/2$) when vertical, between 0 and 90° when tilted to the sun, between 0 and −90° ($-\pi/2$) when tilted away from the sun, and a_s the surface's azimuth (south 0°, east 90°, north 180°, west −90°).

For a horizontal plane facing the sun the formula reduces to:

$$\cos\chi_h = \sin\delta\sin\varphi + \cos\delta\cos\varphi\cos\omega$$

For vertical planes facing the sun it becomes:

South $\cos\chi_{\mathrm{v,south}} = -\sin\delta\cos\varphi + \cos\delta\sin\varphi\cos\omega$

West $\cos\chi_{\mathrm{v,west}} = -\cos\delta\sin\omega$

North $\cos\chi_{\mathrm{v,north}} = -\sin\delta\cos\varphi - \cos\delta\sin\varphi\cos\omega$

East $\cos\chi_{\mathrm{v,east}} = \cos\delta\sin\omega$

Where the beam radiation on a horizontal surface is known ($E_{\mathrm{SD,h}}$), its value on any tilted surface ($E_{\mathrm{SD,s}}$) follows from:

$$E_{\mathrm{SD,s}} = \max\left(0, E_{\mathrm{SD,h}} \cos\chi_s / \cos\chi_h\right) \tag{1.13}$$

Beam radiation seems to be predictable, but the big unknown is the atmospheric turbidity (T_{Atm}). Cloudiness, air pollution and relative humidity all intervene, but their impact is very complex and varies from day to day.

1.2.2.2 Diffuse radiation

Whether the sky is blue or cloudy, diffuse solar radiation reaches the earth from sunrise to sunset. At the earth's surface, it's as if the rays come from all directions. The simplest model considers the sky as a uniformly radiating vault. Any surface, whose slope forms an angle with the horizontal, sees part of it. As a black body at constant temperature, each point on the vault has equal luminosity, turning the surface's view factor into:

$$F_{s,sk} = (1 + \cos s_s)/2 \tag{1.14}$$

If $E_{Sd,h}$ is the diffuse radiation on a horizontal surface, tilted it becomes:

$$E_{Sd,s} = E_{Sd,h}(1 + \cos s_s)/2 \tag{1.15}$$

Better approximating reality is the sky as a vault with the highest luminosity at the solar disk and the lowest at the horizon. With the position of any point P on that vault characterized by its azimuth a_P and height h_P, luminosity there writes as $L(a_P,h_P)$. The angle Γ between the normal on a surface with slope s_s and the line from its centre to this point P now equals:

$$\cos\Gamma = \sin s_s \cos h_P \cos(a_s - a_P) + \cos s_s \sin h_P$$

Diffuse radiation on a tilted surface therefore becomes:

$$E_{Sd,s} = K_D \iint\limits_{a_P,h_P} \left[L_{a_P,h_P} \cos h_P \cos\Gamma\right] dh_P da_P \tag{1.16}$$

with $K_D = 1 + 0.03344\cos[0.017202(d - 2.75)]$, $0 \le a_P \le 2\pi$, $0 \le h_P \le \pi/2$, $\cos\Gamma \ge 0$ and:

$$L_{a_P,h_P} = L_{sd}\underbrace{\frac{\left[0.91 + 10\exp(-3\varepsilon) + 0.45\cos^2\varepsilon\right]\{1 - \exp[-0.32\,\mathrm{cosec}(h_P)]\}}{0.27385\left\{0.91 + 10\exp\left[-3\left(\frac{\pi}{2} - h_S\right)\right] + 0.45\sin^2 h_S\right\}}}_{f} \tag{1.17}$$

where L_{sd} is the luminosity at the solar disk and ε the angle between the line from the surface's centre to P and the normal on the vault in P, which coincides with the solar beam there:

$$\cos\varepsilon = \cos h_S \cos h_P \cos(a_S - a_P) + \sin h_S \cos h_P$$

Entering this formula and the multiplier from Eq. (1.17) in Eq. (1.16), results in:

$$E_{Sd,s} = K_D L_{zenith} \iint\limits_{a_P,h_P} [f \cos h_P \cos\Gamma] dh_P da_P$$

The luminosity at the solar disk then equals:

$$L_{sd} = 0.8785[h_s(°)] - 0.01322[h_s(°)]^2 + 0.003434[h_s(°)]^3 + 0.44347 + 0.0364 T_{Atm}$$

Table 1.3 Uccle, multiplier f_{mo} for the total monthly diffuse radiation.

Azimuth → Slope ↓	0 S	22.5	45	67.5	90 E,W	112.5	135	157.5	180 N
0	1.00	1.00	1.00	1.00	1.00	1.00	1.00	1.00	1.00
22.5	1.03	1.03	1.02	1.01	1.00	0.99	0.98	0.97	0.96
45	1.05	1.04	1.03	1.01	0.99	0.96	0.94	0.92	0.92
67.5	1.06	1.05	1.03	0.99	0.94	0.90	0.86	0.84	0.83
90	1.06	1.04	1.00	0.94	0.87	0.81	0.76	0.73	0.71
112.5	0.98	0.97	0.92	0.85	0.76	0.68	0.63	0.60	0.60
135	0.80	0.78	0.74	0.67	0.59	0.53	0.49	0.47	0.47
157.5	0.58	0.56	0.51	0.48	0.46	0.43	0.41	0.40	0.34
180	0.00	0.00	0.00	0.00	0.00	0.00	0.00	0.00	0.00

with T_{Atm} atmospheric turbidity and $h_s(°)$ solar height in degrees. On a monthly basis, this set of formulas is simplified to:

$$E_{Sd,s} = E_{Sd,h} f_{mo}(1 + \cos s_s)/2$$

with f_{mo} a multiplier that corrects the monthly diffuse radiation, calculated according to (1.15) for the luminosity at the solar disk effect. For Uccle, f_{mo} takes the values given in Table 1.3.

1.2.2.3 Reflected radiation

Surfaces on earth reflect part of the beam and diffuse radiation received. To calculate the intensity, all surroundings are considered to be acting as one horizontal plane with reflectivity 0.2, called the albedo. Every surface then receives reflected radiation proportional to the view factor with that horizontal plane (F_{se}):

$$E_{Sr,s} = 0.2\left(E_{SD,h} + E_{Sd,h}\right)F_{se} = 0.2\left(E_{SD,h} + E_{Sd,h}\right)(1 - \cos s_s)/2 \qquad (1.18)$$

Reflected radiation on a horizontal surface facing the sky ($s_s = 0$) looks to be zero though in reality this is not by definition the case. A low-sloped roof for example gets radiation reflected from surrounding higher buildings. Also, an albedo of 0.2 is too simplistic. White snow gives a higher value.

1.2.2.4 Total radiation

Beam, diffuse and reflected irradiation together give the total solar radiation that a surface receives. The appendix contains tables with values for Uccle, while Table 1.4 here summarizes the average, minimum and maximum monthly totals on a horizontal surface measured there, together with the monthly mean cloudiness, calculated as one minus the ratio between the measured and clear sky total on a horizontal surface. Table 1.5 in turn lists the monthly totals on a horizontal surface for several locations

Table 1.4 Monthly total solar irradiation on a horizontal surface (MJ/(m^2.month)) and cloudiness at Uccle, average, maximum and minimum values for 1958–1975.

	J	F	M	A	M	J	J	A	S	O	N	D
Total solar radiation												
Mean	72	129	247	356	500	538	510	439	327	197	85	56
Min.	61	104	177	263	406	431	408	366	279	145	63	41
Max.	93	188	311	485	589	640	651	497	444	274	112	78
Average cloudiness												
Mean	0.47	0.44	0.42	0.42	0.36	0.35	0.38	0.38	0.34	0.39	0.49	0.50
Min.	0.55	0.55	0.58	0.57	0.48	0.48	0.51	0.48	0.44	0.55	0.62	0.63
Max.	0.31	0.19	0.27	0.20	0.25	0.22	0.21	0.30	0.11	0.15	0.33	0.30

across Europe, while Figure 1.8 shows the annual totals with the ratio between least and most sunny location approaching a value of 2.

How sunny locations are is important when questions are raised such as where to promote photovoltaic cells (PV) as renewable. Within Europe, PV is good in Portugal, Spain, Southern France, Southern Italy and Greece, but of only moderate benefit in the north-west and Scandinavia.

1.2.3 Longwave radiation

Longwave radiation gives extra heat loss as it can chill the outer surface and the layers outside of the insulation to temperatures below those of outdoors, even below the dew point outdoors. Under-cooling, as it is called, turns the outdoor air into a moisture source rather

Table 1.5 Monthly total solar irradiation on a horizontal surface for several locations in Europe (MJ/(m^2.month)).

Location	Month											
	J	F	M	A	M	J	J	A	S	O	N	D
Den Bilt (NL)	72	132	249	381	522	555	509	458	316	193	86	56
Eskdalemuir (UK)	55	112	209	345	458	490	445	370	244	143	70	39
Kew (UK)	67	115	244	355	496	516	501	434	311	182	88	54
Lulea (S)	6	52	182	358	528	612	589	418	211	80	14	1
Oslo (N)	44	110	268	441	616	689	624	490	391	153	57	27
Potsdam (D)	104	137	238	332	498	557	562	412	267	174	88	70
Roma (I)	182	247	404	521	670	700	750	654	498	343	205	166
Torino (I)	171	212	343	474	538	573	621	579	422	281	181	148
Bratislava (Sk)	94	159	300	464	597	635	624	544	389	233	101	72
Copenhagen (Dk)	54	114	244	407	579	622	576	479	308	159	67	38

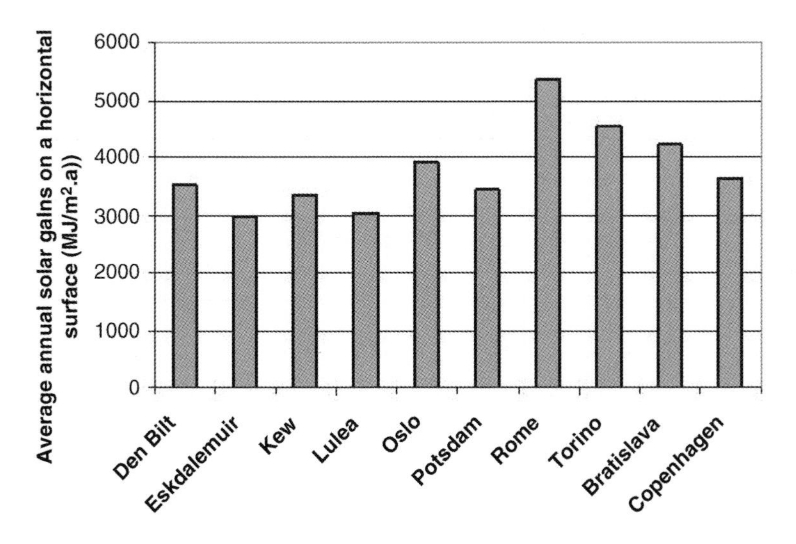

Figure 1.8 Annual solar radiation on a horizontal surface.

than a drying medium with a visible result being condensation on the outer surface of insulating glass, EIFS-stucco and tiled or slated well-insulated pitched roofs (Figure 1.9).

Under-cooling reflects the longwave balance between the atmosphere, represented by the celestial vault, the terrestrial plane and the surface considered, with the vault acting as a selective radiant body, absorbing all incoming radiation but emitting only a fraction. Calculation assumes a vault at air temperature with emissivity below 1, absorptivity 1 and reflectivity 0. Several formulas quantify its emissivity. To give a few (p_e in Pa, θ_e in °C):

Clear sky

(1) $\varepsilon_{L,sky,o} = 0.75 - 0.32 \cdot 10^{-0.051 p_e/1000}$

(2) $\varepsilon_{L,sky,o} = 0.52 + 0.065\sqrt{p_e/1000}$

(3) $\varepsilon_{L,sky,o} = 1.24\left(\dfrac{p_e/1000}{273.15 + \theta_e}\right)^{1/7}$

Cloudy sky

(4) $\varepsilon_{L,sky} = \varepsilon_{L,sky,o}(1 - 0.84c) + 0.84c$

Figure 1.9 Rime formation on a well-insulated pitched roof due to under-cooling.

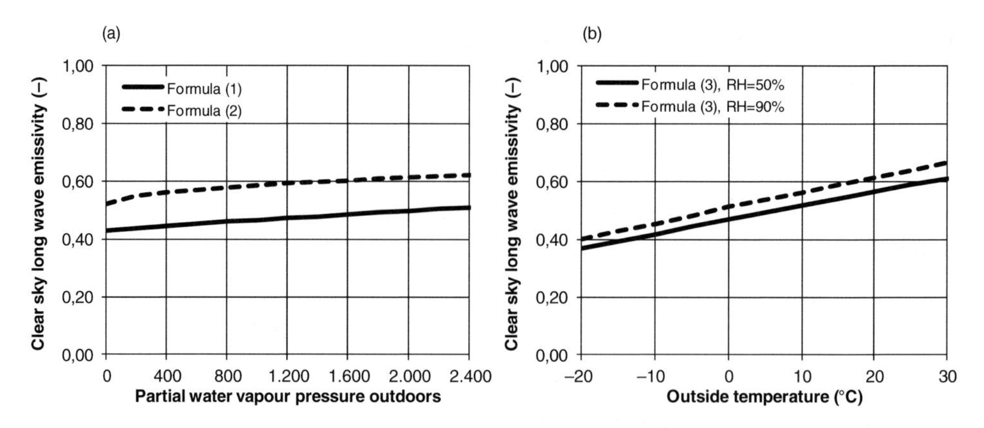

Figure 1.10 Sky emissivity, left according to the clear sky formulas (1) and (2), right according to the clear sky formula (3).

Clear sky emissivity thus drops with air temperature but increases with partial water vapour pressure outdoors as water vapour in fact acts as a strong greenhouse gas. In the cloudy sky formula, c stands for the cloudiness factor, 0 for clear, steps 0.125 for hardly covered to more covered and 1 for a covered sky. According to Figure 1.10(a) left, the formulas (1) and (2) apparently give a different emissivity, whereas Figure 1.10(b), illustrating formula (3), suggests that (1) applies to lower and (2) to somewhat higher temperatures outdoors.

The following two equations for the black body emittance of the surface considered and the terrestrial plane describe the 'celestial vault/surface/terrestrial plane' radiant system:

$$M_{\mathrm{b,s}} = \left[1 + \frac{\rho_{\mathrm{L,s}}}{\varepsilon_{\mathrm{L,s}}}\left(F_{\mathrm{s,t}} + F_{\mathrm{s,sky}}\right)\right] M'_{\mathrm{s}} - \frac{\rho_{\mathrm{L,s}}}{\varepsilon_{\mathrm{L,s}}}\left(F_{\mathrm{s,t}} M'_{\mathrm{t}} + F_{\mathrm{s,sky}} M'_{\mathrm{sky}}\right)$$

$$M_{\mathrm{b,t}} = \left[1 + \frac{\rho_{\mathrm{L,t}}}{\varepsilon_{\mathrm{L,t}}}\left(F_{\mathrm{t,s}} + F_{\mathrm{t,sky}}\right)\right] M'_{\mathrm{t}} - \frac{\rho_{\mathrm{L,t}}}{\varepsilon_{\mathrm{L,t}}}\left(F_{\mathrm{t,s}} M'_{\mathrm{s}} + F_{\mathrm{t,sky}} M'_{\mathrm{sky}}\right)$$

In both, the suffix s stands for the surface, the suffix t for the terrestrial plane, the suffix sky for the celestial vault, M'_{sky} for the radiosity of the celestial vault and $F_{\mathrm{s,t}}$, $F_{\mathrm{s,sky}}$, . . . for the view factors between these radiant bodies. As that plane and celestial vault surround the surface, the sum $F_{\mathrm{s,t}} + F_{\mathrm{s,sky}}$ is 1. The same holds for the view factor between the terrestrial plane and the celestial vault. In fact, the one between that plane and the surface is so small that using zero does not falsify the second equation. Celestial vault radiosity thus simplifies to:

$$M'_{\mathrm{sky}} = \varepsilon_{\mathrm{L,sky}} M_{\mathrm{b,sky},\theta_{\mathrm{e}}}$$

turning the two equations into:

$$M_{\mathrm{b,s}} = \frac{1}{\varepsilon_{\mathrm{L,s}}} M'_{\mathrm{s}} - \frac{\rho_{\mathrm{L,s}}}{\varepsilon_{\mathrm{L,s}}}\left(F_{\mathrm{s,t}} M'_{\mathrm{t}} + F_{\mathrm{s,sky}} \varepsilon_{\mathrm{L,sky}} M_{\mathrm{b,sky},\theta_{\mathrm{e}}}\right) \tag{1.19}$$

$$M_{\mathrm{b,t}} = \frac{1}{\varepsilon_{\mathrm{L,t}}}\left(M'_{\mathrm{t}} - \rho_{\mathrm{L,t}} \varepsilon_{\mathrm{L,sky}} M_{\mathrm{b,sky},\theta_{\mathrm{e}}}\right) \tag{1.20}$$

The radiant heat flow rate at the surface is now:

$$q_R = q_{Rs,t} + q_{Rs,sky} = \frac{\varepsilon_{L,s}}{\rho_{L,s}}\left(M_{b,s} - M'_s\right) = \varepsilon_{L,s}F_{s,t}\left(M_{b,s} - \varepsilon_{L,t}M_{b,t}\right)$$
$$+ \varepsilon_{L,s}F_{s,sky}\left[M_{b,s} - \left(\rho_{L,env}\frac{F_{s,t}}{F_{s,t}} + 1\right)\varepsilon_{L,sky}M_{b,sky,\theta_e}\right]$$

Assuming the terrestrial plane to be a black body at outdoor air temperature further simplifies the formula:

$$q_R = \varepsilon_{L,s}C_b\left[\left(F_{s,t} + F_{s,sky}\right)\left(\frac{T_{se}}{100}\right)^4 - \left(F_{s,t} + F_{s,sky}\varepsilon_{L,sky}\right)\left(\frac{T_e}{100}\right)^4\right] \tag{1.21}$$

Linearization by replacing the celestial vault by a black body at temperature

$$\theta_{sk,e} = \theta_e - (23.8 - 0.2025\,\theta_e)(1 - 0.87c)$$

gives:

$$q_R = \varepsilon_{L,s}C_b\left[\begin{array}{l}\left(F_{s,t}F_{Ts,t} + F_{s,sky}F_{Ts,sky}\right)(\theta_{se} - \theta_e)\\ \qquad + (23.8 - 0.2025\,\theta_e)(1 - 0.87c)F_{s,sky}F_{Ts,sky}\end{array}\right] \tag{1.22}$$

a result that fits with expression (1.21) for a sky emissivity somewhat higher than in Figure 1.10(b). Combining (1.22) with the convective heat exchanged quantifies under-cooling:

$$\begin{aligned}q_{ce} + q_{Re} = {} & \left[h_{ce} + \varepsilon_{L,s}C_b\left(F_{s,t}F_{Ts,t} + F_{s,sky}F_{Ts,sky}\right)\right](\theta_{se} - \theta_e)\\ & + \varepsilon_{L,s}C_bF_{s,sky}F_{Ts,sky}(23.8 - 0.2025\,\theta_e)(1 - 0.87c)\end{aligned} \tag{1.23}$$

$$q_T + q_{ce} + q_{Re} = 0$$

where q_T is the heat flow rate by transmission to or from the exterior surface. Figure 1.11 shows calculated data for a low-slope roof with thermal transmittance 0.2 W/(m^2.K),

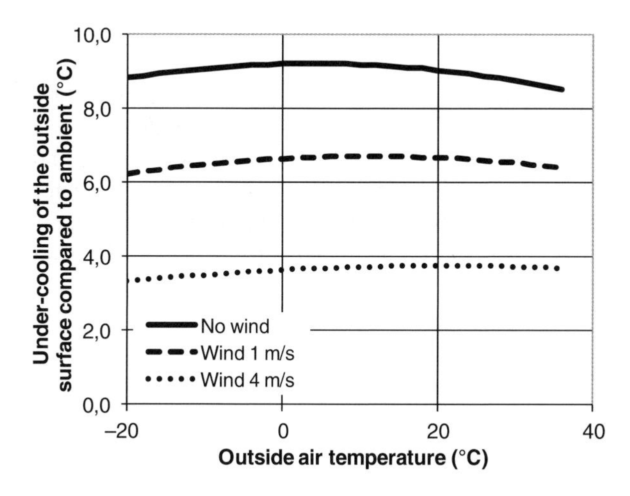

Figure 1.11 Lightweight low-sloped roof, U = 0.2 W/(m^2.K), membrane with $e_L = 0.9$: under-cooling represented by $\theta_e - \theta_{se}$.

a membrane with longwave emissivity 0.9 and very low thermal inertia. Parameters are the air temperature and wind speed. The outside surface temperature can drop substantially below ambient.

1.2.4 Relative humidity and (partial water) vapour pressure

Both relative humidity (ϕ_e) and (partial water) vapour pressure (p_e) impact the moisture tolerance of building enclosures and buildings in a straightforward way. Table 1.6 summarizes monthly means for several locations across Europe.

On average, relative humidity hardly changes between winter and summer. Vapour pressure, however, does. In temperate climates the inverse often holds between day and night: large differences in relative humidity and fairly constant vapour pressures. A sudden temperature rise lowers the relative humidity while a sudden drop may push it up to a misty 100%. During rainy weather the wet bulb temperature closely follows the raindrop temperature. When it is as warm as the air, relative humidity will near 100%. Also, the environment strongly influences relative humidity and vapour pressure, with higher values in forests and river valleys than in cities. Again, the annual variation is often written in terms of a Fourier series with one harmonic, though, as Figure 1.12

Table 1.6 Monthly mean relative humidity (%) and vapour pressure (Pa, bold) for several locations all over Europe.

Location	Month											
	J	F	M	A	M	J	J	A	S	O	N	D
Uccle (B)	89.6	89.0	84.0	78.5	77.8	78.9	79.9	79.8	84.2	88.3	91.2	92.7
	663	**681**	**757**	**854**	**1151**	**1334**	**1529**	**1489**	**1346**	**1084**	**806**	**780**
Aberdeen (UK)	81.5	80.3	74.9	72.3	75.0	78.5	74.1	79.1	80.6	81.6	78.1	77.1
	596	**596**	**631**	**714**	**860**	**1107**	**1161**	**1207**	**1122**	**955**	**695**	**614**
Catania (I)	66.5	72.4	68.8	69.8	71.2	70.0	62.0	68.6	69.4	69.6	68.5	65.9
	816	**912**	**964**	**1114**	**1469**	**1849**	**1985**	**2252**	**1972**	**1471**	**1183**	**900**
Den Bilt (NL)	86.1	82.1	76.0	75.9	72.9	72.7	77.1	78.9	80.7	84.4	85.5	86.8
	578	**596**	**631**	**811**	**1028**	**1263**	**1411**	**1443**	**1306**	**1086**	**772**	**587**
Kiruna (S)	83.0	82.0	77.0	71.0	64.0	61.0	68.0	72.0	77.0	81.0	85.0	85.0
	177	**171**	**221**	**324**	**476**	**710**	**1011**	**914**	**676**	**436**	**292**	**219**
Malmö (S)	87.0	86.0	83.0	76.0	73.0	74.0	78.0	77.0	82.0	85.0	87.0	89.0
	510	**496**	**561**	**711**	**958**	**1262**	**1531**	**1464**	**1269**	**969**	**753**	**627**
Munich (G)	83.7	81.9	76.8	72.3	74.9	76.8	74.2	76.1	79.2	82.9	83.5	85.7
	451	**484**	**598**	**780**	**1043**	**1361**	**1483**	**1446**	**1267**	**938**	**660**	**515**
Rome (I)	76.1	71.3	66.5	69.1	71.4	71.3	61.6	68.5	72.3	74.1	78.1	79.4
	794	**764**	**816**	**1048**	**1400**	**1795**	**1881**	**2042**	**1776**	**1346**	**1124**	**874**
Västerås (S)	84.0	82.0	74.0	66.0	62.0	65.0	69.0	74.0	81.0	83.0	86.0	86.0
	364	**355**	**402**	**540**	**766**	**1081**	**1354**	**1328**	**1085**	**793**	**602**	**483**

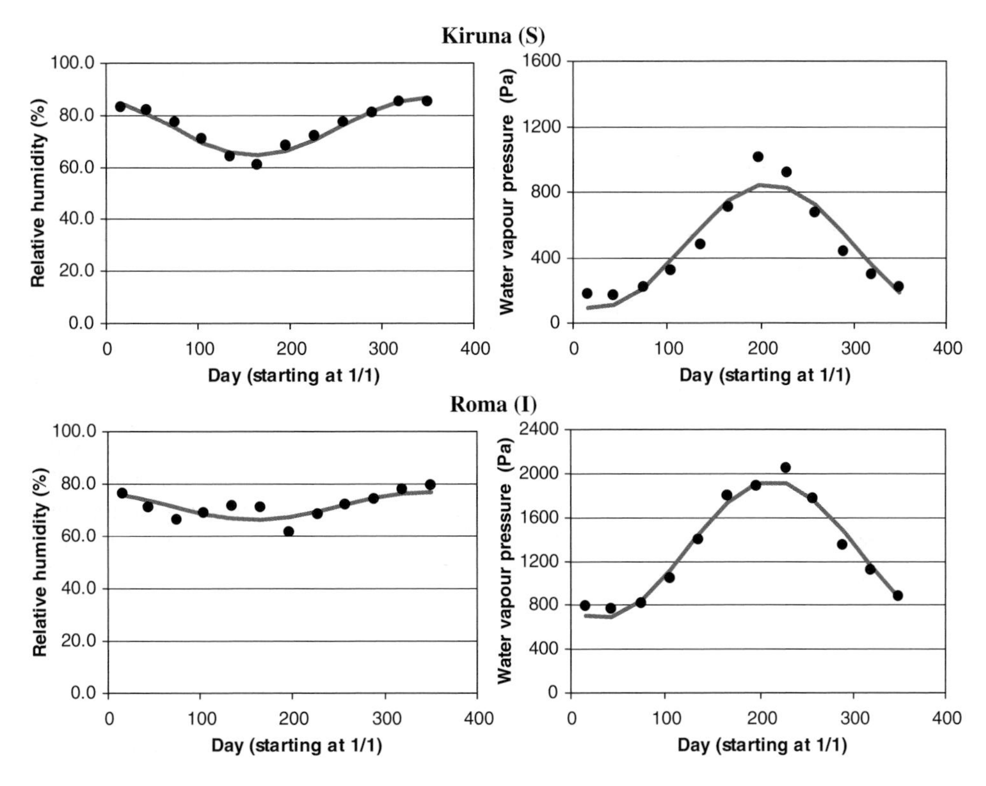

Figure 1.12 Monthly mean relative humidity and vapour pressure in Kiruna (Sweden), a cold climate, and in Rome (Italy), a rather warm climate – harmonic fit.

illustrates, deviation from the monthly averages can be significant:

$$\phi_e = \overline{\phi}_e + \hat{\phi}_e \cos\left[\frac{2\pi(t-d)}{365.25}\right] \quad p_e = \overline{p}_e + \hat{p}_e \cos\left[\frac{2\pi(t-d)}{365.25}\right] \tag{1.24}$$

with:

	Relative humidity (*RH*), %			Vapour pressure (*p*), Pa			Quality of the fit
	Average	Amplitude	*d* Days	Average	Amplitude	*d* Days	
Kiruna	75.5	11.1	346	469	380	209	*RH*±, *p*−
Roma	71.6	5.3	342	1305	627	214	*RH*−, *p*±

1.2.5 Wind

Wind impacts the hygrothermal response of building enclosures. A higher speed hastens the initial drying period but lowers the sol-air temperature and counteracts under-cooling. The thermal transmittance of single glass is increased as wind speed increases. Also infiltration, exfiltration and wind washing of poorly mounted insulation layers becomes more importance. Wind combined with precipitation gives wind-driven rain. Wind also governs comfort conditions outdoors. High speeds exert unpleasant dynamic forces on people, chills them and makes it harder to open and close doors.

1.2.5.1 Wind speed

In meteorology, wind speed is seen as a horizontal vector, whose amplitude and direction are measured in the open field at a height of 10 m. The average per 3 seconds is referred to as the immediate, and the average per 10 minutes as the mean speed. The spectrum shows various harmonics, while the vector changes direction constantly. Locally, the environment impacts the amplitude and direction. Venturi effects in small passages increase speed, while alongside buildings eddies develop and the windward and leeward side see zones of still air created.

As following formula shows, due to the friction that terrain roughness causes, wind speed increases from 0 m/s at grade up to a fixed value in the 500–2000 m thick atmospheric boundary layer in contact with the earth:

$$v_h = v_{10}\, K \ln(h/n) \tag{1.25}$$

v_{10} being the average wind speed in flat open terrain, 10 metres above grade, v_h average wind speed h metres above grade, n terrain roughness and K a friction-related factor:

Terrain upwind	K	n
Open sea	0.128	0.0002
Coastal plain	0.166	0.005
Flat grassland, runway area at airports	0.190	0.03
Farmland with low crops	0.209	0.10
Farmland with tall crops, vineyards	0.225	0.25
Open landscape with larger obstacles, forests	0.237	0.50
Old forests, homogeneous villages and cities	0.251	1.00
City centres with high-rises, industrial developments	≥0.265	≥2

Figure 1.13 and Table 1.7 show the annual distribution of the wind vector over all directions at Uccle. For almost half the year the wind comes from the west through south-west to south. Of course, these distributions differ between locations.

1.2.5.2 Wind pressure

The fact that wind exerts a pressure against any obstacle follows from Bernoulli's law. Without friction the sum of kinetic and potential energy in a moving fluid must remain constant. In a horizontal flow, potential energy is linked to pressure. When an obstacle stops the flow, all kinetic becomes potential, which gives the following as pressure (p_w in Pa):

$$p_w = \rho_a v_w^2/2 \tag{1.26}$$

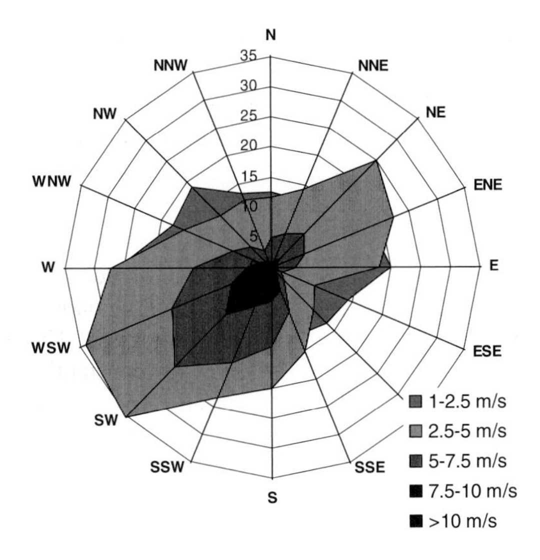

Figure 1.13 Typical wind rose for Uccle, Belgium.

Table 1.7 Uccle, ‰ of the time wind comes from the different directions (monthly means, 1931–1960).

Orientation	Month											
	J	F	M	A	M	J	J	A	S	O	N	D
N	18	22	43	54	59	65	57	37	54	22	15	14
NNE	20	32	48	59	62	60	57	38	47	33	30	13
NE	41	65	79	103	85	72	64	44	67	65	58	47
ENE	54	65	68	68	75	56	40	38	62	63	56	56
E	56	61	48	53	72	45	34	41	63	63	64	41
ESE	28	32	30	28	39	25	26	25	34	38	32	29
SE	36	38	34	33	31	25	25	28	45	50	39	35
SSE	60	53	57	40	38	29	26	36	55	70	60	65
S	90	98	81	49	52	44	44	53	62	100	92	109
SSW	120	109	86	62	70	55	64	88	71	118	105	132
SW	163	130	113	105	92	103	131	152	104	124	142	147
WSW	125	115	100	97	92	100	120	145	103	85	119	125
W	92	73	79	78	71	95	109	121	94	77	89	91
WNW	50	49	54	59	49	74	68	64	55	46	51	46
NW	28	35	73	62	59	84	76	50	44	26	29	32
NNW	19	23	37	50	54	68	59	40	40	20	19	18

Table 1.8 Wind pressure coefficients. Exposed building, up to three storeys, rectangular floor plan, length-to-width ratio 2 to 1 (reference speed: local, at building height).

Location		Wind angle							
		0	45	90	135	180	225	270	315
Face 1	(Wind side)	0.5	0.25	−0.5	−0.8	−0.7	−0.8	−0.5	0.25
Face 2	(Rear side)	−0.7	−0.8	−0.5	0.25	0.5	0.25	−0.5	−0.8
Face 3	(Side wall)	−0.9	0.2	0.6	0.2	−0.9	−0.6	−0.35	−0.6
Face 4	(Side wall)	−0.9	−0.6	−0.35	−0.6	−0.9	0.2	0.6	0.2
Roof Pitches	Wind side	−0.7	−0.7	−0.8	−0.7	−0.7	−0.7	−0.8	−0.7
<10°	Rear side	−0.7	−0.7	−0.8	−0.7	−0.7	−0.7	−0.8	−0.7
Average		−0.7	−0.7	−0.8	−0.7	−0.7	−0.7	−0.8	−0.7
Pitches	Wind side	−0.7	−0.7	−0.7	−0.6	−0.5	−0.6	−0.7	−0.7
10–30°	Rear side	−0.5	−0.6	−0.7	−0.7	−0.7	−0.7	−0.7	−0.6
Average		−0.6	−0.65	−0.7	−0.65	−0.6	−0.65	−0.7	−0.65
Pitches	Wind side	0.25	0	−0.6	−0.9	−0.8	−0.9	−0.6	0
>30°	Rear side	−0.8	−0.9	−0.6	0	0.25	0	−0.6	−0.9
Average		−0.28	−0.45	−0.6	−0.45	−0.28	−0.45	−0.6	−0.45

with ρ_a the air density in kg/m^3 ($\approx$1.2 kg/m^3) and v_w the wind speed in m/s. Yet, since no obstacle stretches to infinity, the flow is only impeded and redirected. That makes wind pressure on buildings different from the stop-the-flow value, a fact accounted for by the pressure coefficient (C_p):

$$p_w = \rho_a C_p v_w^2 / 2 \tag{1.27}$$

where v_w is a reference wind speed measured at the nearest weather station or at 1 m above the roof ridge. Field measurements, wind tunnel experiments and CFD have shown that pressure coefficients change depending on the reference and the up- and downwind environment. On buildings, their value varies from place to place, being highest at the edges and upper corners, lowest down in the middle. At the wind side, values are positive, at the rear side and sides parallel or nearly parallel to the wind negative. Values are found in the literature, see Table 1.8.

1.2.6 Precipitation and wind-driven rain

In humid climates, rain is the largest moisture load on buildings. The term 'wind-driven' applies to the horizontal component, and 'precipitation' to the vertical one. While in windless weather the horizontal component remains zero, higher wind speeds increase it. Horizontal and angled outside surfaces collect rain under all circumstances whereas

Table 1.9 Precipitation: duration and amounts per month, means for Uccle (1961–1970).

Precipitation	Month											
	J	F	M	A	M	J	J	A	S	O	N	D
Duration (h)	53.6	57.3	51.9	51.5	41.9	36.0	42.9	38.8	36.5	52.5	67.3	72.7
Amount (l/m^2)	57.4	61.4	56.8	63.6	63.1	74.9	92.3	78.6	54.6	74.5	76.5	84.5

vertical surfaces are struck by wind-driven rain only, albeit that they also suffer from run-off coming from angled and sometimes horizontal surfaces. In countries where outside wall buffering ensures rain-tightness, wind-driven rain to some extent impacts end energy use for heating and cooling.

1.2.6.1 Precipitation

Precipitation is as variable as the wind. Table 1.9 and Figure 1.14 give mean durations (hours) and amounts per month for Uccle. As with wind speed, no annual cycle appears, though slightly more rain falls in summer. The absolute maxima noted between 1956 and 1970 were: 4 l/m^2 in 1 minute, 15 l/m^2 in 10 minutes and 42.8 l/m^2 in 1 hour.

Showers are characterized by a droplet distribution $F_{\text{d,precip}}$. According to Best (1950), the relation with rain intensity is:

$$F_{\text{d,precip}} = 1 - \exp\left[-\left(\frac{d}{1.3 g_{\text{r,h}}^{0.232}}\right)^{2.25}\right] \tag{1.28}$$

with d the droplet size in m and $g_{\text{r,h}}$ the precipitation in l/(m^2.h). In windless weather, droplets have a final speed (d droplet size in mm, see also Figure 1.15) given by:

$$v_\infty = -0.166033 + 4.91844d - 0.888016d^2 + 0.054888d^3 \tag{1.29}$$

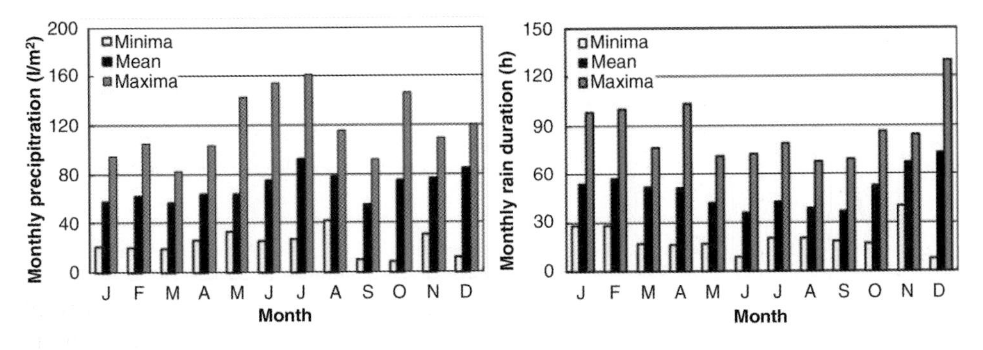

Figure 1.14 Precipitation: monthly amounts and duration for 1961–1970 at Uccle (averages, maxima and minima).

Depending on the mean droplet size, following terms describe precipitation:

Mean droplet size, mm	Precipitation
0.25	Drizzle
0.50	Normal
0.75	Strong
1.00	Heavy
1.50	Downpour

1.2.6.2 Wind-driven rain

The drag force that wind exerts inclines droplet trajectories. Final tilt angle depends on droplet size, wind direction compared to horizontal and speed increase with height. A constant horizontal speed keeps the inclined droplet trajectories straight. Wind-driven rain intensity ($g_{r,v}$ in kg/(m^2.s)) then becomes:

$$g_{r,v} = 0.222 v_w g_{r,h}^{0.88} \tag{1.30}$$

with $g_{r,h}$ the precipitation in kg/(m^2.s). This equation is simplified to:

$$g_{r,v} = 0.2 v_w g_{r,h} \tag{1.31}$$

a relation that fits well in open field. For wind speeds beyond 5 m/s, wind-driven rain exceeds precipitation there. In built environments, the equation fails. Intensity on a facade in fact depends on horizontal rain intensity, raindrop size distribution, building volumes up and down wind, building geometry, building orientation compared to wind direction, where on the enclosure and local detailing. Buildings in an open neighbourhood may catch 40 times as much wind-driven rain as buildings in a densely built environment. Local amounts on the facade are linked to the wind flow pattern around the building. For low-rises, the upper corners see the highest values. Under drizzle,

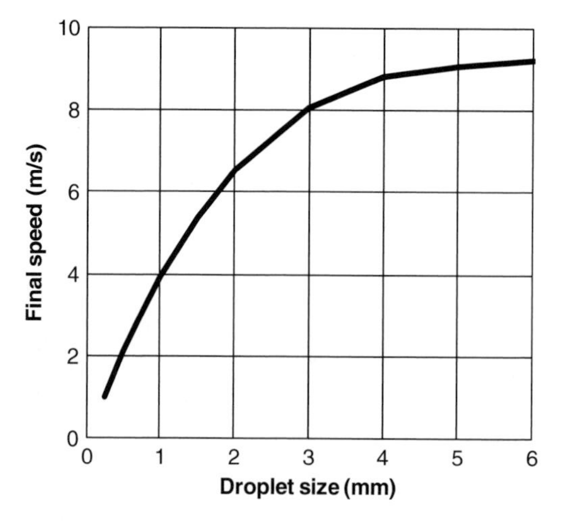

Figure 1.15 Final vertical speed of a raindrop in windless weather.

Table 1.10 Wind-driven rain on an 18-storey block of flats in Munich, measured by the Fraunhofer Institut für Bauphysik, May–November 1972.

Spot	Wind-driven rain, kg/m^2
Facade west, middle 3th floor	29
Facade west, middle 9th floor	55
Facade west, middle 16th floor	65
Roof edge north	115
Roof edge south	130

high-rises catch most at the highest floors, the higher edges and the upper corners. The differences diminish with higher rain intensity and wind speed. Measured deposits on an 18-storey block of flats, listed in Table 1.10, underline these trends.

The formula used to describe local wind-driven rain intensity on building enclosures is:

$$g_{r,v} = (0.2 C_r v_w \cos\theta) g_{r,h} \tag{1.32}$$

with θ the angle between wind direction and normal to the surface and C_r the wind-driven rain factor, a function of the type of precipitation, the surroundings, the facade spot in question, local detailing, and so on. On average, C_r is between 0.25 and 2. The product $0.2 C_r v_w \cos\theta$ is called the 'catch ratio', the amount of wind-driven rain hitting a facade spot as a fraction of the precipitation in the open field during the same rain event. The formula is based on experimental evidence. For that, catch ratios on the south-west facade of a test building, calculated using CFD for the wind field and droplet tracing for the rain trajectories, were compared with measured data, see Figure 1.16. Both fit reasonably well.

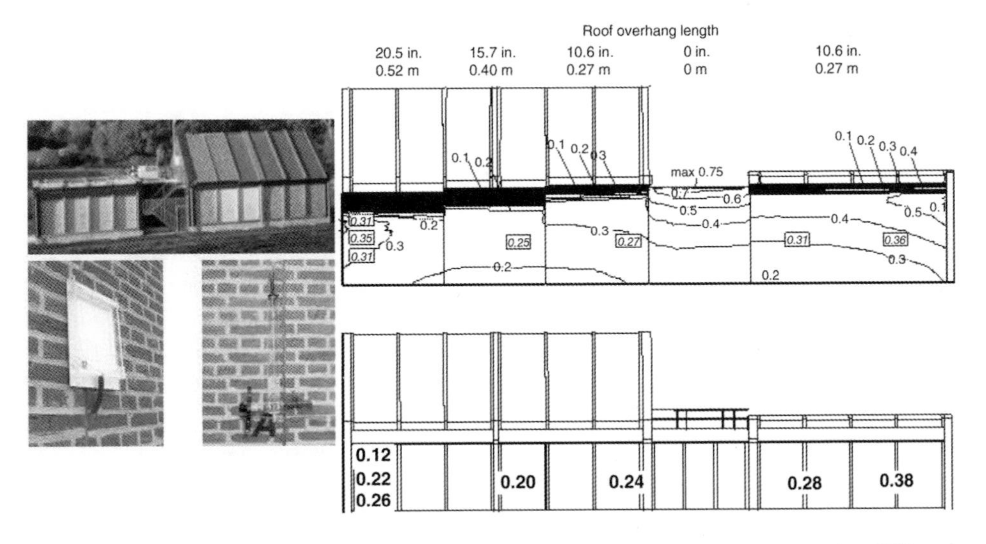

Figure 1.16 SW facade of a test building: calculated lines (top) of equal catch ratio using CFD and droplet tracing compared to measured catches (below).

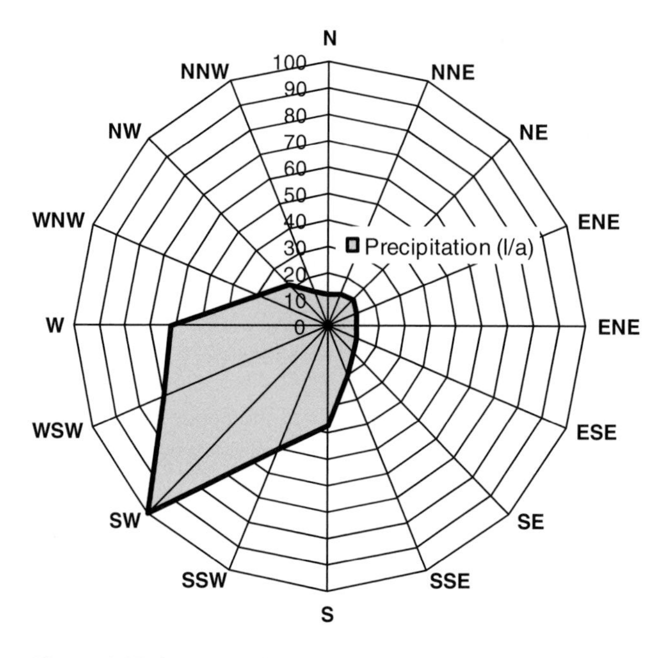

Figure 1.17 Average rain rose for Uccle.

In north-western Europe and as Figure 1.17 confirms for Uccle, wind-driven rain mainly comes from the south-west. Happily, strong winds and heavy precipitation rarely coincide. Between 1956 and 1970 the largest hourly means measured at Uccle were:

Precipitation, l/h	Mean wind speed, m/s
15	5.8
42	2.2

1.2.7 Microclimates around buildings

The microclimate around buildings is a complex reality. The immediate environment has an important impact, as it affects shading, wind direction and speed, wind-driven rain patterns, temperatures and so on. But also each part of the enclosure experiences differences, depending on factors such as slope and orientation, its position, whether or not it is shaded and/or sheltered by protrusions and volumes above. All these have impacts on the damage tolerance of a building.

This complexity, including the ever-changing, time-dependent character of the weather, forces us to simplify, using design temperatures, reference days, reference years, indoor climate classes, and so on. It must therefore be admitted that any evaluation using these, results in an approximate picture of reality, loaded with uncertainty.

1.2.8 Standardized outdoor climate data

1.2.8.1 Design temperature

In north-west European countries with massive building tradition, the design temperature (θ_{ed}, °C) for heating is either the lowest running two-day mean over a period of 20 or the lowest daily mean over a period of 10 years. Many countries have maps giving values. In the USA and Canada, two countries with a timber-frame tradition, heating designers can choose between two hourly values: one exceeded 99.6% or 99% of the time during $8790 \times T$ hours with T the years considered, actually 1986–2010. For cooling, the hourly mean exceeded 0.4% or 1% of the time is used. The same holds for the wet bulb temperatures, needed to calculate the power that (de)humidification requires.

Both approaches remain quite close. Take for example Uccle. The two-day mean for the period 1971–1990 is −8 °C. The 99.6% and 99% hourly means for the period 1986–2010 are −6.6 °C and −4.4 °C, respectively. The fact that the period used by the USA and Canada misses the cold winter of 1979 explains most of the difference.

1.2.8.2 Reference years

Air temperature, solar radiation, longwave radiation and wind are of decisive importance for the net heating and cooling demand and overheating risk. To allow the use of building energy simulation tools (BES), hourly thermal reference years (TRYs) were constructed. Mean ones give the same annual net heating and cooling demand as found on average over, for example, 30 successive years. Warm ones help in evaluating overheating and allow a better guess of the cooling load. Cold ones do the same for the heating load. TRYs exist for several locations worldwide. For steady state calculations, monthly TRYs are used. The Belgian three used the 1931–1960 averages and standard deviations. For the mean, cold and warm year, see Table 1.11. Cloudiness J in the table stands for the ratio between the actual and clear sky solar radiation on a horizontal surface. The energy performance of buildings TRY in that country however took the 1961–1990 averages with temperature, total ($E_{ST,hor}$) and diffuse solar radiation ($E_{sd,hor}$) on a horizontal surface as parameters, see Table 1.12.

1.2.8.3 Very hot summer, very cold winter day

A simpler choice than using TRYs is looking for a hot summer's day and a cold winter's day with probability once in, for example, 20 years, and considering both as being the last of a spell of the same hot and cold days. This allows analytic solutions to be used for the transient response. Table 1.13 lists the two for Uccle. They are slightly reframed to fit the spell assumption.

1.2.8.4 Moisture reference years

Since the end of the 1960s, calculation tools have allowed us to predict the moisture response of assemblies. They require, among others, knowledge of the indoor and outdoor ambient conditions. So, the 1990s saw the introduction of the concept of moisture reference year, the year with return rate one in every ten years giving the

Table 1.11 Monthly mean TRYs for Belgium (mean, *cold*, **warm**).

Month	θ_e (°C)	$E_{ST,hor}$(W/m^2)	Wind (m/s)	RH (%)	p (Pa)	J (–)
January	3.8	26.6	3.8	91	730	0.52
February	3.2	55.2	5.1	87	668	0.58
March	7.1	88.5	3.9	85	857	0.56
April	9.1	131.4	4.7	81	935	0.56
May	11.9	172.6	3.4	77	1072	0.59
June	16.7	216.9	3.2	80	1520	0.68
July	16.1	162.5	3.0	79	1445	0.53
August	17.1	158.0	3.5	83	1618	0.60
September	14.3	128.7	3.4	83	1352	0.67
October	11.0	64.9	3.6	91	1194	0.54
November	6.1	29.6	3.8	91	856	0.45
December	3.1	19.3	3.6	91	694	0.46
January	*–3.9*	*33.2*	*4.1*	*86*	*379*	*0.65*
February	*–1.4*	*63.1*	*3.3*	*85*	*462*	*0.66*
March	*3.1*	*94.3*	*3.6*	*81*	*618*	*0.59*
April	*6.7*	*101.5*	*4.0*	*87*	*853*	*0.43*
May	*10.8*	*151.5*	*3.7*	*84*	*1087*	*0.51*
June	*14.2*	*178.7*	*3.1*	*79*	*1278*	*0.56*
July	*16.1*	*171.4*	*3.7*	*80*	*1463*	*0.55*
August	*15.9*	*136.7*	*3.5*	*86*	*1552*	*0.52*
September	*12.8*	*123.0*	*2.9*	*82*	*1211*	*0.64*
October	*7.5*	*53.9*	*3.3*	*90*	*932*	*0.44*
November	*4.2*	*40.2*	*4.6*	*88*	*725*	*0.62*
December	*0.2*	*25.3*	*3.2*	*90*	*558*	*0.61*
January	**6.8**	**25.3**	**5.2**	**87**	**859**	**0.50**
February	**7.6**	**55.4**	**4.4**	**89**	**928**	**0.58**
March	**8.5**	**109.4**	**3.5**	**85**	**942**	**0.69**
April	**12.3**	**130.7**	**3.0**	**86**	**1229**	**0.55**
May	**15.4**	**206.8**	**2.8**	**72**	**1258**	**0.70**
June	**18.8**	**235.2**	**3.1**	**74**	**1605**	**0.74**
July	**20.0**	**242.9**	**3.5**	**73**	**1706**	**0.79**
August	**19.9**	**181.7**	**2.9**	**74**	**1718**	**0.69**
September	**16.5**	**171.0**	**3.5**	**70**	**1313**	**0.89**
October	**13.8**	**86.2**	**2.6**	**86**	**1356**	**0.71**
November	**9.4**	**35.1**	**5.1**	**88**	**1036**	**0.54**
December	**7.4**	**15.4**	**5.3**	**87**	**895**	**0.37**

Table 1.12 Energy performance legislation, reference year for Belgium.

Month	θ_e (°C)	$E_{ST,hor}$ (MJ/m^2)	$E_{sd,hor}$ (MJ/m^2)
January	3.2	71.4	51.3
February	3.9	127.0	82.7
March	5.9	245.5	155.1
April	9.2	371.5	219.2
May	13.3	510.0	293.5
June	16.3	532.4	298.1
July	17.6	517.8	305.8
August	17.6	456.4	266.7
September	15.2	326.2	183.6
October	11.2	194.2	118.3
November	6.3	89.6	60.5
December	3.5	54.7	40.2

highest wetness in an assembly. Each moisture event, however, demanded another year. A wet and windy one best serves for rain leakage. Solar-driven vapour flow scores highest in years with sunny summers, interrupted by regular rainy spells. Interstitial condensation is worst in cold, humid ones with little sun in summer. Initially, the exercise ran for interstitial condensation only, see Table 1.14.

An alternative consists of using a so-called moisture index. As more wetting and less drying decreases moisture tolerance, this index must combine the two. In Canada, the exercise bases on normalized wetting (WI_{norm}) and drying (DI_{norm}) functions:

$$WI_{norm} = \frac{WI - WI_{min}}{WI_{max} - WI_{min}} \quad DI_{norm} = \frac{DI - DI_{min}}{DI_{max} - DI_{min}} \tag{1.33}$$

where WI equals the annual amount of wind-driven rain hitting a facade with given orientation at the location considered and DI is:

$$DI = \sum_{h=1}^{k} \overline{x}_{sat}\left(1 - \overline{\phi}\right)$$

with $\overline{x}_{sat}$ the hourly mean vapour saturation ratio, $\overline{\phi}$ the hourly mean relative humidity outdoors at that location and Σ denoting the sum of all hourly means over a year. The minimums WI_{min} and DI_{min} and maximums WI_{max} and DI_{max} represent the most and least severe year in Ottawa out of a file of 30. For any other location, WI and DI are taken from the third year in terms of severity over the 30 years of hourly data there. The moisture index then is:

$$MI = \sqrt{WI_{norm}^2 - (1 - DI_{norm})^2} \tag{1.34}$$

Table 1.13 Uccle: cold winter (declination −17.7°) and hot summer day (declination 17.7°).

Hour↓	Cold winter day			Hot summer day		
	Temp. °C	Direct sun, ⊥, W/m^2	Diffuse sun, horizontal, W/m^2	Temp. °C	Direct sun, ⊥, W/m^2	Diffuse sun, horizontal, W/m^2
0	−13.3	0	0	20.2	0	0
1	−13.6	0	0	19.8	0	0
2	−13.8	0	0	19.5	0	0
3	−14.2	0	0	19.0	0	0
4	−14.6	0	0	18.5	0	0
5	−15.0	0	0	18.0	136	6
6	−15.4	0	0	17.8	301	50
7	−15.5	0	0	17.8	440	87
8	−15.5	42	8	20.0	555	121
9	−14.9	336	47	22.0	645	146
10	−14.3	553	72	25.5	711	165
11	−13.1	636	89	28.0	752	180
12	−11.8	669	100	30.0	768	187
13	−11.7	656	105	30.8	759	188
14	−11.5	611	94	30.5	726	179
15	−11.9	564	72	30.0	669	165
16	−12.3	375	39	29.0	587	147
17	−12.8	47	11	27.8	480	121
18	−13.3	0	0	26.5	348	90
19	−13.7	0	0	25.5	192	49
20	−14.0	0	0	24.2	11	7
21	−14.5	0	0	23.0	0	0
22	−15.0	0	0	22.0	0	0
23	−15.1	0	0	20.5	0	0
24	−15.2	0	0	20.2	0	0

This index has two drawbacks. First it only considers problems caused by wind-driven rain, overlooking problematic moisture deposits due to interstitial condensation. Secondly, it excludes the effects of solar radiation, although drying primarily depends on the difference between the vapour saturation ratio at the surface and the vapour ratio in the air, a reality that the sun heavily impacts.

1.2.8.5 Equivalent temperature for condensation and drying

In cold and temperate climates interfaces where interstitial condensate deposits sit at the outside of the thermal insulation, where the remaining thermal resistance to outdoors is

Table 1.14 Moisture reference years for interstitial condensation, four European cities.

Month	Uccle			Rome			Copenhagen			Luleå		
	θ_e °C	p_e Pa	$E_{ST,h}$ W/m^2	θ_e °C	p_e Pa	$E_{ST,h}$ W/m^2	θ_e °C	p_e Pa	$E_{ST,h}$ W/m^2	θ_e °C	p_e Pa	$E_{ST,h}$ W/m^2
J	2.7	675	24.0	8.0	747	73.3	−0.1	569	18.2	−12.6	190	2.4
F	1.7	587	50.4	10.1	877	205.3	−0.4	507	53.1	−13.2	183	22.9
M	5.9	724	99.0	10.2	811	257.8	−0.3	476	83.4	−5.2	357	54.8
A	8.5	932	115.5	13.2	997	289.2	7.4	817	159.6	−2.4	415	109.8
M	12.6	1123	180.0	18.3	1323	243.6	12.0	1080	227.5	3.6	538	179.6
J	15.5	1374	215.0	22.3	1733	257.0	14.7	1355	229.4	10.8	894	234.4
J	14.9	1423	148.2	24.5	1976	271.1	15.3	1474	186.3	15.4	1365	203.9
A	16.2	1492	172.8	22.3	1747	224.0	15.1	1435	178.3	12.3	1144	122.6
S	13.5	1300	120.6	19.4	1524	187.8	12.6	1275	125.4	7.7	883	71.2
O	11.1	1097	102.1	17.1	1446	125.5	7.7	961	54.9	−2.4	430	28.4
N	4.0	716	40.1	11.5	1029	74.9	5.0	793	26.6	−4.1	401	4.3
D	4.8	783	15.6	9.6	907	51.2	1.1	614	15.9	−17.1	123	0.5

mostly small. In these interfaces, vapour saturation pressure fluctuates with temperature. As the relation between the two is exponential, higher interface temperatures have a greater weight on the mean vapour saturation pressure. So, over a month, that mean will differ from the value at the mean interface temperature. It is therefore sensible to calculate it beforehand and link its value to a 'fictive' outdoor temperature, called the 'equivalent temperature for condensation and drying', which works due to the low remaining thermal resistance to outdoors.

Quantification goes as follows. First, the sol-air temperature per orientation and slope is calculated for each month, using hourly temperature, solar and longwave radiation data. Then, the hourly temperatures in the condensation interface are fixed. As it sits close to outdoors, a steady-state approach suffices:

$$\theta_j = \theta_e^* + R_e^j\left(\theta_i - \theta_e^*\right)/R_a \tag{1.35}$$

with θ_i the temperature indoors, θ_e^* the sol-air temperature, R_e^j the thermal resistance from the condensation interface to outdoors and R_a the thermal resistance of the assembly environment to environment. Next, vapour saturation pressure is linked to these hourly interface temperatures and the mean for the month considered is calculated as:

$$\overline{p}_{sat,j} = \frac{1}{T}\int_o^T p_{sat,j}\,dt \tag{1.36}$$

Table 1.15 Uccle, equivalent temperature for condensation and drying ($a_K = 1$, $e_L = 0.9$).

Slope Az.	Annual mean $\overline{\theta}^*_{ce}$ (°C)					Annual amplitude $\hat{\theta}^*_{ce}$ (°C)				
	N	NW/NE	W/E	SW/SE	S	N	NW/NE	W/E	SW/SE	S
0	14.4	14.4	14.4	14.4	14.4	12.6	12.6	12.6	12.6	12.6
15	13.7	13.9	14.4	14.7	14.8	12.4	12.5	12.6	12.7	12.7
30	12.9	13.4	14.2	14.9	16.0	11.7	12.0	12.3	12.4	12.4
45	12.2	12.8	13.9	14.7	14.9	10.9	11.3	11.6	12.0	11.8
60	11.9	12.6	13.7	14.6	14.7	9.8	10.6	11.2	11.3	11.1
75	11.8	12.4	13.5	14.3	14.6	9.3	10.0	10.7	10.6	10.3
90	12.1	12.6	13.6	14.2	14.4	9.1	9.5	9.9	9.7	9.5

Related interface temperature is then:

$$\overline{\theta}^*_{\overline{p}_{sat,j}} = F(\overline{p}_{sat,j}) \tag{1.37}$$

From it, the monthly mean equivalent temperature for condensation and drying follows:

$$\overline{\theta}^*_{ce} = \frac{\overline{\theta}_{\overline{p}_{sat,j}} + \overline{\theta}_i (R^j_e / R_a)}{1 - R^j_e / R_a} \approx \frac{\overline{\theta}_{\overline{p}_{sat,j}}}{1 - R^j_e / R_a} \tag{1.38}$$

The almost equal to sign ≈ indicates that the ratio R^j_e / R_a is so small that neglecting its product with the indoor temperature has virtually no impact, a fact that anyhow introduces some uncertainty.

Once this is done for all months, the annual mean is fixed and the first and second harmonic transposed into an annual amplitude, and a time function $C(t)$. Table 1.15 gives the mean and amplitude for a non-shaded, sun-radiated surface with shortwave absorptivity 1 and longwave emissivity 0.9 for Uccle, while Table 1.16 lists the time function.

Correction for non-shaded, sun-radiated surfaces with different shortwave absorptivity first requires combining the annual mean and amplitude into a relation of the form:

$$\theta^*_{ce} = \overline{\theta}^*_{ce} + \hat{\theta}^*_{ce} C(t) \tag{1.39}$$

Table 1.16 Uccle, month-based time function.

Month	$C(t)$	Month	$C(t)$	Month	$C(t)$	Month	$C(t)$
January	−0.98	April	−0.10	July	+1.00	October	−0.10
February	−0.85	May	+0.55	August	+0.85	November	−0.55
March	−0.50	June	+0.90	September	+0.55	December	−0.90

Then the mean and amplitude are adapted to account for the actual shortwave absorptivity:

$$\left[\overline{\theta}^*_{\text{ce}}\right] = \alpha_{\text{K}}\left(\overline{\theta}^*_{\text{ce}} - \overline{\theta}'_{\text{e}}\right) + \overline{\theta}'_{\text{e}} \quad \left[\hat{\theta}^*_{\text{ce}}\right] = \alpha_{\text{K}}\left(\hat{\theta}^*_{\text{ce}} - \hat{\theta}'_{\text{e}}\right) + \hat{\theta}'_{\text{e}} \tag{1.40}$$

In these equations $\overline{\theta}'_{\text{e}}$ and $\hat{\theta}'_{\text{e}}$ are the annual mean and amplitude of the equivalent temperature for condensation and drying on a north-facing surface with slope of 45°, shortwave absorptivity 0 and longwave emissivity 0.9, equal to 8.5 and 7.1 °C at Uccle. If the longwave emissivity is also 0 or the surface stays shaded, the mean and amplitude of the air temperature take over, 9.8 and 6.9 °C at Uccle. For non-shaded, sun-radiated surfaces with longwave emissivity between 0 and 0.9, a linear interpolation applies for $\overline{\theta}'_{\text{e}}$ and $\hat{\theta}'_{\text{e}}$:

$$\overline{\theta}'_{\text{e}} = 8.5 + 1.3\left(\frac{0.9 - e_{\text{L}}}{0.9}\right) \quad \hat{\theta}'_{\text{e}} = 7.1 - 0.2\left(\frac{0.9 - e_{\text{L}}}{0.9}\right) \tag{1.41}$$

The effective monthly mean equivalent temperature for condensation and drying on a non-shaded, sun-radiated surface then becomes:

$$\left[\theta^*_{\text{ce}}\right] = \left[\overline{\theta}^*_{\text{ce}}\right] + \left[\hat{\theta}^*_{\text{ce}}\right] C(t) \tag{1.42}$$

An estimate for sun-radiated surfaces that get shade during part of the day consists of adapting Eq. (1.40):

$$\left[\overline{\theta}^*_{\text{ce}}\right] = \alpha_{\text{K}}\left(\overline{\theta}^*_{\text{ce}} - \overline{\theta}'_{\text{e}}\right)\frac{E_{\text{s,T,real}}}{E_{\text{s,T}}} + \overline{\theta}'_{\text{e}} \quad \left[\hat{\theta}^*_{\text{ce}}\right] = \alpha_{\text{K}}\left(\hat{\theta}^*_{\text{ce}} - \hat{\theta}'_{\text{e}}\right)\frac{E_{\text{s,T,real}}}{E_{\text{s,T}}} + \hat{\theta}'_{\text{e}} \tag{1.43}$$

with $E_{\text{s,T}}$ the total monthly solar radiation that the surface should receive, if non-shaded, and $E_{\text{s,T,real}}$ the total monthly solar radiation actually recorded.

Calculating the equivalent temperature for condensation and drying for any other cold or temperate climate location proceeds as explained.

1.2.8.6 Monthly mean vapour pressure outdoors

The monthly mean vapour pressure outdoors is quantified using the same time function C(t):

$$p_{\text{e}} = \overline{p}_{\text{e}} + \hat{p}_{\text{e}} C(t) \tag{1.44}$$

with $\overline{p}_{\text{e}}$ the annual mean and $\hat{p}_{\text{e}}$ the amplitude, in Uccle 1042 and 430 Pa.

1.3 Indoors

Building use presumes comfortable temperatures. Whether or not relative humidity demands control depends on climate and building use. Those two and the air pressure differences between rooms and outdoors fix the environmental load that the enclosure has to endure.

1.3.1 Air temperatures

1.3.1.1 In general

The air temperature not only affects thermal comfort but also impacts the end energy use for heating and cooling plus the enclosure's moisture response. Standards focus

Table 1.17 Operative temperatures needed according to DIN 4701 and EN 12 831.

Building type	Room	Temperature	
		DIN 4701	EN 12831
Dwelling	Daytime room and bedroom	20	20
	Bathroom	24	24
Hospital	Nursery	22	20
	Surgery, premature births	25	
Office buildings	All rooms	20	20
	Corridors, rest rooms	15	
Indoor swimming pool	Natatorium	28	
	Showers	24	
	Cabins	22	
Schools	Classroom		20
Department store			20
Church			15
Museum, gallery			16

on the operative temperature, see Table 1.17, but in most buildings, air temperature is the factor that is controlled, although a thermostat somewhere on a wall senses a mixture of the local air and surface temperatures, the two impacted by the little heat that the device produces. In insulated buildings, operative and air temperature hardly differ.

1.3.1.2 Measured data

Residential buildings – Figure 1.18 shows weekly mean air temperatures measured between 1972 and 2008 in 283 daytime rooms, 338 bedrooms and 37 bathrooms of poorly insulated homes. The figures picture the data as a function of the mean outdoor temperature for the same week. Between 2002 and 2005, data was also collected in 39 well-insulated homes. Table 1.18 lists the regression line constants and correlation coefficient for the two sets.

In the poorly insulated dwellings, the weekly daytime room means are near to the comfort value with only a limited impact of the mean outdoors, proving these rooms are on average well heated. Bedrooms and bathrooms instead show significantly lower

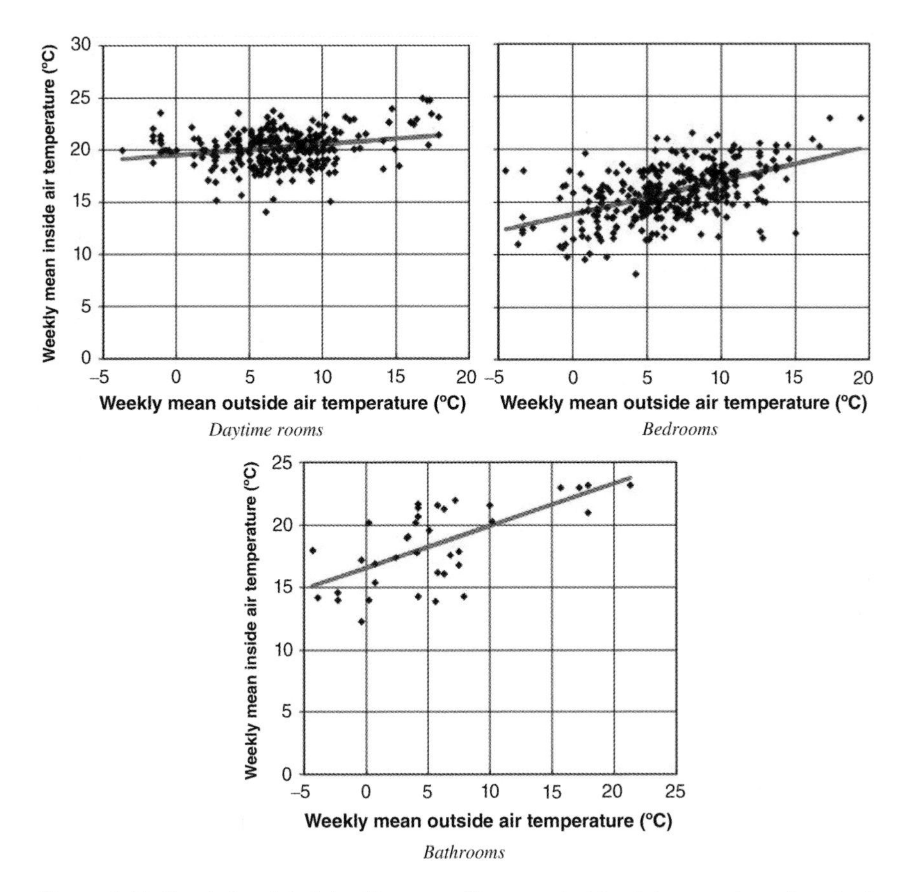

Figure 1.18 Poorly insulated dwellings: weekly mean inside air temperatures in a daytime rooms (top left), bedrooms (top right) and bathrooms (below).

Table 1.18 Measured weekly mean air temperatures.

	Number of weekly records	$\overline{\theta}_i = a + b\overline{\theta}_e$		Correlation coefficient
		a (°C)	b (–)	
Poorly insulated homes				
Daytime rooms	283	19.5	0.11	0.06
Bedrooms	338	13.8	0.32	0.26
Bathrooms	37	16.5	0.34	0.43
Insulated homes				
Daytime rooms	39	19.3	0.22	0.89
Bedrooms	78	15.6	0.42	0.95
Bathrooms	39	19.9	0.20	0.82

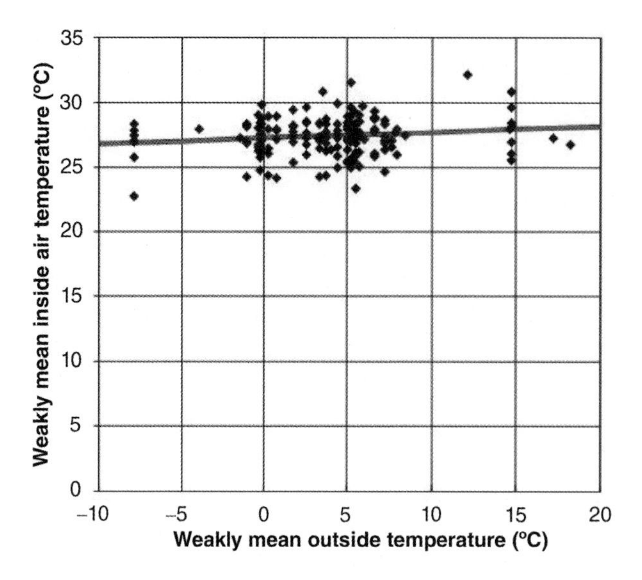

Figure 1.19 Natatoriums: weekly mean inside temperature.

weekly means that change strongly with the mean outdoors. While bathrooms look intermittently heated, bedrooms apparently lack heating. Their indoor temperature merely reflects the balance between transmission and ventilation gains from adjacent rooms, transmission and ventilation losses to outdoors, solar gains through the window and internal gains. Insulated, the conclusions differ. Daytime rooms remain well heated but bedrooms look warmer, probably due to the better insulation, although still, they are either intermittently or poorly heated. Bathrooms are also warmer but the outdoor temperature still intervenes.

Poorly heating the sleeping rooms seems a widespread dweller's habit, which has its advantages. For people, facing temperature differences favours health. After all, the blankets ensure comfort while sleeping.

Natatoriums – Figure 1.19 groups 162 weekly mean air temperatures logged in 16 natatoriums, together with the corresponding weekly mean outdoors. The least square line and correlation coefficient are given in Table 1.19.

Table 1.19 Natatoriums, measured weekly mean air temperatures.

	$\theta_i = a + b\theta_e$		
Number of weekly records	a (°C)	b (–)	Correlation coefficient
162	27.3	0.045	0.02

On average, the value measured fits well with the preferred 28 °C operative temperature. Nonetheless, the lowest (22.2 °C) and highest value (32.2 °C) deviate substantially. Well known is that leisure swimmers like higher temperatures, and competition swimmers prefer lower temperatures.

1.3.2 Relative humidity and vapour pressure

Relative humidity and vapour pressure are key factors impacting moisture tolerance. Both also influence the indoor environmental quality. In temperate climates the two are mostly allowed to vary freely. Values depend on the whole building heat, air, moisture balance. Only museums, archives, cleanroom spaces, computer rooms, surgical units and intensive care units require control, as do dwellings in hot and humid or really cold climates.

1.3.2.1 Vapour release indoors

Tables 1.20–1.23 list literature data about the overall daily vapour release, the vapour release per activity and a more detailed overview of the hourly releases by families as function of their living patterns.

Table 1.24 suggests the following relation between vapour release and number of children:

$$G_{v,P} = 3.2 + 2.8 n_{\text{children}} \qquad r^2 = 0.28(\text{kg}/(\text{day.child}))$$

This 2.8 kg looks high. A more modest value is 1 kg extra per day per child listed in the final column of the table.

Table 1.20 Daily vapour release, depending on the number of family members.

Family members	Average water vapour release in kg/day		
	Low water usage[a)]	Average water usage[b)]	High water usage[c)]
1	3 to 4	6	9
2	4	6	11
3	4	9	12
4	5	10	14
5	6	11	15
6	7	12	16

a) Dwelling frequently unoccupied.
b) Families with children.
c) Teenage children, frequent showers, etc.

Table 1.21 Vapour release linked to metabolism, activity and others.

	Activity	Release, g/h
Adults, metabolism	Sleeping	30
	At rest (depends on temperature)	33–70
	Light physical activity	50–120
	Moderate physical activity	120–200
	Heavy physical activity	200–470
Bathroom	Bath (15′)	60–700
	Shower (15′)	≤660
Kitchen	Breakfast preparation (four people)	160–270
	Lunch preparation (four people)	250–320
	Dinner preparation (four people)	550–720
	Dish cleaning by hand	480
	Daily average release	100
Laundry drying	After dry spinning	50–500
	Starting from wet	100–1500
Plants (per pot)		5–20
Young trees		2000–4000
Full-grown trees		$2–4 \times 10^6$

Table 1.22 Vapour release by a family of two.

Hour	Persons present	Release in g/h				Total
		Persons	Cooking	Hygiene	Laundry	
Both working, weekday						
1–5	2	120				120
6	2	120	240	720		1080
7	2	120	240			360
8–17	0					0
18	2	120				120
19–20	2	120	480			600
21–24	2	120				120
Total		1680	1440	1200		4320

Table 1.23 Vapour release by a family of four.

Hour	Persons present	Release in g/h				Total
		Persons	Cooking	Hygiene	Laundry	
Father working, oldest child at school, mother and one child at home, weekday						
1–5	4	240				240
6–7	4	240	480	720		1440
8	2	120			120	240
9	1	120			180	300
10	1	120	720		180	1020
11–12	2	120	1200	120		1440
13–14	2	120	480	120		720
15	2	120			120	240
16–17	2	180			120	300
18	4	240				240
19	4	240	480			720
20	4	240	480	240		960
21–22	4	240				240
23	4	240		240		480
24	4	240				240
Total	4680	6000	2400	840	13 920	

Table 1.24 Daily vapour release in relation to the number of family members (several sources).

Number of family members			
2 (no children)	3 (one child)	4 (two children)	>4 (more children)
8	12	14	>14
			+1 kg/(day.child)
	10		
7	20		
		14.6	
13.2	19.9	23.1	
	11.5		
	5–12		
	6–10.5		
4.3		13.7	
8.2	12.1	14.1	14.4
Mean: 8.14	Mean: 11.9	Mean: 15.9	

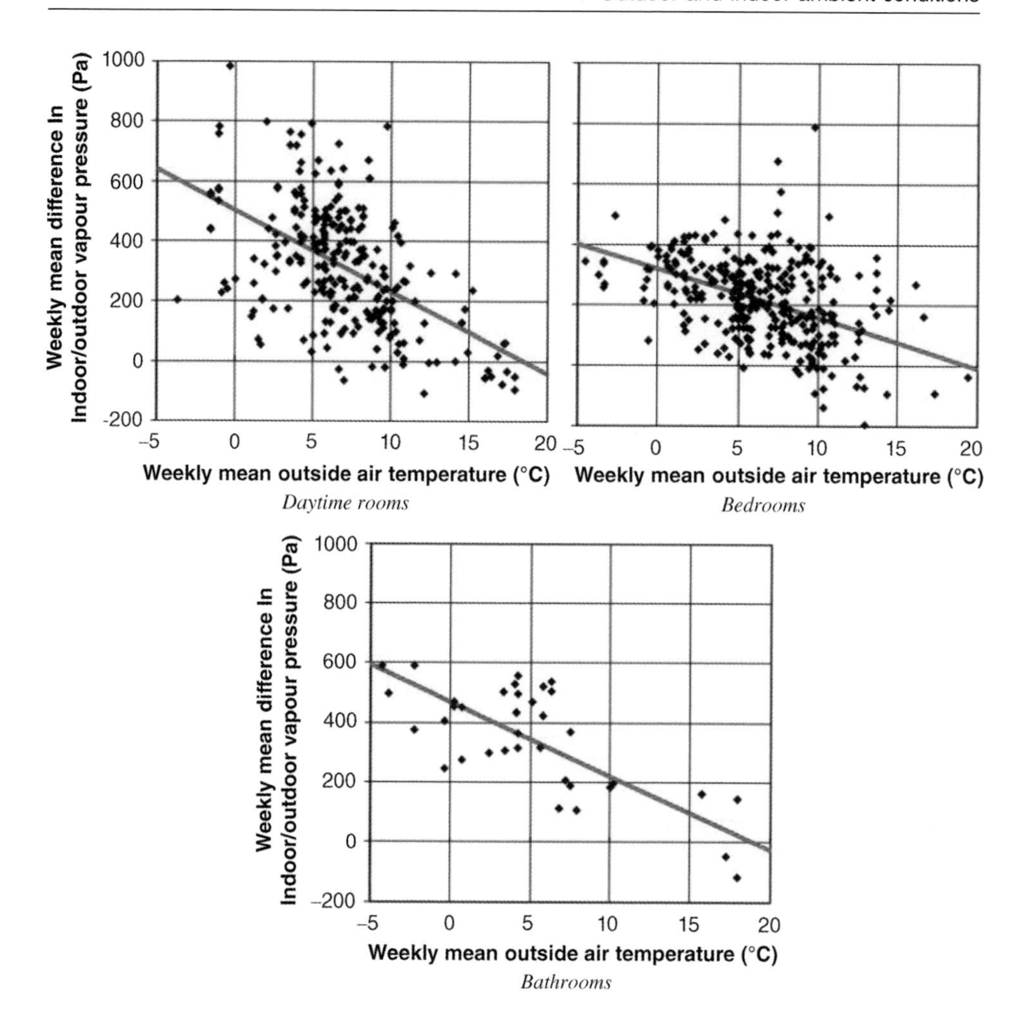

Figure 1.20 Poorly insulated homes: difference in weekly mean inside/outside vapour pressure in daytime rooms (top left), bedrooms (top right) and bathrooms (below).

1.3.2.2 Measured data

Residential buildings – Figure 1.20 displays the weekly mean differences in indoor/outdoor vapour pressure measured between 1972 and 2008 in the 261 daytime room, 315 bedrooms and 37 bathrooms of a number of poorly insulated homes. The figures present the data as a function of the mean air temperature outdoors for the same week. Between 2002 and 2005, data also came from 39 well-insulated homes.

Table 1.25 lists the least square lines, the 95% line and the correlation coefficients. For the poorly insulated dwellings, also a multiple regression between relative humidity and the air temperature indoors and outdoors is included. In general, the difference decreases when it is warmer outdoors. Reasons are more intense ventilation in summer and the

Table 1.25 Measured weekly indoor/outdoor vapour pressure difference and relative humidity.

	Number of records	$\Delta p_{ie} = a + b\theta_e$		Correlation r^2	$\varphi_i = a + b\theta_i + c\theta_e$			Correlation r^2
		a, Pa	b, Pa/°C		a, %	b,%/°C	c, %/°C	
Poorly insulated homes								
Daytime rooms	261	504	−27.2	0.28	99.6	−2.6	−0.87	0.40
95% line		704	−27.2					
Bedrooms	351	319	−16.3	0.19	94.5	−3.4	−1.9	0.47
95% line		465	−16.3					
Bathrooms	37	469	−24.9	0.61	94.5	−2.5	−1.2	0.58
95% line		575	−24.9					
Insulated homes								
Daytime rooms	39	299	−12.4	0.53				
95% line		594	−18.0					
Bedrooms	39	261	−10.7	0.18				
95% line		497	−13.1					
Bathrooms	39	338	−11.3	0.64				
95% line		658	−16.1					

long-term hygric inertia of fabric, furniture and furnishings that dampens and shifts the annual amplitude indoors compared to outdoors.

Most humid are the daytime rooms. Relative humidity in all rooms decreases with higher indoor and increases with higher outdoor temperatures. A comparison with the insulated dwellings shows that bedrooms and bathrooms remain quite close but that daytime rooms look much dryer when insulated. Why this should be is unclear. Probably kitchen stove hoods have become standard, or these more recent dwellings have fewer inhabitants, more volume per inhabitant and are better ventilated.

Natatoriums – Figure 1.21 shows 162 records of weekly mean differences in indoor/outdoor vapour pressure in the 16 natatoriums mentioned, together with the mean outdoor temperature during the same week. The least square analysis and the correlation coefficients are given in Table 1.26.

Again, the weekly means drop when it is warmer outdoors. Also notable is the large difference between the mean and the 95% line, indicating that natatoriums range from humid to very humid.

Nevertheless, as Figure 1.22 underlines, related relative humidity scores far below the 60–70% that wet swimmers need to feel comfortable when drying. The majority in fact have an enclosure, which was not designed to tolerate the comfortable relative humidity needed without suffering from severe moisture damage. However, keeping relative humidity as low as measured, demands intense ventilation, resulting in far too high end energy use for heating.

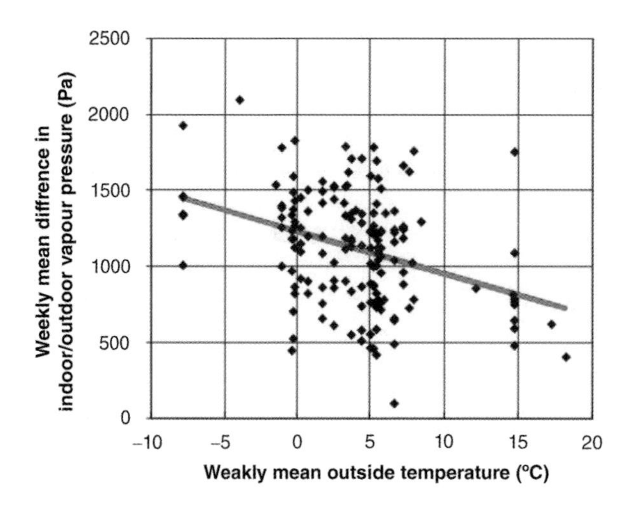

Figure 1.21 Natatoriums: weekly mean indoor/outdoor vapour pressure difference.

Table 1.26 Natatoriums, weekly mean indoor/outdoor vapour pressure difference, relative humidity.

Number of records	Line	$\Delta p_{ie} = a + b\theta_e$		Correlation coefficient r^2	$\phi_i = a + b\theta_i + c\theta_e$			Correlation coefficient r^2
		a, Pa	b, Pa/°C		a %	b %/°C	c %/°C	
162	Mean	1229	−27.7	0.13	125	−2.8	0.36	0.17
	95%	1793	−27.7					

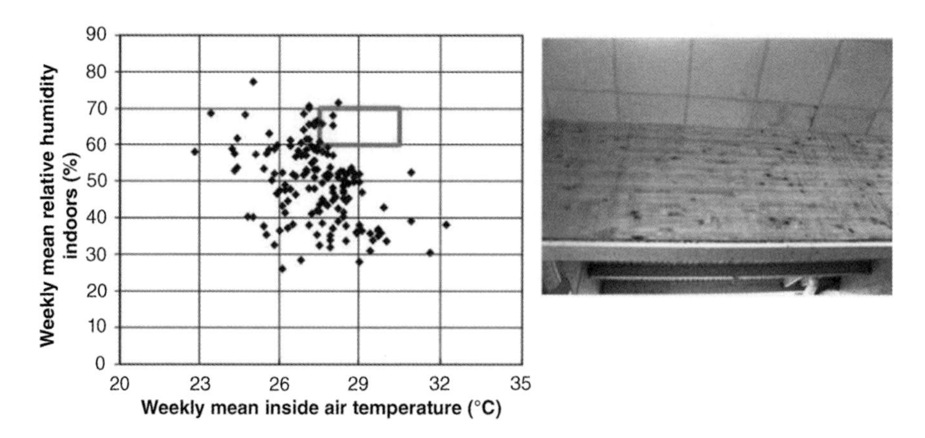

Figure 1.22 Natatoriums: left weekly mean relative humidity indoors in relation to the weekly mean indoor temperature with the rectangle giving the comfort zone, right interstitial condensate dripping off and humidifying the acoustical ceiling in a natatorium.

Table 1.27 Belgium, indoor climate classes ($\overline{\theta}_{e,m}$: gliding monthly mean outdoor temperature).

Indoor climate class		Upper pivot value $\Delta\overline{p}_{ie}$ in Pa $\overline{\theta}_{e,m}$<0? Set 0	Applies to:
1	$\overline{\theta}_{e,m} = 0$ $\overline{\theta}_{e,m} \geq 0$	150 $150 - 8.9\,\overline{\theta}_{e,m}$	All buildings with hardly any vapour release, except for short periods of time (dry storage rooms, sport arenas, garages, etc.)
2	$\overline{\theta}_{e,m} = 0$ $\overline{\theta}_{e,m} \geq 0$	540 $540 - 29\,\overline{\theta}_{e,m}$	Buildings with limited vapour release per m^3 of air volume and appropriate ventilation (offices, schools, shops, large dwellings, apartments)
3	$\overline{\theta}_{e,m} = 0$ $\overline{\theta}_{e,m} \geq 0$	670 $670 - 29\,\overline{\theta}_{e,m}$	Buildings with higher vapour release per m^3 of air volume, but still appropriately ventilated (small dwellings, hospitals, pubs, restaurants)
4		$>670 - 29\,\overline{\theta}_{e,m}$	Buildings with high vapour release per m^3 of air volume (natatoriums, breweries, several industrial complexes, hydrotherapy spaces)

1.3.2.3 Indoor climate classes

The difference in indoor/outdoor vapour pressure is such an important parameter for moisture tolerance that buildings have been grouped into indoor climate classes, using that difference as pivot. Table 1.27 and Figure 1.23 summarize the classes used in Belgium since 1982. The pivot values reflect following situations:

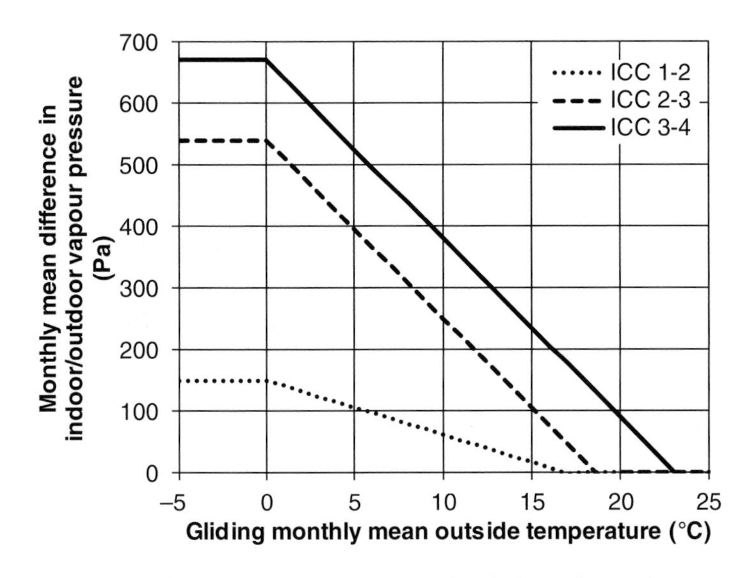

Figure 1.23 Belgium, pivots between the four indoor climate classes.

Table 1.28 Belgium, pivot values between the indoor climate classes recalculated (if $\Delta\overline{p}_{ie} < 0$, set 0).

Indoor climate class		Pivot (Pa)		Pivot (Pa)
1–2	$\overline{\theta}_{e,m} = 0$	87	$\overline{\theta}_{e,m} \geq 0$	$87 - 8.9\,\overline{\theta}_{e,m}$
2–3	$\overline{\theta}_{e,m} = 0$	550	$\overline{\theta}_{e,m} \geq 0$	$550 - 29\,\overline{\theta}_{e,m}$
3–4	$\overline{\theta}_{e,m} = 0$	1030	$\overline{\theta}_{e,m} \geq 0$	$1030 - 29\,\overline{\theta}_{e,m}$

Class 1/2 On a monthly mean basis vapour diffusion does not give interstitial condensation in a continuously shaded, airtight low-slope roof without vapour retarder, composed of non-hygroscopic materials and covered with a vapour tight membrane.

Class 2/3 On a monthly mean basis vapour diffusion does not give accumulating condensate in a north-orientated, non-hygroscopic, airtight wall without vapour retarder at the inside, finished at the outside with a vapour tight cladding.

Class 3/4 On a monthly mean basis vapour diffusion results in accumulating condensate in the roof, used for the class 1/2 pivot but now radiated by the sun.

In indoor climate class 4 a correct heat, air, moisture design of the enclosure is mandatory. In classes 1, 2 and 3, respecting simple design rules suffices. In the 1990s, the pivots were recalculated using transient modelling. The main result was an important shift of the pivot between indoor climate classes 3 and 4, see Table 1.28.

Other countries did the same exercise. Table 1.29 and Figure 1.24 summarize the results produced by Finland and Estonia. They used data collected in 101 dwellings spread over the two countries to come up with pivot values.

A Europe-wide adoption of the concept came with EN ISO 13788, introduced in 2001. The standard grouped buildings not into three or four but into five classes with pivot values as listed in Table 1.30 and shown in Figure 1.25.

Indoor climate classes lose usability as soon as airflow-driven vapour movement intervenes. The concept in fact was developed to evaluate the slow interstitial condensate deposit that diffusion induces, not to judge the much faster built-up convective vapour flow causes. Even indoor climate class 1 buildings may experience problems then.

Table 1.29 Finland and Estonia, indoor climate, pivot values (Δp_{ie}, $\overline{\theta}_{e,w}$: weekly means).

Indoor climate	Difference in indoor/outdoor vapour pressure (Pa)		
	$\overline{\theta}_{e,w} < 5\,°C$	$5 \leq \overline{\theta}_{e,w} \leq 15\,°C$	$\overline{\theta}_{e,w} > 15\,°C$
Low humidity load	540	$540 - 34\,\overline{\theta}_{e,w}$	200
Average humidity load	680	$680 - 41\,\overline{\theta}_{e,w}$	270
High humidity load	810	$810 - 47\,\overline{\theta}_{e,w}$	340

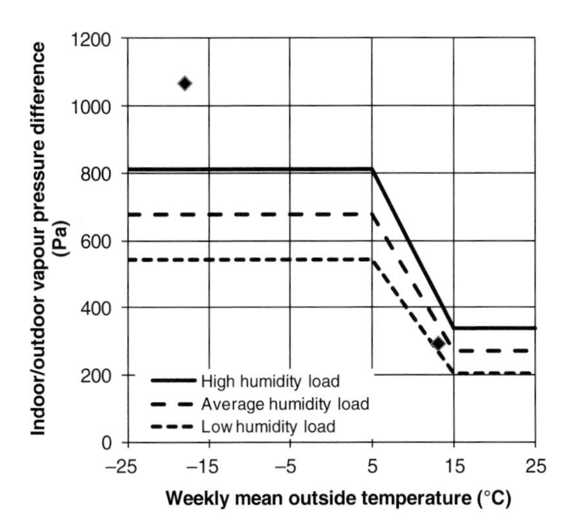

Figure 1.24 Finland and Estonia, indoor climate: differences in indoor/outdoor vapour pressure (the two individual dots represent averages measured in northern Canada).

Table 1.30 EN ISO 13788, indoor climate classes, pivot values.

Indoor climate classes	Annual mean $\Delta\overline{p}_{ie}$(Pa)	
	$\overline{\theta}_{e,m} < 0$	$\overline{\theta}_{e,m} \geq 0$
1–2	270	$270 - 13.5\,\overline{\theta}_{e,m}$
2–3	540	$540 - 27.0\,\overline{\theta}_{e,m}$
3–4	810	$810 - 40.5\,\overline{\theta}_{e,m}$
4–5	1080	$1080 - 54.0\,\overline{\theta}_{e,m}$

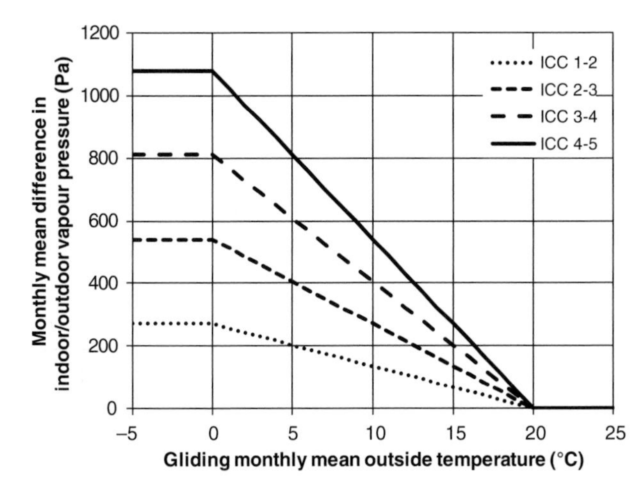

Figure 1.25 The indoor climate classes according to EN ISO 13788.

1.3.3 Indoor/outdoor air pressure differentials

Thermal stack, wind and fans induce air pressure differentials that act as driving force for air infiltration and exfiltration. Wind in turn also favours wind washing, while thermal stack is responsible for indoor air washing and air looping in and across envelope assemblies. Together with the air, non-negligible amounts of water vapour and enthalpy are displaced, causing the thermal transmittances and transient thermal response properties to lose significance; interstitial condensation to become more likely and bulky; end energy consumption to increase; draft complaints to emerge; and sound insulation to degrade.

In winter, thermal stack gives overpressure indoors. In warm weather, that becomes under-pressure when the indoor temperatures drop below that outdoors. Pressure differentials from wind change with wind speed and direction. They also depend on the pressure coefficients and leak distribution over the enclosure and the leak distribution indoors. When, for example, the enclosure is most permeable at the windward side, overpressure indoors will follow with the air leaving at the leeward side and at the sides more or less parallel to the wind. The pressures that forced air heating and mechanical ventilation give differ between spaces; some industrial buildings experience extreme values. In the supply plenum above the HEPA filter ceiling of a clean room laboratory an overpressure up to 150 Pa was measured. An overpressure of 300 Pa was noted in the top-floor of a brewery's hop kiln.

Where wind mainly induces transient differentials and those caused by mechanical ventilation are often negligible, except in very airtight buildings and in some industrial buildings, thermal stack is always active. Pressure differentials induced and related inflows and outflows may reach high values, as Figure 1.26 illustrates for a medium-rise

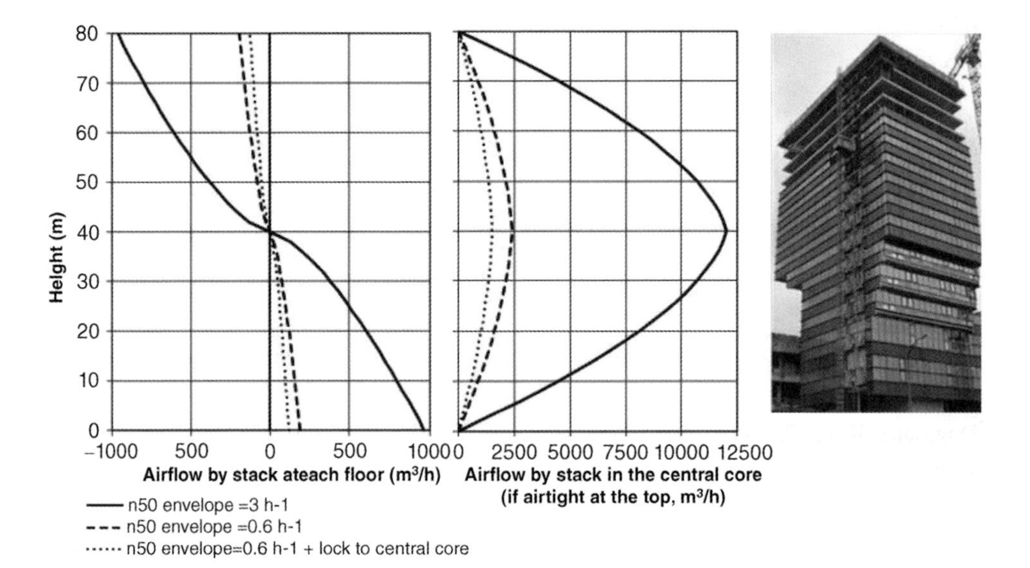

Figure 1.26 Medium-rise office building, air flow caused by thermal stack.

office building. Minimization demands specific measures, such as extreme airtightness of the envelope and an airtight lock around the central core.

Further reading

Anon (1983) *Draft European Passive Solar Handbook,* Section 2 on Climate, CEC.

Flemish government (2006) *Energy Performance Decree,* Text and Addenda (in Dutch).

ASHRAE (2009) *Handbook of Fundamentals*, Chapter 27, Climatic Design Information.

ASHRAE (2013) *Handbook of Fundamentals*, Chapter 27, Climatic Design Information.

Berdahl, P. and Fromberg, R. (1982) The thermal radiance of clear skies. *Solar Energy*, **29**(4), 299–314.

Best, A.C. (1950) The size distribution of rain drops. *Quarterly Journal of the Royal Meteorological Society*, **76**, 16–36.

Blocken, B. (2004) *Wind-driven rain on buildings measurement. Numerical modelling and applications*, PhD thesis, KU Leuven 323 pp.

Blocken, B., Carmeliet, J. and Hens, H. (1999) *Preliminary Results on Estimating Wind Driven Rain Loads on Building Envelopes: A Numerical and Experimental Approach*, Laboratory of Building Physics, KU Leuven, 78 pp.

Blocken, B., Carmeliet, J. and Hens, H. (2000) *On the use of microclimatic data for estimating driving rain on buildings*, Proceedings of the International Building Physics Conference, TU/e, pp. 581–588.

BRE (1971) *An index of exposure to driving rain*, digest 127.

CD with weather data from KU Leuven (1996 –2008) Laboratory of Building Physics.

Cornick, S. and Rousseau, M. (2003) *Understanding the Severity of Climate Loads for Moisture-related Design of Walls*, IRC Building Science Insight, 13 pp.

Cornick, S., Djebbar, R. and Dalgliesh, W. (2003) *Selecting Moisture Reference Years Using a Moisture Index Approach,* IRC Research Paper, 19 pp.

Costrop, D. (1985) *The use of climate data for simulating of thermal systems*, Rapport KU Leuven RD Energie (in Dutch).

Davies, J. and McKay, D. (1982) Estimating solar irradiance and components. *Solar Energy*, **29**(1), 55–64.

Dogniaux, R. (1978) *Recueil des données climatologiques exigentielles pour le calcul des gains solaires dans l'habitat et l'estimation de la consommation de l' énergie pour le chauffage des bâtiments*, KMI (Contract EG) (in French).

EN ISO (2001) 13788. *Hygrothermal performance of building components and building elements – Internal surface temperature to avoid critical surface humidity and interstitial condensation*, calculation method.

EN ISO (2002) 15927. *Hygrothermal performance of buildings*. Calculation and presentation of climatic data.

Geving, S. and Holme, J. (2012) Mean and diurnal indoor air humidity loads in residential buildings. *Journal of Building Physics*, **35**(2), 392–421.

Hartmann, T., Reichel, D. and Richter, W. (2001) Feuchteabgabe in Wohnungen – alles gesagt? *Gesundheits-Ingenieur*, **122**(4) 189–195 (in German).

Hens, H. (1975) *Theoretical and experimental study of the hygrothermal behaviour of building- and insulating materials by interstitial condensation and drying, with application on low-sloped roofs*, doctoraal proefschrift, KU Leuven (in Dutch).

IEA-Annex 14 (1991) *Condensation and Energy*, Sourcebook, ACCO Leuven.

ISO (1980) *Climatologie et industrie du bâtiment*, Rapport, Genève (in French).

Kalamees, T., Vinha, J. and Kurnitski, J. (2006) Indoor humidity loads in lightweight timber-framed detached houses. *Journal of Building Physics*, **29**(3), 219–246.

Kumaran, K. and Sanders, C. (2008) *Boundary conditions and whole building HAM analysis,* Final report IEA-ECBCS Annex 41 Whole Building Heat, Air, Moisture Response, Vol. **3**, ACCO, Leuven, 235 pp.

Künzel, H.M. (2010) Indoor Relative Humidity in Residential Buildings – A Necessary Boundary Condition to Assess the Moisture Performance of Building Envelope Systems.

Künzel, H.M., Zirkelbach, D. and Sedlbauer, K. (2003) *Predicting Indoor Temperature and Humidity Conditions including Hygrothermal Interactions with the Building Envelope*, Proceedings of the First Intenational Conference on Sustainable Energy and Green Architecture, Bangkok, 8–10 October 2003.

Laverge, J., Delgust, M. and Janssens, A. (2014) Moisture in bathrooms and kitchens: the impact of ventilation, *Proceedings of NSB 2014 Lund*, pp. 998–1005 (data stick).

Liddament, M.W. (1986) *Air Infiltration Calculation Techniques*, AIVC.

Poncelet, L. and Martin, H. (1947) *Main characteristics of the Belgian climate*, Verhandelingen van het KMI (in Dutch).

Sanders, C. (1996) *Environmental Conditions,* Final report IEA-ECBCS Annex 24 'Heat, Air and Moisture Transfer in Insulated Envelope Parts', Vol. 2, ACCO, Leuven, 96 pp.

Straube, J. (1998) *Moisture control and enclosure wall systems*, PhD thesis, University of Waterloo (Canada).

Straube, J. and Burnett, E. (2000) *Simplified prediction of driving rain on buildings*, Proceedings of the International Building Physics Conference, TU/e, pp. 375–382.

Ten Wolde, A. and Pilon, C. (2007) The Effect of Indoor Humidity on Water Vapor Release in Homes, *Proceedings of the Buildings X Conference, Clearwater Beach* (CD-ROM).

Uyttenbroeck, J. (1985) *Exterior climate data for energy consumption and load calculations*, Rapport WTCB RD Energie (in Dutch).

Uyttenbroeck, J. and en Carpentier, G. (1985) *Exterior climate data for building physics applications*, Rapport WTCB RD Energie (in Dutch).

Vaes, F. (1984) *Hygrothermal behaviour of lightweight ventilated roofs*, onderzoek IWONL KULeuven TCHN, eindrapport (in Dutch).

Van Mook, F. (2002) *Driving rain on building envelopes*, PhD thesis, TU/e, 198 p.

Wouters, P., L'Heureux, D. and Voordecker, P. (1987) *The wind as parameter in ventilation studies, a short description*, rapport WTCB (in Dutch).

WTCB (1969) *Study of the heat gains by natural illumination of buildings*, Part 1, Solar irradiation, Researchrapport 10 (in Dutch).

WTCB tijdschrift (1982) *Moisture response of building components*, 1 (52 pp.) (in Dutch and French).

2 Performance metrics and arrays

2.1 Definitions

The term 'performance metrics' applies to all physical qualities and properties of the built environment, whole buildings and building assemblies that are expressible in an engineering way, predictable at the design stage and controllable during and after construction. The difference between properties and qualities is subtle. While the first do not reflect a graded judgement, the second do. Area and thickness of an assembly are properties, its thermal transmittance a quality. A low value in fact is seen as beneficial, a high value as undesirable. The phrase 'in an engineering way' underlines the necessity of arriving at specific requirements and reference values. These can be imposed by law or put forward by the principal. Some are so obvious that they are considered as essential.

Characteristic of performances is their hierarchical structure, top down nature, with at the highest level the built environment (level 0), followed by the whole building (level 1), then all building assemblies – also called building parts – (level 2) and finally the materials and material layers composing the assemblies (level 3). A correct rationale starts with requirements and metrics at the levels 0 and 1, before descending to levels 2 and 3. This track gives designers the greatest freedom in making choices and presenting solutions.

2.2 Functional demands

A whole set of functional demands helps in establishing a fitness-for-purpose approach, from which performance formulations emanate. At the built environment level, the demands include accessibility for public transport, quality of the urban environment, proximity of shops and the parking infrastructure. At the whole building level we must consider flexibility, communication infrastructure, sanitary facilities, maintenance, source energy consumption, durability and sustainability, while in the direct environment factors such as ergonomics, thermal, acoustic, visual and olfactory comfort, indoor air quality and functionality come into play. Of course, form, space and architectural quality are also important assets that give buildings added value, but these should not offer designers reasons to deny the importance of a high performance.

2.3 Performance requirements

Requirements turn the sets of performance formulations that emanate from the functional demands into the metrics, expressed in an engineering way. At the design stage, built environments and buildings only exist on computer screens and drawings. The only means of ensuring that the performance requirements imposed are met, are past experience, calculation, computer simulation and sometimes model testing at reduced scale. Or, a performance-based design only makes sense to the extent that calculation tools and computer models exist, which allow predicting and evaluating the quality levels demanded.

Controlling, during and after construction, whether the requirements imposed are met, in turn requires well established measuring methods and protocols. For some things, such as floor or beam bending, airtightness, reverberation times, air-based sound and contact noise reduction and visual comfort, testing is quite simple and not very time consuming.

Applied Building Physics: Ambient Conditions, Building Performance and Material Properties, Second Edition. Hugo Hens.

Others, such as measuring thermal transmittances, end energy consumption for heating and cooling, testing several aspects related to moisture tolerance, demand time and often require complex methodologies. Prototype evaluation and quality guarantee have therefore replaced such controls. In any case, this presumes that a well-established agreement policy exists. For that purpose, certification institutes were set up at the national level, Europe-wide (e.g. EOTA, European Organisation for Technical Approvals) and worldwide (WFTAO, World Federation of Technical Assessment Organisations). EOTA delivers the Euromark for free trade of building materials and components between member states. WFTAO in turn fosters worldwide cooperation between continents, such as Europe, North America and Oceania.

2.4 A short history

Interest in performance-based design dates back to the 1970s. Driving forces were better knowledge and tools available in fields as structural mechanics, material science, building physics and building services, but it was also spurred on by a series of events, including the energy crises of that decade, complaints about sick buildings, the growing interest in sustainable construction, the search for more safety and a slowly growing concern about a global warming. Quality consciousness and insurance demands accelerated that evolution. But a few barriers still continue to hamper this process. Even today, specifications often remain too descriptive, simply imposing which materials to use and how to apply them. Several designers still only use the performance rationale of 'according to the art of construction'. This was acceptable in times when building was based on tradition and the range of different materials was limited, but today, with waves of new materials, new components, new construction systems and alternative solutions flooding onto the markets, nobody really knows what that art looks like now, neither does such a simplistic approach allow to enforce correct quality metrics by contract.

In 1979, an inter-industrial research team was the first to publish a 'Performance Guide for Buildings' in Dutch, which today can be seen as an early trial to produce coherent sets of performance requirements. In nine volumes, requirements were formulated at the whole building level, followed by requirements for facades, roofs, partition walls, floors, staircases and water distribution, waste water, heating, ventilation, electrics, transport and communication systems. The metrics applied were based on the functional demands advanced by ISO DP 6241:

1. Structural integrity
2. Fire safety
3. Safety in use
4. Tightness for water and air
5. Thermal comfort
6. Indoor air quality
7. Acoustic comfort
8. Visual comfort
9. Contact comfort
10. Vibratory comfort
11. Hygiene
12. Functionality
13. Durability
14. Economy

Missing from that list are sustainability and source energy use. The first was not really a pressing issue in the 1970s, while energy was considered an economic reality rather than a performance on its own with considerable environmental impact.

The international interest in a performance-based approach grew from the 1980s on. Within CIB (the International Council for Research and Innovation in Buildings) a working group W60, Performance Concept in Buildings, was established. A real breakthrough came with the European Construction Products Directive (CPD), which delineated six groups of functional demands:

1. Structural integrity
2. Fire safety
3. Health, hygiene, environment
4. Safety in use
5. Acoustic comfort
6. Energy efficiency

Building systems, components and materials have to comply with these demands and with related performance requirements before they can be traded freely between member states. In 2011, the CPD was reviewed, now under the name CPR.

A full application of a performance-based approach was first introduced in the Netherlands with their Building Decree, first published in 1992 and updated in 2000 and 2010. Meanwhile, Canada, Australia, New Zealand and other countries also redrafted their building regulations in a performance-based way.

2.5 Performance arrays

2.5.1 Overview

2.5.1.1 The built environment

Level 0, the built environment, with urban physics as one of the leading disciplines, has gained quite some interest in recent decades. Research has started into fields such as wind comfort, thermal comfort outdoors, including the impact of trees along streets, fountains on squares, shading and the kind of street paving, the urban heat island effect and how to neutralize its impact, air quality outdoors and the energy infrastructure in estates and cities – see Table 2.1 for an introductory array. Today, urban physics is a full-grown segment of building physics, taught at architectural, civil and building engineering departments worldwide.

Table 2.1 Level 0, the built environment, performance array.

Field	Performances
Functionality	Zoning, traffic organization and regulation (cars, public transport, bikes, pedestrians), traffic-free zones, parks and squares
Heat and mass	Urban heat island effect Thermal comfort outdoors during the cold half of the year Thermal comfort outdoors during the warm half of the year Wind comfort Air quality outdoors (pollution and the spread of contaminants and dust) Energy infrastructure and end energy use at district and urban level
Sound	Sound experience in the urban environment
Sustainability	CO_2 emissions, total GWG emissions, climate neutrality

2.5.1.2 Whole buildings and building assemblies

In 2000, the Annex 32 'Integral Building Envelope Performance Assessment' of the International Energy Agency (IEA), Executive Committee on Energy in Buildings and Communities, put forward quite complete performance arrays, covering level 1, whole buildings, as well as level 2, building assemblies, see Tables 2.2 and 2.3. Both arrays are based on the knowledge fields that shape building engineering.

Some of these performances are discussed in detail in Chapter 3. All have helped to establish the global evaluation schemes that have been worked out over the past few decades and are applied in many countries. Examples include LEED in the USA, BREEAM in the UK, DNGB in Germany, Miljöbyggnad in Sweden, CASBEE in Japan, IGBC in India, the 'Assessment of Office Buildings' tool in Belgium and, internationally, the Active House Alliance. Besides the performances in the table, most schemes also consider issues such as water usage, the materials used with recycling as an important aspect, waste policies, building location including accessibility by public transport and by bike, building efficacy and level of innovation. What importance each of these aspects are given varies between schemes. Their use has had a growing impact on the real estate market in the sense that buildings that score higher gain in financial value.

2.5.2 In detail

2.5.2.1 Functionality

Functionality coincides with what the CPD calls safety of use. Functionality has a major impact on floor layout, all room and auxiliary space dimensions and on the finishes used. A bedroom must be high enough to accommodate a wardrobe with an upper shelf. Balustrades should be 0.9 m high or more. A doorway must be at least 0.9 m wide and 2.1 m high – it can be wider and higher but that is left to the design intent of the architect. However, 1.2 m width is needed in patient rooms because beds have to be able to pass through. If needed, enough height must be left below windows to install radiators. A kitchen floor finish should not absorb fat and should remain non-slip even when wet. Trickle vents have to be burglar-safe and cleanable. For a detailed discussion on functional requirements, see the literature.

2.5.2.2 Structural adequacy

Buildings bear mechanical loads, which may cause problems, unless the three conditions below are met – and this applies to the foundations, the load bearing structure, all infill, the separate layers and all details. Examples of the last two are the wind resistance that a roof membrane needs and the strength that a point of suspension must have.

Constructions have to remain stable. Foundations and retaining walls must ensure sliding and settling equilibrium. A building must not slip or turn over and a facade panel has to remain vertical under extreme wind load. Roof girders should not bend excessively when bearing the design snow load, and so forth.

Table 2.2 Level 1, whole buildings, performance array.

Field		Performances
Functionality		Safety of use Adaptation to usage Flexibility
Structural adequacy		Global stability Strength and stiffness for vertical loads Strength and stiffness for horizontal loads Dynamic response
Building physics related	Heat and health	Thermal comfort year round, included avoiding overheating Indoor air quality
	Sound	Acoustic comfort *Room acoustics (reverberation time, speech intelligence, others)* *Overall sound insulation with emphasis on flanking transmission*
	Light	Visual comfort *Day lighting* *Artificial lighting*
	Heat and mass	Energy efficiency Moisture tolerance
	Fire[a)]	Fire containment by compartmentalisation and the right fire resistance of all enclosing floors and partition walls Escape routing Active fire fighting
Durability		Functional service life Economic service life Technical service life
Maintenance		Accessibility Ease of cleaning
Costs		Total and net present value Life cycle costs
Sustainability		Life cycle inventory and assessment Global evaluation schemes

a) In the Netherlands and Germany, fire safety is seen as a part of building physics. Not so in Belgium and the Scandinavian countries, where fire safety is considered a separate knowledge field. Even more, as is the case for what the Anglo-Saxon world calls ‘building science’ the term ‘building physics’ often contains the heat and mass part only, with energy as a logical appendix. Acoustics and lighting then are dealt with totally separately.

Stresses should remain below allowed and fracture safety should be as imposed. Own weight, dead weight, realistic combinations of live load, static and dynamic wind load, snow load and accidental loads are the external forces intervening. Control starts with calculating the internal normal and shear forces, the bending and torsion moments. These are then

Table 2.3 Level 2, building assemblies, performance array.

Field		Performance
Structural adequacy		Strength and stiffness against vertical loads Strength and stiffness against horizontal loads Resistance against buckling Dynamic response
Building physics related	Heat and mass	Airtightness *Infiltration, exfiltration* *Wind washing, indoor air washing* *Venting and ventilation* *Air looping* Thermal insulation *Thermal transmittance (U) (inclusive thermal bridging)* *Thermal transmittance of transparent parts* *Average thermal transmittance of the envelope* Transient response *Dynamic thermal resistance,* *temperature damping, admittance* *Solar transmittance* Moisture tolerance *Construction moisture* *Rain-tightness* *Rising damp* *Hygroscopic loading* *Surface condensation* *Interstitial condensation* *Random leaks* Thermal bridging *Temperature factor* Other (i.e. the contact coefficient)
	Sound	Sound attenuation index Sound insulation of the envelope against noise from outside Sound absorption
	Lighting	Light transmittance of transparent parts Glass percentage in the envelope
	Fire[a)]	Fire reaction of the materials used Fire resistance
Durability		Resistance against physical attack (mechanical load, moisture, temperature, frost, UV-radiation, salt, and others) Resistance against chemical attack (corrosion, etc.) Resistance against biological attack (moss and algae growth)
Maintenance		Resistance against soiling, easiness of cleaning
Costs		Total and net present value
Sustainability		Life cycle profiles

a) See the remark under Table 2.2.

transposed into stresses and compared with what is allowable or judged whether the values found are safe looking to the overall fracture equilibrium.

Buildings must keep their serviceability. This concerns the compressibility of the soil with consequent differential settling and the overall loadbearing structure's stiffness. Allowable deformations in buildings are usually small and depend among others on the type and function of the finishes, which are stricter, for example, for a floor with plastered rather than hung ceiling.

Apart from this, there are also a few other structural adequacy requirements that have to be met. The dynamic acceleration, for example, has to remain acceptable, which defines among others the usability of high-rises.

2.5.2.3 Building physics related quality

This fits within the CPD criteria about health, hygiene, environment, acoustic comfort and energy efficiency. Building physics manages the hygrothermal, acoustic and visual performances and related metrics and requirements – see the Tables 2.2 and 2.3 – and for the discussion in detail of the hygrothermal package, see Chapter 3 on whole buildings and Chapter 4 on building assemblies. For acoustics and light refer to the literature and to volumes 3 and 4 on performance-based building design.

2.5.2.4 Fire safety

Fire has such detrimental consequences for buildings and their users that everyone hopes such events will never happen during the lifetime of a building but for which at design and construction a probability 1 is imposed by law. So, all precautions must be taken to minimise the possible consequences, although eliminating all risk is neither doable nor affordable. The requirements to consider are at three levels: materials and movables, building parts and the whole building. It is the materials and movables that burn. Correctly designed building parts have to keep the fire localized. Issues of structural safety need extra attention. Adequate compartmentalisation and means for active firefighting must help to curtail the fire, allow safe evacuation and facilitate firefighters' access. All have as primary objective guaranteeing a long enough evacuation time and as a second, limiting damage.

For materials, basic is their fire behaviour with a distinction between burnable and non-burnable. As burning is an exothermal oxidation reaction, non-burnable materials don't oxidize or at least do it so slowly that the heat flow released is too small to cause problems. In the overall classification the following aspects must be considered: flammability, flame spread, heat release, smoke formed and burning dripping. The European standard has seven classes, A1 and A2 for non-burnable, B, C, D and E for stepwise increasingly burnable and F for non-tested.

For building parts, fire resistance is the defining property, a number expressing the time in minutes between the start of a standardized fire and the moment that one of the following happens: temperature at the part's side away from the fire becomes high enough to allow fire propagation by radiation; the part is no longer smoke- and flame-tight; the part losing structural integrity. The three criteria are controlled according to the governing standards, and they result in a specific fire resistance value for each part. For non-loadbearing walls, structural integrity is skipped, whereas for columns temperature and tightness demands are not included.

At the whole building level, it is important to guarantee overall stability for a long enough period of time to assure that building users can leave along fire-safe routes, to limit flame and smoke development, to prevent fire-spread to neighbouring buildings and to allow safe firefighting. The metrics include fire compartmentalisation per floor with areas not more than 2500 m^2 connected to safe escape routes and having partition walls and floors in between with fire resistance 90–120′. They also fix fire-free distances to windows on the next floor, impose the widths of staircases along escape routes and specify where to hang fire extinguishers and to install sprinklers.

The severity of the requirements depends on the building height, the fire load per square metre, the users' familiarity with the building and the ease of evacuation. Today, fires and their consequences are often simulated numerically during design, using CFD-based computer tools. For this one has to know the fire load, its division, the dimensions of the compartments, all the properties of the partitions, the mechanical characteristics of the structure and the ventilation flows.

2.5.2.5 Durability

The basic requirement is correct moisture tolerance. Of course, also other elements come into play:

Material	Other elements
Concrete	Correct water/cement factor, well compacted, steel bars correctly concrete covered, limiting chloral ions present
Stucco	Sufficient ductility, fatigue resistant
Stone	Salt attack resistant
Bituminous membranes	UV resistant, fatigue resistant
Synthetics	UV resistant

Some assemblies have specific requirements. For operable windows and doors, for example, the hinges and locks must go on functioning properly even after ten thousands times being opened and closed, while door leafs must not start protruding.

2.5.2.6 Maintenance

No one doubts that easy maintenance is important, but 'easy' involves accessibility, design and finishes chosen. Examples are spaces where strict hygienic and immunologic rules prevail, such as operating theatres, clean rooms in the pharmaceutical industry and haematological centres. Surface layers should be hard with high resistance to puncturing. Finishes must get easily disinfected, so are watertight and not attacked by disinfectants. Other issues include corners rounded off, electrical switches and plug sockets in watertight mounts and pipes passing through shafts.

Further reading

ASHRAE (2011) *Handbook of HVAC Applications*, chapter 43.

ASHRAE (2014) *Handbook of HVAC Applications*, chapter 43.

Becker, R. (2008) *Fundamentals of Performance-based Building Design, Building Simulation*, Volume 1, Number 4.

Blocken, B. (2015) *Computational Fluid Dynamics for Urban Physics: Importance, possibilities, limitations and ten tips and tricks towards accurate and reliable simulations, Buildings and Environment* Golden Issue to celebrate 50 years of the journal.

Bouwbesluit (1991) *Staatsblad, 680, Staatsdrukkerij*, Den Haag (aangepaste versie: 2000) (in Dutch).

CERF (1996) *Assessing Global Research Needs*, Report #96-5061, Washington DC.

CERF (1996) *Construction Industry: Research Prospectuses for the 21th Century*, Report #96-5016, Washington DC.

CIB, *WG 'Performance Concept in Buildings'*, Congress Proceedings.

Hendriks, L. and Hens, H. (2000) *Building Envelopes in a Holistic Perspective*, Final Report IEA, EXCO ECBCS Annex 32, ACCO, Leuven.

Hens, H. (1996) *The performance concept: a way to innovation in construction*, Proceedings of the CIB-ASTM-ISO-RILEM 3rd International Symposium on Application of the Performance Concept in Buildings, Tel-Aviv, December 9–12, pp. 5-1–12.

Hens, H. and Rose, W. (2008) *The Erlanger House at the University of Illinois – A Performance-based evaluation*, Proceedings of the Building Physics Symposium, Leuven, October 29–31, pp. 227–236.

IC-IB (1979) *Prestatiegids voor gebouwen*, 9 delen (in Dutch).

Lovegrove, K., Borthwick, L. and Bowen, N. (1995) *New Draft Performance Based Building Code Based on a Pyramid of Principles*, Australian Building Codes Board News, September, pp. 6–9.

Lstiburek, J. and Bomberg, M. (1996) *The performance linkage approach to the environmental control of buildings*, Part I, construction today. *Journal of Thermal Insulation and Building Envelopes*, **19**, 244–278.

Lstiburek, J. and Bomberg, M. (1996) *The performance linkage approach to the environmental control of buildings*, Part II, construction tomorrow. *Journal of Thermal Insulation and Building Envelopes*, **19**, 386–402.

Ollson, D. (2013) Wide variation in how parameters are regarded in environmental certification systems. *The REHVA European HVAC Journal*, **50/3**, 92–93.

Vitse, P., Vandevelde, P. and Thylde, J. (2003) *Europese testmethoden en classificatie van de brandreactie van bouwproducten*, WTCB-tijdschrift, 2^{e} trimester, pp. 27–36 (in Dutch).

3 Whole building level

Chapter 1 analysed the ambient loads that buildings experience. In Chapter 2 the focus was on performances, with the advancement of three arrays, a first at the built environment level (0), a second at the whole building level (1), and a third at the building parts and assemblies level (2). This present chapter concentrates on six level 1 performances: thermal comfort, health and indoor air quality, energy efficiency, durability, economics and sustainability.

3.1 Thermal comfort

3.1.1 General concepts

Feeling comfortable can be defined as a state of mind expressing satisfaction with the ambient. A prerequisite is healthiness and satisfaction with the living, housing and working conditions. Then, a series of parameters must have values that satisfy the individual. These are air temperature, radiant temperature, air speed and relative humidity for thermal comfort, sound pressure levels for acoustic comfort, luminance, glare and light colour for visual comfort and odours for olfactory comfort. Evaluating an indoor space in terms of comfort presumes knowledge of the impact of each of these parameters. Relationships are well established for thermal, acoustic and visual comfort, but less so for olfactory comfort, and most research indicates that dissatisfaction with one is hardly compensated by adjusting the conditions for the other three.

Of the four mentioned, thermal comfort is the most critical, as satisfaction is directly linked to human physiology, although adaptation plays an important role. Each of the other three involves one of the senses. Although lacking one due to long-lasting overload or injury is highly undesirable, it is less threatening than sustained adverse thermal conditions.

3.1.2 Physiological basis

Human sensitivity to the thermal environment originates from two physiological realities shared by all warm-blooded creatures: being exothermic and homoeothermic.

3.1.2.1 Exothermic

Humans need energy for their metabolism (M), the sum of all chemical activities enabling cell growth, dead cell replacement, respiration, heart and liver activity, digestion, brain activity and others. This forces humans to eat regularly. If not, the body sustains metabolism by first consuming its fat reserves and then the muscle tissue. The blood takes up the heat that these activities produce (Φ_M in W) and helps to transmit it to the environment. What's called basic metabolism ($M_o \approx 73$ W) relates to the energy that a 35-year-old male, 1.7 m tall, weighting 70 kg, needs when sleeping in a thermally neutral environment, 10 h after his last meal. A metabolic rate (M_A) of 58 W per unit body area is named 1 met. On waking up, metabolism rises. Becoming active gives a further increase with peaks at high activity, such as when playing sports. For an overview of the rates linked to different activities, see Table 3.1.

Applied Building Physics: Ambient Conditions, Building Performance and Material Properties, Second Edition. Hugo Hens.

Table 3.1 Metabolic rates per unit body area (W/m^2).

Activity	Air speed m/s	Metabolic rate (M_A), W/m^2	Heat (q_M) W/m^2	Labour (p_M) W/m^2
Rest				
Sleeping	0	41	41	
Lying	0	46	46	
Sitting	0	58 (= 1 Met)	58	
Standing	0	70	70	
Light activity				
Laboratory	0	93	93	
Teaching	0	93	93	
Car driving	0	58–116	58–116	
Cooking	0	96–116	96–116	
Typing	0	62–68	62–68	
Studying	0	78	78	
Normal activity				
Walking on flat terrain				
3.2 km/h	0.9	116	116	
4.8 km/h	1.3	150	150	
Climbing				
3.2 km/h	0.9	174	156	18
4.8 km/h	1.3	232	206	26
Assembling	0.05	128	128	0
Dancing	0.2-2.0	139–255	139–255	0
Heavy activity and sports				
Spadework	0.5	348	279	69
Tennis	0.5–2.0	267	240–267	0-40
Basketball	1.0–3.0	441	397–441	0-44

As indicated in the table, physical activity turns part of the metabolism into labour (P_M in W), so the metabolic equilibrium can be written as:

$$M = P_M + \Phi_M \qquad \text{(W)}$$

With η_{mech} the mechanical efficiency of the body, the labour P_M is also written as:

$$P_M = \eta_{mech} M$$

The result is a metabolic heat flux (q_M), given by:

$$q_M = (1 - \eta_{mech})M/A_{body} = (1 - \eta_{mech})M_A \qquad (W/m^2) \tag{3.1}$$

where A_{body} is the body area in m^2, a function of body mass (m_{body} in kg) and stature (L_{body} in m):

$$A_{body} = 0.202 m_{body}^{0.425} L_{body}^{0.725} \tag{3.2}$$

For a male adult, 1.7 m tall and weighing 70 kg, the formula gives as area some 1.8 m^2.

Between metabolic heat flux and labour the relations are:

$$q_M = \frac{1}{A_{body}}\left(\frac{1 - \eta_{mech}}{\eta_{mech}}\right)P_M \qquad P_M = A_{body}\left(\frac{\eta_{mech}}{1 - \eta_{mech}}\right)q_M \tag{3.3}$$

3.1.2.2 Homoeothermic

The body core temperature must stay close to 36.8 °C, a value that rises a little after eating, excessive alcohol consumption or heavy activity and is subject to a daily cycle, lowest in the morning and highest in the afternoon. It will only keep that 'set' value when on average the metabolic heat equals the heat lost, or:

$$(1 - \eta_{mech})M = \sum \Phi_j \tag{3.4}$$

with $\Sigma\Phi_j$ the sum of all heat flows among body and ambient.

3.1.2.3 Autonomic control system

Without feedback between metabolism and heat loss, equilibrium would be purely accidental, which is why humans possess an autonomic control system with the hypothalamus as relay, which balances metabolism and heat loss. For this, the hypothalamus knows the set temperature and receives temperature signals from the skin, the spinal cord and the mucous membranes in nose and bronchi. From these it distils a real temperature, which can be measured at the tympanum in the ear. The difference between set and real then stimulates a series of physiologic actuators. If positive, meaning that the ambient is too cold for a given metabolism, the subcutaneous blood stream slows down (it may drop from 25 ml/(m^2.s) to 1.7 ml/(m^2.s)), thus diminishing the heat loss through the skin, while shivering and chattering will increase metabolism. The subcutaneous blood stream, of course, cannot decrease indefinitely. Once the real core value is below 28 °C, progressive hypothermia develops, first in the feet, hands, ears and nose, but finally attacking all the vital organs and the brain. If instead the difference is negative, meaning that the ambient is too hot for a given metabolism, the subcutaneous blood stream is boosted, while sweating increases the sensible and latent heat loss. The larger the difference between real and set value, the more the skin is covered with sweat. People also feel languid. Again, there is a limit. Once the real core value passes 46 °C, hyperthermia develops with irreversible brain damage as the probable consequence.

Thanks to this autonomic control system, humans survive under thermally quite extreme conditions. Too intense control interventions, however, are not perceived as comfortable. Absolute thermal comfort therefore is limited to situations where metabolism and heat loss are equilibrated without control action, which is only achieved at rest. Each activity stimulates control, resulting in a relative thermal comfort, characterized by limited subcutaneous blood stream and sweating changes. Nonetheless, this will be felt as satisfactory.

3.1.3 Steady state thermal comfort, the physiology based approach

The steady state comfort model, which assumes constant ambient conditions, was developed by the late P.O. Fanger. The basic assumption is that the clothed body behaves as if at uniform surface temperature in a thermally homogeneous environment, where air temperature and relative humidity maintain the same value everywhere, where all surfaces have the same radiant temperature and where air speed remains the same all around. All this allows human physiology to be translated into a system of algebraic equations with activity and clothing as parameters.

3.1.3.1 Clothing

Clothing increases the thermal resistance between skin and environment, thus limiting heat loss or gain, keeping skin temperature up and allowing us to stay comfortable in too cold an environment for the activities being engaged in. The units used for its thermal resistance are clos, symbol I_{clo}, the ratio between the thermal resistance of any garment and a typical business dress (underwear, shoes, socks, straight trousers, shirt, tie and vest), where 1 clo equals 0.155 m^2.K/W. Table 3.2 gives values for various garments and types of dress The numbers listed could differ from those given by other sources because of the way clos are measured. Testing requires hours in a climate chamber using thermal manikins, which differ slightly between institutes and are never a perfect replica of humans. The clo-value of any dress is assumed equal to the sum of the clo-values of the separate garments:

$$I_{clo,T} = \sum I_{clo,j} \tag{3.5}$$

Sitting in a chair or seat counts as an additional garment with clo-value:

$$I_{clo} = 0.748\, A_{ch} - 0.1 \tag{3.6}$$

where A_{ch} is the contact area in m^2 between the chair or seat and the human body.

3.1.3.2 Heat flow between body and ambient

Going back to Eq. (3.4), the question is what heat flows (Φ) are there between the body and ambient? First comes sensible heat lost or gained by convection and radiation. Things start with the heat transmitted across the clothing:

$$q_{cl} = \frac{\theta_{skin} - \theta_{cl}}{0.155\, I_{clo,T}} \tag{3.7}$$

Table 3.2 Clo-values.

Garments and clothing	I_{Clo}	f_{cl}
Nude	0	1
Briefs	0.14	1
Long underwear top, long underwear bottom	0.35	1
Bra	0.01	1
Undershirt	0.08	1
Socks	0.02	1
Tights	0.03	1
Short or swimming trunk	0.1	1
Office chair (is an extra, add to the garment)	0.15	1
Bed clothes (mattress, sheets and blankets)	>1.6	1
Short, summer shirt, socks, sandals	0.3–0.4	1.05
Light summer clothing (straight trousers, shirt)	0.5	1.1
Female dress (bra, slip, tights, skirt, shirt,)	0.7–0.84	1.1
Typical business dress	1.0	1.15
Typical business dress with raincoat	1.5	1.16
Polar outfit	>4	—

where θ_{skin} is the uniform skin and θ_{cl} the uniform surface temperature of the clothed body. In practice, average temperatures replace uniformity. At the surface of the cloths, transmission is divided into convection to the air (q_c) and radiant exchanges with all surrounding surfaces (q_R). The convective part is:

$$q_c = h_c f_{cl}(\theta_{cl} - \theta_a) \tag{3.8}$$

with θ_a the uniform air temperature and h_c the mean convective surface film coefficient along the clothed body. The multiplier f_{cl} accounts for the larger surface with clothing compared to nude (see Table 3.2, third column). For natural convection the surface film coefficient equals:

$$h_c = 2.4(\theta_{cl} - \theta_a)^{0.25} \tag{3.9}$$

With forced convection the expression becomes:

$$h_c = 12\sqrt{v_a} \tag{3.10}$$

with v_a the mean relative air velocity along the clothed body including any body movement. The radiant part is written as:

$$q_R = 5.67 e_L F_T f_R f_{cl}(\theta_{cl} - \theta_r) \tag{3.11}$$

with f_R a reduction factor as parts of the clothed body see each other, e_L the average longwave emissivity of the clothed body and θ_r the radiant temperature representing the ambient. For a standing individual f_R is 0.73, for a sitting one 0.7.

Convection, radiation and linked transmission give gains or losses at the skin: gains when the difference between the weighted average of air and radiant temperature $(a\bar{\theta}_a + (1-a)\theta_R)$ and the mean skin temperature is positive, losses when negative. For thermal comfort outdoors, solar radiation and longwave losses to the sky must be accounted for.

Second is sensible loss or gain by respiration. The sensible part of the heat per m^2 of body exchanged in this way starts from the relation between exhaled air ($G_{a,b}$ in kg/s) and metabolism:

$$G_{a,b} = 1.43 \times 10^{-6} M_A A_{skin}$$

This gives as heat flux (c_a specific heat capacity of air):

$$q_{bs} = \frac{\Phi_{a,b}}{A_{skin}} = c_a G_{a,b}(34 - \theta_a) = 1.43 \times 10^{-6} c_a M_A (34 - \theta_a) \tag{3.12}$$

Turning to latent heat loss, first come perspiration and sweating, both of which induce a latent heat exchange:

$$q_{PS} = l_b(0.06 + 0.94 f_w) \underbrace{\left(\frac{p_{sat,skin} - p}{\frac{1}{f_{cl}\beta} + Z_{clo,T}} \right)}_{B} \tag{3.13}$$

with f_w a factor quantifying the skin fraction moistened by sweat, zero at a metabolic flux below 58 W/m^2 and a value proportional to the ratio between the amount of sweat and the term B beyond 58 W/m^2, l_b heat of evaporation for water, ≈2500 kJ/kg, p the vapour pressure in the surrounding air in Pa, $p_{sat,skin}$ the mean vapour saturation pressure at the skin in Pa, β the surface film coefficient for diffusion at the clothed body surface in s/m and $Z_{clo,T}$ the average diffusion resistance of the clothing in m/s, zero where sweat wets the garments. The vapour flux in kg/(m^2.s) lost by sweating follows from the ratio between the latent heat exchanged and the heat of evaporation.

Then comes the latent part in the heat exchanged by respiration (W/m^2):

$$q_{bl} = 3.575\, M_A \left(0.033 - 6.21 \times 10^{-6} p\right) \tag{3.14}$$

with 0.033 the water vapour ratio in the exhaled air in kg/kg.

3.1.3.3 Comfort equations

Looking again at Eq. (3.4), dividing the metabolic heat by the skin area, transposing all heat fluxes just discussed into the formula, moving the latent and sensible heat fluxes by respiration to the left because they do not link to the garments worn, and equalizing the transmission flux to convection and radiation gives:

$$\begin{aligned}(1-\eta_{\text{mech}})M_{\text{A}} \quad & -0.143\times 10^{-2}M_{\text{A}}(34-\theta_{\text{a}})-3.575\,M_{\text{A}}\left(0.033-6.21\times 10^{-6}p\right)\\ & -l_{\text{b}}(0.06+0.94f_{\text{w}})\left(\frac{p_{\text{sat,skin}}-p}{\dfrac{1}{f_{\text{cl}}\beta}+Z_{\text{clo,T}}}\right)=\frac{\theta_{\text{skin}}-\theta_{\text{cl}}}{0.155\,I_{\text{clo,T}}}\\ & =f_{\text{cl}}\left[h_{\text{c}}(\theta_{\text{cl}}-\theta_{\text{a}})+e_{\text{L}}5.67\,F_{\text{T}}f_{\text{R}}(\theta_{\text{cl}}-\theta_{\text{r}})\right]\end{aligned} \tag{3.15}$$

Obeying this equation is a prerequisite for a comfortable steady state ambient, though it also requires acceptable sweating fluxes (g_{sw} in kg/(m^2.s)) and an allowable average skin temperature:

$$g_{\text{sw}}=\max\left\{1.68\times 10^{-7}[(1-\eta)M_{\text{A}}-58],0\right\} \qquad \theta_{\text{skin}}=35.7-0.0276(1-\eta)M_{\text{A}}$$

The three equations together allow to evaluate overall steady state thermal comfort.

3.1.3.4 Comfort parameters and variables

The comfort equation above contains two human parameters, activity and clothing, the latter with its clo-value, longwave emissivity and vapour resistance, and four ambient variables: air temperature, radiant temperature, relative humidity and air velocity relative to the moving human body. Actually, it is not relative humidity but vapour pressure in the air that is the third variable, although as air temperature already has an effect, relative humidity is a perfect substitute. Air velocity fixes the convective surface film coefficient for forced convection, while for natural convection temperature difference between clothed body and surrounding air suffices.

In theory, all combinations of these four environmental variables that satisfy the comfort equations define environments that should be experienced as comfortable for given activity and clothing. Since the system only has three equations for four variables, the number of comfortable environments for any combination of clothing and activity looks infinite.

3.1.3.5 Thermally equivalent environments and comfort temperatures

Manipulating four variables in three equations is complex. Therefore, since thermal comfort research started, people involved have tried to replace the four by one, using the concept of 'thermally equivalent environment'. As mentioned, for a given activity and clothing, an infinite number of comfort situations exist. All are equivalent, though infinite must include environments where some variables have a given value and the others a fixed value. Assume an ambient with relative air velocity zero, radiant and air temperature equal and relative humidity 100%, or the wet bulb equal to the dry bulb temperature. The only unknown then is that common temperature value, called the operative temperature, symbol θ_o. Equal operative temperature thus characterizes all environments perceived as comfortable at given activity and clothing. Its numerical value follows from solving the system of comfort equations for given metabolism and clothing. Of course, other rules could be put forward. If for example relative humidity is set at 50%, the effective temperature, symbol θ_{eff}, comes out as the comfort characteristic.

In practice, thermal comfort is most critical in environments where people perform light manual or intellectual labour. Temperature fluctuations, if any in such ambient, remain moderate, and slower than the time needed for the body to adapt, while in temperate climates clothing with longwave emissivity around 0.9 hardly reaches a clo-value 1.2. The air temperature is typically 15–27 °C, whereas the radiant temperature is mostly less the 8 °C lower. Relative humidity is 30–70%, while the relative air velocity remains below 0.2 m/s. In such cases, respiration, perspiration and sweating dissipate some 30% of the metabolic heat. This turns convection and radiation into the only adjustable heat fluxes, which simplifies the main comfort equation to ($F_T = 0.98$, $f_R = 0.7$):

$$0.7(1 - \eta_{mech})M_A = f_{cl}[h_c(\theta_{cl} - \theta_a) + 3.5(\theta_{cl} - \theta_r)] \tag{3.16}$$

Assuming the air and radiant temperature are equal to the operative one, the expression reduces to:

$$0.7(1 - \eta_{mech})M_A = f_{cl}[h_c + 3.5](\theta_{cl} - \theta_o) \tag{3.17}$$

Both together give as operative temperature:

$$\theta_o = \frac{h_c\theta_a + 3.5\theta_r}{h_c + 3.5} \tag{3.18}$$

Or its value is given by the respective surface film coefficients weighted average of the air and radiant temperature. With a convective surface film coefficient between $2.4(6)^{0.25}$ and $2.4(14)^{0.25}$ for a 6–14 °C difference between clothed body and air, that weighted average hardly differs from the arithmetic one:

$$\theta_o = \frac{(3.8 \text{ to } 4.6)\theta_a + 3.5\theta_r}{(3.8 \text{ to } 4.6) + 3.5} \approx 0.55\theta_a + 0.45\theta_r \approx 0.5(\theta_a + \theta_r) \tag{3.19}$$

The air and radiant temperature are thus of equal importance. A higher value of the one compensates a too low value of the other. What comfort value the operative temperature should have follows from the equations above. In winter, 21 °C seems optimal for an individual who is moderately active and whose dress has a clo-value 1.2, whereas in summer 26 °C remains comfortable for a clo-value down to 0.6–0.8. Of course, the operative temperature can only be used as comfort parameter in environments where the conditions listed above are met. If not, one has to return to the complete set of comfort equations. Such exercises show that relative humidity is the least important ambient variable. Only beyond 25 °C does a too high value feel sultry. Of course, its impact is broader than thermal comfort. Humans perceive humid air as less fresh. Another fact that such exercises show is why, in hot climates, ceiling fans improve thermal comfort. Increased air velocity compensates for a too high operative temperature.

3.1.3.6 Comfort appreciation

The physiological thermal comfort approach suggests that individuals should react equally. Not so. Several studies suggest that women are less tolerant to cold than men. They express more dissatisfaction – a difference manifest in overcooled offices in summer. Several factors are involved here, among them clothing. Compared with men's button-down shirts and jackets, the clothing that women wear in summer tends to be lighter and often exposes cold-sensitive areas, such as the back of the neck and the ankles. Also children, adults and the elderly appreciate comfort differently. That some are skinny and others thin, that a percentage have an ideal body mass index, that quite a number are corpulent, produces variations, as do subtle differences in core temperature, a more or less sensitive autonomic control system and past experiences. Illness even impairs the hypothalamus, while extreme activity restrains its functionality.

These differences emerge when comfort appreciation is at stake. For a given activity and dress, ambients that are 'comfortable' according to the mathematical model are not perceived as such by individuals. On the contrary, a percentage judge them too cold, much too cold, too warm or much too warm. This turns the appreciation into a statistical reality. An approach, developed by P.O. Fanger, combines three concepts.

The load (L) stands for the difference between metabolic heat and heat lost:

$$L = (1 - \eta_{\mathrm{mech}})M - \sum \Phi_j \tag{3.20}$$

The higher it is, the more the conditions drift away from comfortable. A negative load marks environments colder than, a positive one warmer than desired, while zero load fits with the comfort equations for the activity and dress given.

The predicted mean vote (PMV) represents the perceived thermal comfort according to the sensation scale of Table 3.3. The concept has been evaluated through laboratory experiments involving thousands of individuals working in simulated office environments. The results enabled the following statistic relationship between load and vote, now called predicted mean, to be advanced:

$$PMV = [0.303\,\exp(-0.036\,M_{\mathrm{A}}) + 0.28]L \tag{3.21}$$

If calculation gives a $PMV > 3$ or $PMV < -3$, then the value is set at 3 or -3, respectively.

Table 3.3 Thermal sensation scale.

Vote	Thermal sensation
3	Much too warm
2	Too warm
1	Somewhat too warm
0	Neutral
−1	Somewhat too cold
−2	Too cold
−3	Much too cold

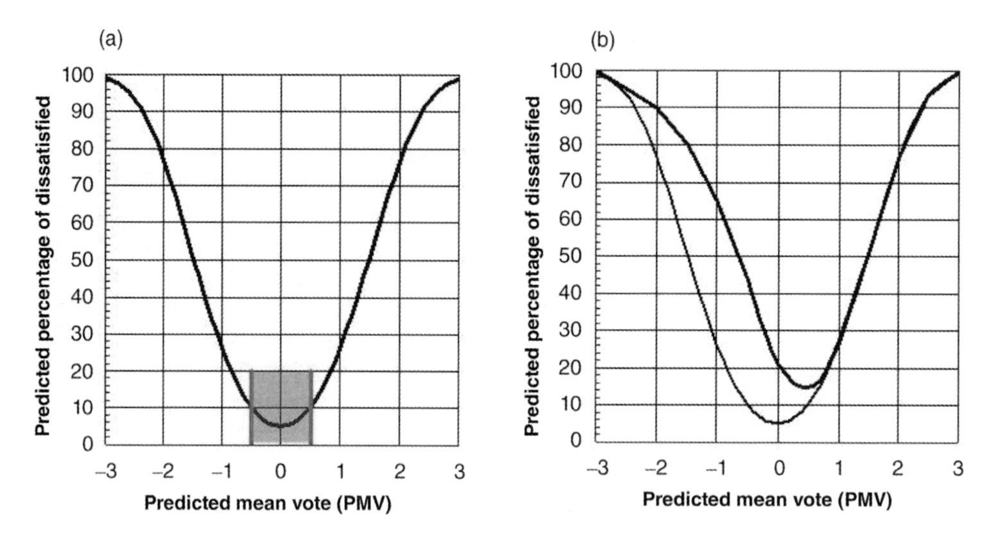

Figure 3.1 (a) PPD versus PMV as proposed by P.O. Fanger and included in ISO 7730. (b) PPD versus PMV according to Mayer (Fraunhofer Institut für Bauphysik)).

Finally, using the test results, a relation was formed between the predicted mean vote and the percentage of dissatisfied, again referred to as predicted. Dissatisfied were those wearing a standard dress with known clo-value, doing office work and judging the conditions too warm, much too warm, too cold or much too cold (Figure 3.1):

$$PPD = 100 - 95\exp\left[-\left(0.03353\, PMV^4 + 0.2179\, PMV^2\right)\right] \tag{3.22}$$

Even at *PMV* zero, 5% remain statistically dissatisfied, 2.5% quoting the conditions as too warm or much too warm and 2.5% as too cold or much too cold. The curve looks symmetric, meaning that people react equally when warmer or colder. In any case, as well, the 5% dissatisfied at *PMV* zero as the symmetry is questioned. Some researchers claim higher percentages with *PPD* 15% at *PMV* 0.5 and more negative than positive dissatisfied (Figure 3.1).

PPD below 10% is judged acceptable. According to Fanger, this correlates with *PMV* –0.5 to +0.5. For Mayer, less than 10% dissatisfied is fiction. The author's own measurements and comfort enquiries in air-conditioned office buildings tend to support the Mayer results. As Figure 3.2 illustrates, for one of the buildings surveyed, perceived comfort gave even more dissatisfied for a mean vote close to zero than Mayer found.

3.1.4 Steady state thermal comfort, the adaptive model

Fanger put forward the PMV/PPD model in 1970, but by the early 1980s questions arose. The model hardly considered psychological and adaptive aspects that impact perceived thermal comfort. Instead, it overestimated the impact of the ambient. Expectations in fact have an effect, as do adaptations such as changing posture, slowing down activity or varying clothing. Together, they widen the comfort band beyond the 10% *PPD*.

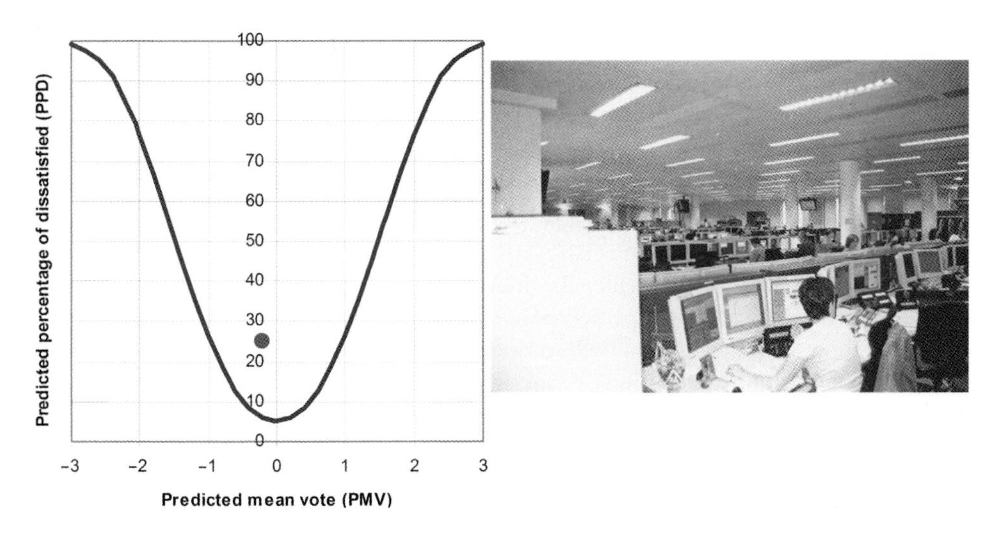

Figure 3.2 Comfort enquiry in an air conditioned office building: mean vote close to zero. The dot gives the number of dissatisfied counted.

The PMV/PPD model also overlooks motivations such as energy saving, which enable people to accommodate to thermal environments that should be uncomfortable. The importance of adaptation was clearly seen in the comfort response of employees, who could adjust their thermal environment by opening windows. On average, they complained less than employees who could not, even when the conditions the latter experienced were comfortable according to the PMV/PPD model.

Field studies in naturally ventilated buildings revealed a relationship between the operative temperature that employees prefer and the temperature outdoors, with comfort values different from those the PMV/PPD-model proposed. Each author, however, advanced a different equation, some coupling the preferred operative temperature to the monthly mean outdoors:

$$\theta_{o,comfort} = 11.9 + 0.534\overline{\theta}_{e.ref}\ (10\,°C \leq \theta_e \leq 34\,°C) \tag{3.23}$$

while others used a weighted four- or seven-day value:

$$\overline{\theta}_{e,ref} = \left(\theta_{e,today} + 0.8\theta_{e,today-1} + 0.4\theta_{e,today-2} + 0.2\theta_{e,today-3}\right)/2.4$$

$$\overline{\theta}_{e,rref} = \left(\theta_{e,-1} + 0.7\theta_{e,-2} + 0.6\theta_{e,-3} + 0.5\theta_{e,-4} + 0.4\theta_{e,-5} + 0.3\theta_{e,-6} + 0.2\theta_{e,-7}\right)/3.8$$

The results were comfort equations, such as:

$$\theta_{o,comfort} = 20.4 + 0.06\overline{\theta}_{e,ref} \quad \left(\overline{\theta}_{e,ref} \leq 12.5\,°C\right) \tag{3.24}$$

Usually, scatter around the least square lines was large, turning the equations above into statistical means that did not perfectly reflect individual preferences.

Recent studies confirm that the PPD/PMV model does well in air-conditioned but less well in naturally ventilated buildings. A German project on adaptive comfort refined that statement by giving following equations for the measured operative temperatures:

Office buildings	$\theta_e \leq 10\,°C$	$\theta_e > 10\,°C$
Heated only, natural ventilation	$\theta_o = 22.5 - 0.2\theta_e$	$\theta_o = 17.9 - 0.42\theta_e$
Heated, balanced ventilation	$\theta_o = 23.1 - 0.05\theta_e$	$\theta_o = 20.0 - 0.23\theta_e$
Air conditioning, window airing impossible	$\theta_o = 22.2 - 0.1\theta_e$	$\theta_o = 22.2 - 0.1\theta_e$
Air conditioning, window airing possible	$\theta_o = 23.4 - 0.05\theta_e$	$\theta_o = 23.4 - 0.05\theta_e$

According to the ASHRAE scale, the indoor climate in the heated only, naturally ventilated office buildings was perceived as too warm in summer but pleasing in winter. The other three scored neutral in both seasons. Unfortunately, the report did not specify the outside temperature used (daily mean, weighted four-day mean, monthly mean). In later publications, Fanger adapted the PMV/PPD model by multiplying *PMV* with an expectancy factor (*e*):

Expectancy	Boundary conditions	*e*
High	Naturally ventilated office building in regions where air-conditioning is the norm Regions where heat waves are limited in duration	0.9–1.0
Moderate	Naturally ventilated office building in regions where air-conditioning is less common Regions where the summers are warm	0.7–0.9
Low	Naturally ventilated office building in regions where air conditioning is the exception Regions where it is hot the whole year	0.5–0.7

3.1.5 Thermal comfort under non-uniform and under transient conditions

Aside of being time independent, steady state included that the human body has equal skin temperature everywhere, and the environment is characterized by one air temperature, the same radiant temperature all around, one relative humidity and one relative air velocity, all of which misrepresents reality. The human skin is not at equal temperature and the clothed body does not sense the same ambient conditions all around. Instead, it acts as a complex thermal system: inert, partly clothed, blood pumped around continuously, different skin temperatures, and so on, while air temperature can differ between head and ankles, each body part senses other radiant temperatures, relative humidity changes a little along the clothed body and relative air velocity varies considerably, depending on how the body and its parts move.

3.1.5.1 Refined body model

Studying thermal comfort under spatially non-uniform, transient conditions often starts from a body split into 15 parts (Figure 3.3), each having a bone core surrounded by muscles, fat and skin, with the central part 3 housing the respiratory tract, the heart and the internal organs. A specific thermal conductivity, a given mass and different clo-values characterize each part, while blood flow and heat conduction connect them.

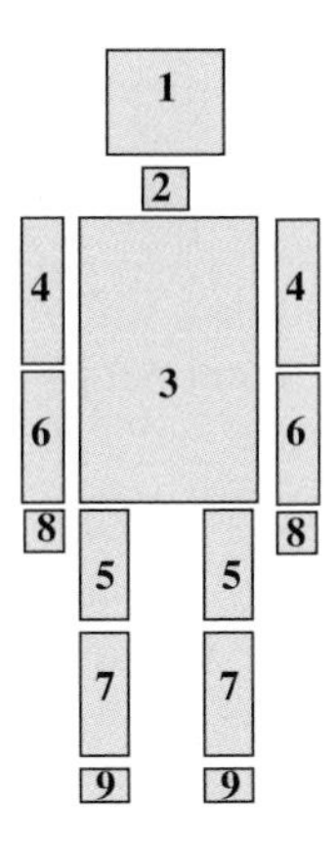

Figure 3.3 The 15 body parts.

Table 3.4 lists the masses for a human weighing 71 kg. Of course, each individual's weight will be different, partly explaining why humans react differently under non-uniform, transient thermal conditions.

Local skin and core temperature noted by the control system lets the relay determine the thermoregulatory responses required: sweating, shivering and more or less blood pumped around, thanks to vasodilatation or constriction. A core temperature of 36.8 °C and average skin temperature of 33.7 °C define the basal situation. Maximum vasodilatation occurs at 37.2 °C core temperature with the total blood flow then equalling seven times the basal value. Maximum constriction coincides with an average skin temperature of 10.7 °C, dropping blood flow to one eighth of the basal value. Each cardiac output between these limits distributes the blood over the 15 parts proportionally to their skin area, with the core temperature above 36.8 °C guiding vasodilatation and the average skin temperatures below 33.7°C vasoconstriction. Calculation of course requires a refined grid per part with local thermal sensation (S_{local}) function of the

Table 3.4 Body part masses.

	Mass (kg)								
Body part→	1	2	3	4	5	6	7	8	9
Brain	1.398								
Abdomen			11.14						
Lung			2.919						
Bone	1.777	0.233	4.169	0.473	1.482	0.266	0.655	0.164	0.412
Muscles	0.452	0.581	11.98	1.580	2.539	0.883	1.120	0.141	0.215
Fat	0.282	0.071	9.014	0.240	1.134	0.134	0.508	0.149	0.347
Skin	0.187	0.031	1.231	0.181	0.341	0.101	0.153	0.093	0.127

skin and overall core temperature and their time derivatives:

$$S_{\text{local}} = F\left(\theta_{\text{skin}}, \frac{\mathrm{d}\theta_{\text{skin}}}{\mathrm{d}t}, \theta_{\text{core}}, \frac{\mathrm{d}\theta_{\text{core}}}{\mathrm{d}t}\right) \tag{3.25}$$

Although the model helps to master non-uniform, transient thermal conditions, still, in practice, the consequences are bundled under 'local discomfort' and 'drifts and ramps'.

3.1.5.2 Local discomfort

Complaints about draughts are common in buildings with air conditioning. The term reflects unwanted local cooling by air movement. It is especially the neck, the lower back and the ankles that are sensitive. The reason is the high surface film coefficient by natural convection (h_{co}) there. Testing has shown that the value not only depends on the mean relative air velocity (v in m/s), but also on the ratio between that mean and its standard deviation, called the turbulence intensity (Tu in %):

$$h_{\text{c}} = h_{\text{co}} + 0.27\sqrt{vTu} \qquad (\text{W}/(\text{m}^2.\text{K})) \tag{3.26}$$

Flow direction is also a factor, with air striking the back perceived as worse than air on the chest, and air from above as worse than air from below. However, the equation for the draft rate (DR) does not distinguish between these four:

$$\begin{aligned} &DR = (34 - \theta_{\text{a}})(\overline{v} - 0.05)^{0.62}(0.37\overline{v}Tu + 3.14) \\ &\overline{v} < 0.05\ \text{m/s} \rightarrow \overline{v} = 0.05\ \text{m/s}, \qquad DR > 100 \rightarrow DR = 100 \end{aligned} \tag{3.27}$$

Vertical differences in air temperature may cause complaints. Figure 3.4 shows the percentage of dissatisfied (PD) for seated people as a function of the difference in air temperature between head and ankles.

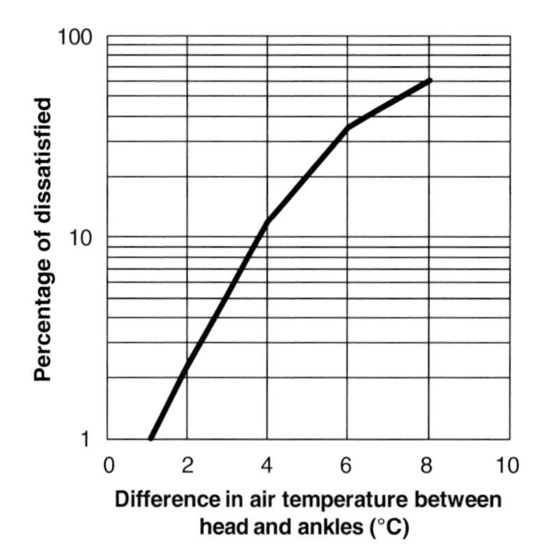

Figure 3.4 Seated people, air temperature difference between head and ankles, percentage of dissatisfied (PD).

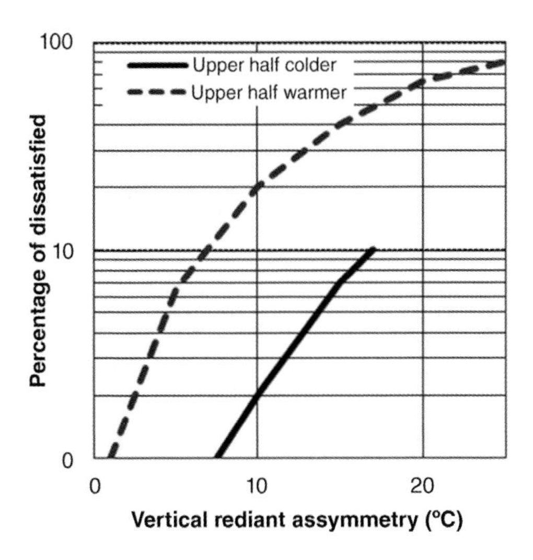

Figure 3.5 Vertical radiant asymmetry, percentage of dissatisfied (PD).

Also radiant asymmetry matters. The human body faces two half-spaces, each with its own radiant temperature. This asymmetry gives problems when the difference between the two passes certain values. Figure 3.5 gives the *PD* for vertical asymmetry.

It reaches 5% when the upper half is 14 °C colder. When warmer, the 5% limit is 4.5 °C, a figure that fixes the allowable mean temperature of a heated ceiling, see Table 3.5, albeit the acceptable temperature of the heating elements still depends on the view factor ($F_{h.c}$) with the head (i.e. the view factor between a dot and a surface). *PD* values above 5% are sometimes allowed. For example, a heated ceiling can have a temperature at design conditions (θ_{eb}):

$$\theta_{ceil,max} \leq 1.67\theta_{ceil} - 0.67\theta_{o,comfort} \quad (3.28)$$

Figure 3.6 shows the percentage of dissatisfied for horizontal asymmetry.

If caused by a cold surface, the *PD* becomes 5% for a difference 10 °C. A single glazed wall could be responsible for that, which is why the figure is referred to as the glass effect. A warmer surface rarely generates complaints.

The floor temperature is another critical factor. Transmission gains from, or losses to were not included in the steady state comfort balance. Related heat exchange is too

Table 3.5 Permitted mean temperature of a radiant ceiling.

$\mathbf{F_{h,c}}$	0.06	0.08	0.10	0.12	0.14	0.16	0.18	0.20	0.22	0.24	0.26	0.28
$\boldsymbol{\theta_{ceil}}$ (°C)	50	43	38	35.2	33	31.3	30	28.6	27.5	26.8	26.0	25.8

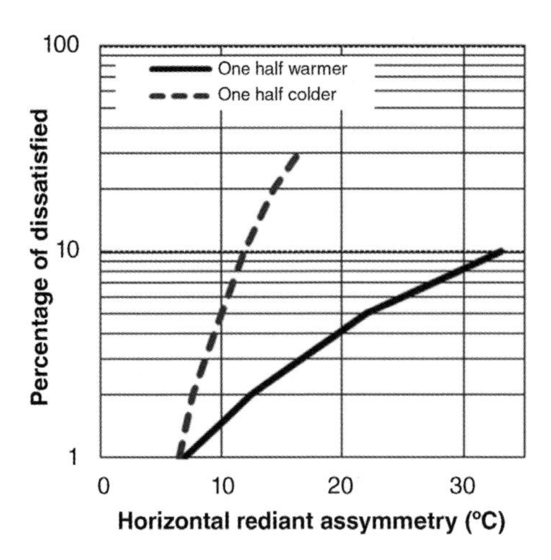

Figure 3.6 Horizontal radiant asymmetry, percentage of dissatisfied (PD).

limited for that. Nonetheless, people perceive foot comfort as critical. Influencing parameters are the floor temperature (θ_{fl}) and its contact coefficient (b_{fl}). Their importance depends on how long the soles of the feet contact the floor, and whether shoes are worn or not. As an array:

Contact duration	Very short	Short	Long
Bare foot	$\theta_{fl} = F(b_{fl})$	$\theta_{fl} = F(b_{fl})$	$\theta_{fl} = F(b_{fl})$
Wearing shoes	θ_{fl}	θ_{fl}, (b_{fl})	θ_{fl}

Barefoot, comfortable floor temperatures depend on the contact coefficient of the floor cover, see Table 3.6. Figure 3.7 gives the *PD* for long-lasting floor contact when wearing shoes. A *PD* of 10% relates to a floor temperature of 19–28 °C, with 28 °C the limit temperature for floor heating and 19 °C for floor cooling. In circulation zones, where floor contact is short, the limit values are less severe, 17 and 30 °C.

3.1.5.3 Drifts and ramps

The temperature transients indoors mainly consist of drifts and ramps. Both refer to non-cyclic monotonic increases and decreases of the operative temperature with the passive ones called drifts and the HVAC controlled ones ramps. Cyclic variations instead refer to situations where the operative temperature repeatedly rises and drops. Acceptability depends on three factors: whether or not it is under control of the occupant; temperature rise or drop rate (°C/h); and peak-to-peak difference for cycling. There are no negative impacts on thermal comfort when drifts, ramps and cycles result from actions by

Table 3.6 Floor temperature and contact duration, covers from low to high contact coefficient.

Floor cover:	Floor temperature (°C)		
Contact:	Very short	Short	Long
Carpet	21	24.5	21-28
Cork	24	26	23–28
Parquet (Nordic pine)	25	25	22.5–28
Parquet (oak)	26	26	24.5–28
Vinyl tiles	30	28.5	27.5–29
Vinyl on felt	28	28	27.5–28
Linoleum on planks	28	26	24–28
Linoleum on concrete	28	27	26–28.5
Concrete	30	28.5	27.5–29
Marmora, tiles	30	29	28–29.5

individuals, such as opening or closing windows. Not so for very abrupt and important changes beyond individual control. For an average operative temperature at comfort level, drifts and ramps with periods between 15′ and 4 h should not exceed the following values.

Time period	15′	30′	1 h	2 h	4 h
Operative temperature step allowed in degrees C (+/−)	1.1	1.7	2.2	2.8	3.3

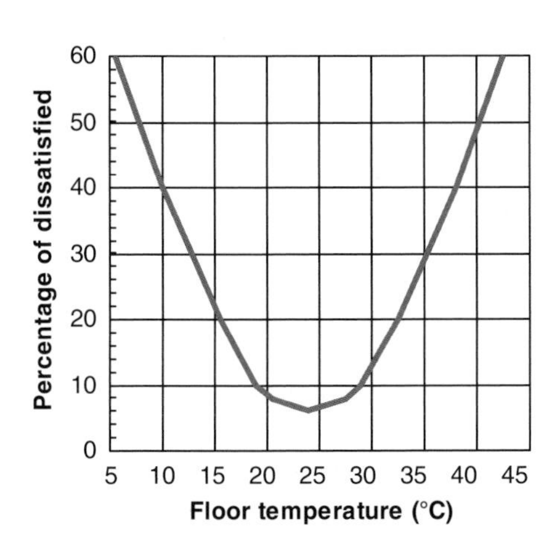

Figure 3.7 Wearing shoes, long floor contact, percentage of dissatisfied (PD).

Where cyclic variations prevail and periods are shorter than 15 minutes, peak-to-peak differences should not exceed 1.1 °C. If superimposed on longer cyclic variations, the less than 1.1 °C restriction combines with the drift and ramp restriction for the longer period.

3.1.6 Standard-based comfort requirements

ASHRAE 55-2013 and ISO EN 7730-2005 contain thermal comfort requirements, see Tables 3.7 and 3.8. While for conditioned spaces ASHRAE restricts the requirements to one series of predicted *PD*s, the EN fixes four categories, allowing designers to choose between high (I), good (II), acceptable (III) and low (IV) expectations.

Table 3.7 ASHRAE 55-2013.

Comfort		P PD≤, %
Overall		10
Local	Draft	20
	Vertical air temperature difference	5
	Radiant asymmetry	5
	Feet	10

Table 3.8 ISO EN 7730-2005.

Overall						
Category	Overall comfort		Operat. temperature (°C)		Maximum air speed (m/s)	
	PPD (%)	PMV	Summer	Winter	Summer, 0.6 clo, cooling	Winter, 1 clo, heating
I	≤6	−0.2 to 0.2	23.5–22.5	21.0–23.0	0.18	0.157
II	6–10	−0.5 to 0.5	23.0–26.0	20.0–24.0	0.22	0.18
III	10–15	−0.7 to 0.7	22.0–27.0	19.0–25.0	0.25	0.21
IV	15–25	−1 to 1	21.0–28.0	17.0–26.0	0.28	0.24

Local						
Category	Vertical temp. gradient (°C)	Floor temperature°C	Radiant asymmetry (°C)			
			Warm ceiling	Cold ceiling	Warm wall	Cold wall
I	≤2	19–29	≤5	≤14	≤23	≤10
II	2–3	19–29	≤5	≤14	≤23	≤10
III	3–4	17–31	5–7	14–18	23–35	10–13
IV	—	—	—	—	—	—

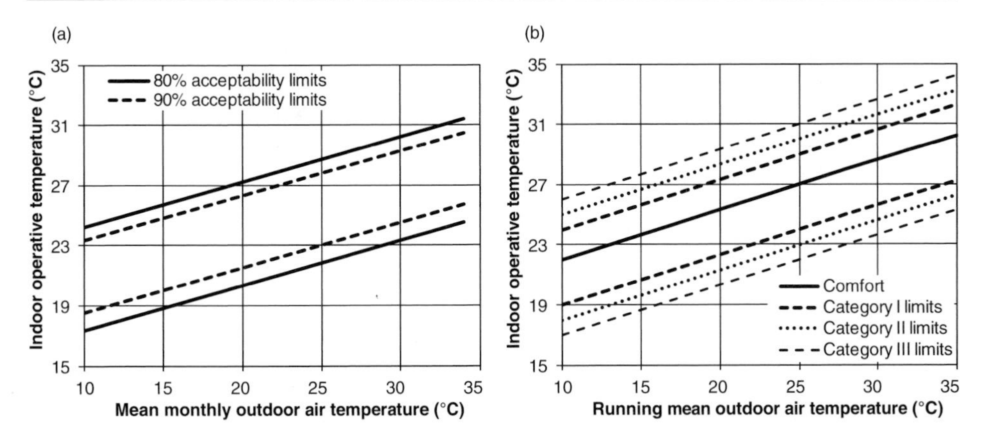

Figure 3.8 Adaptation, (a) according to ASHRAE 55-2013, (b) to ISO EN 7730-2005.

Both standards also consider adaptation in naturally ventilated spaces, see Figure 3.8. Related formulas are:

ANSI/ASHRAE 55-2013

80% acceptability	lower limit: $\theta_o = 14.4 + 0.3\overline{\theta}_{e,month}$	upper limit: $\theta_o = 21.2 + 0.3\overline{\theta}_{e,month}$
90% acceptability	lower limit: $\theta_o = 15.5 + 0.3\overline{\theta}_{e,month}$	upper limit: $\theta_o = 20.3 + 0.3\overline{\theta}_{e,month}$

ISO EN 7730-2005

Outdoors:

$$\overline{\theta}_{e,ref} = \left(\theta_{e,-1} + 0.7\theta_{e,-2} + 0.6\theta_{e,-3} + 0.5\theta_{e,-4} + 0.4\theta_{e,-5} + 0.3\theta_{e,-6} + 0.2\theta_{e,-7}\right)/3.8$$

Comfort	$\theta_o = 18.7 + 0.33\overline{\theta}_e$	
Category I	lower limit: $\theta_o = 15.7 + 0.33\overline{\theta}_{e,ref}$	upper limit: $\theta_o = 20.7 + 0.33\overline{\theta}_{e,ref}$
Category II	lower limit: $\theta_o = 14.7 + 0.33\overline{\theta}_{e,ref}$	upper limit: $\theta_o = 21.7 + 0.33\overline{\theta}_{eref}$
Category III	lower limit: $\theta_o = 13.7 + 0.33\overline{\theta}_{e.ref}$	upper limit: $\theta_o = 22.7 + 0.33\overline{\theta}_{e,ref}$

3.1.7 Comfort related enclosure performance

The operative temperature with weighting factor 0.5 for air and radiant allows to calculate the mean thermal transmittance of the room enclosure needed to ensure acceptable thermal comfort in heating climates. For this, all surfaces are considered grey. Radiant temperature in the room's centre then approaches:

$$\theta_r = \sum(A_j\theta_{s,j})/\sum A_j \tag{3.29}$$

with A_j the area and $\theta_{s,j}$ the temperature of each surface (j). The steady state relation between surface temperature and thermal transmittance is:

$$\theta_{s,j} = \theta_o - (U_j/h_i)(\theta_o - \theta_k) \tag{3.30}$$

where, for partitions, θ_k is the operative temperature in the neighbouring room and for envelope assemblies the sol-air temperature outdoors. Per envelope assembly this equation becomes:

$$\theta_{s,l} = (1 - U_l/h_i)\theta_o + (U_l/h_i)\theta_e^*$$

For a partition with room k reshuffling gives:

$$\theta_{s,m} = (1 - a_k U_k/h_i)\theta_o + (a_k U_k/h_i)\theta_{o,k} \quad \text{with} \quad a_k = (\theta_o - \theta_{o,k})/(\theta_o - \theta_e^*)$$

The ratio a_k becomes 0 when both rooms are at the same operative temperature, and approaches 1 when the partition is well insulated and the neighbouring room is not heated but intensely ventilated. Substituting both relations in Eq. (3.30) results in:

$$\theta_r = \theta_o - (U_m/h_i)(\theta_o - \theta_e^*) \tag{3.31}$$

where U_m is the mean thermal transmittance of the room enclosing surfaces:

$$U_m = \frac{\sum (A_l U_l) + \sum (a_k A_k U_k)}{\sum A_l + \sum A_k}$$

$\sum A_l$ refers to all envelope assemblies and $\sum A_k$ to all partitions. Introducing (3.31) in $0.5(\theta_a + \theta_r)$ gives:

$$\theta_o = \frac{h_i \theta_a + U_m \theta_e^*}{h_i + U_m} \tag{3.32}$$

The lower the mean thermal transmittance of the room enclosing surfaces, that is the better insulated the envelope assemblies and the larger the partition's share in the enclosing area, the smaller the difference between operative and air temperature. Writing the formula as:

$$\theta_a = (1 + U_m/h_i)\theta_o - (U_m/h_i)\theta_e^*$$

shows that a better thermal insulation allows the air temperature to be lowered without jeopardizing thermal comfort. An additional advantage is that less end energy will be needed to warm the ventilation air.

A related envelope performance metric starts by imposing a maximum gap between air and radiant temperature, thus fixing an upper limit for the mean thermal transmittance. Contrary to what passive building adepts claim, this does not require the extremely high thermal resistance values they impose. Figure 3.9, for example, underlines that the difference in central operative temperature is negligible above a mean thermal resistance of the envelope, equal to $3\,m^2.K/W$.

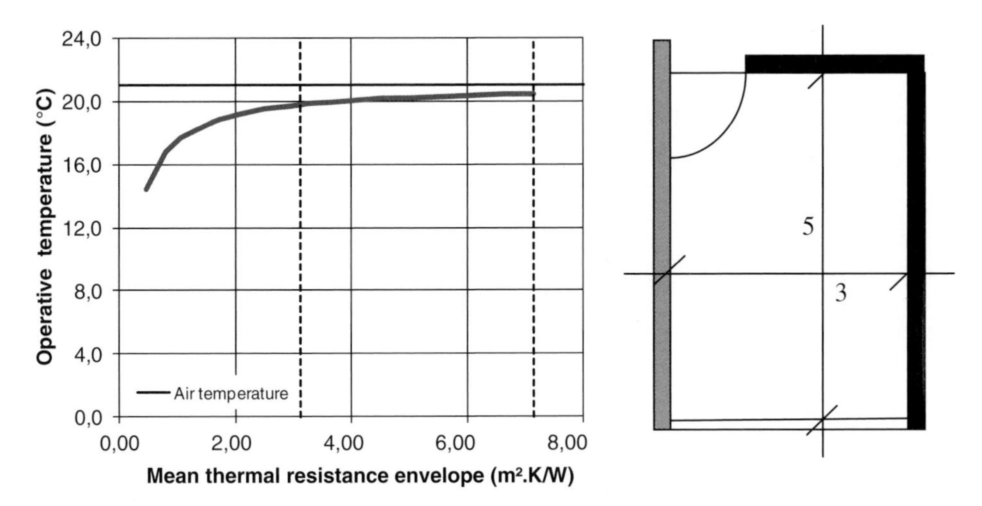

Figure 3.9 Single upper corner office, area 15 m^2, volume 40.5 m^3, window 8.1 m^2. Operative temperature for an air temperature of 21 °C depending on the mean thermal resistance of the envelope. The two black coloured partition walls and the floor separate the office from equally heated offices and corridor.

3.2 Health and indoor environmental quality

3.2.1 In general

Dwelling and health are so basic that their conjunction has inspired many phantom theories. Popular was bio-ecology, a belief based on paradigms. Everyone is subjected to terrestrial radiation. Lines of equal radiation form quadrants that local water arteries disturb. A dowsing rod reveals the disturbances. Living in a dwelling whose footprint stays in a quadrant ensures healthiness. Inhabiting one that stands on the quadrant lines causes sleeplessness, headaches and sometimes cancer. Static electrical fields are curative, while low-frequency alternating fields are unhealthy. One's psychological and physiological equilibrium benefits from negative ions in the air, which the magic pyramid produces. And, receiving cosmic radiation ensures well-being.

These paradigms shaped sets of construction rules. Use healthy building, insulating and finishing materials, the adjective 'healthy' referring to materials where the basic substances are still recognizable, examples being loam, lime, brick (if hand-moulded), straw, timber, cork, cellulose and wax. Materials manufactured chemically are unhealthy. They shield the curative cosmic rays, act as electrostatic bodies and emit all kind of noxious substances. Examples include concrete, polymers, most metals and bitumen. Walls must breath. Only healthy materials ensure that this can happen. Vapour-retarding materials and vapour barriers should be excluded. Radiant heat is healthier than convective heat. As heavy tile stoves ensure a perfect balance between both, they figure as the best heating choice.

Figure 3.10 Straw bale construction.

Happily, the aversion towards thermal insulation and airtightness that spoiled discussions have faded away. Today, bio-ecologists defend both, although only healthy insulation materials and only air barriers not made of synthetics should be used. Favoured are cellulose, sheep's wool, sea grass and straw bales (Figure 3.10). That most behave badly when humid, that some attract rodents and that all lack fire resistance if not treated chemically, seems of no concern.

Yet, not one of these bio-ecological paradigms and construction rules withstands the laws of physics. But still they infect the public discussion on indoor environment and health. Many people in fact simply prefer fairy tales instead of a sober-minded scientific approach.

That there is a relation between habiting and health is not denied by any scientist active in the field. Pollutants in the air people breath are found outdoors and indoors. These include viruses, bacteria, fungal spores, house dust, volatile and semi-volatile organic compounds, bio-odours, tobacco smoke, radon, nitrous oxides, carbon monoxide, vapours, fine dust and fibres. Moisture has a special status as it often acts as a necessary condition for health-threatening species to become activated. Complaints about sick buildings and so-called sick building syndrome have even brought indoor air quality to the forefront of people's concern.

As people in temperate climates spend, on average, more than 80% of their lifetime indoors, maintaining healthy buildings assumes that emission of harmful contaminants should be minimized, that dilution by ventilation should keep concentrations acceptable and that, if this is impossible, air disinfection or personal protective measures should be considered.

3.2.2 Health

The World Health Organization (WHO) defines health as physical, psychological and social well-being rather than just an absence of disease or handicap. Indoors, this involves uneasiness, stress and various kinds of discomfort.

3.2.3 Definitions

With pollutants and their effects, two notions surface: concentration and exposure time. Although not a paradigm, harmfulness generally aggravates with higher concentration and longer exposure times. Definitions therefore link both:

Emission	Pollutant release – typically a function of time
Immission	Pollutant load – immissions outdoors and indoors depend on pollutant distribution in the air, on where the receiver stands and on their breathing rhythm and volume
ppm	Parts per million – the number of molecules of a component per million molecules in the mixture
Radioactive radiation	Radiation emitted by decaying atom nuclei – radioactivity consists of α-rays (helium atoms), β-rays (electrons) and electromagnetic γ-rays
Becquerel (Bq)	Unit of radioactive decay – one becquerel means a decay of one nucleus per second
Sievert (Sv)	Unit of radioactive dose equivalent – the risk of developing biological effects is proportional to the sievert value received, where 1 sievert is equivalent to an effective dose of 1 joule per kg of recipient mass
AMP (USA)	Peak concentration allowed during 15′
ACC (USA)	Highest concentration allowed 15′ per 8 hours a day (excluding the periods when AMP applies)
MAC, TWA8 (USA)	Mean concentration allowed 8 hours a day, 5 days a week
DNEL (EU REACH)	Maximum concentration allowed during a given exposure time (from peak to 24 hours a day)
AIC, NOAEL (EU REACH)	Maximum concentration without negative health effects

3.2.4 Relation between pollution outdoor and indoors

Pollution indoors depends partially on what happens outdoors. No building is perfectly airtight and, even if it were, people entering and leaving permit unclean outdoor air to enter and mix with the air indoors. Also ventilation lets outdoor air in. As a result, outdoor pollutants will contaminate the indoors, though not to the same concentration as outdoors. Indeed, on its way inside, concentration decreases thanks to filters in ventilation systems and due to the deposit and side adsorption in the many enclosure leaks. Once indoors, house dust and the surfaces of partition walls, furniture and furnishings adsorb a proportion of the pollutants, while others react to form new compounds. The pollutants that enter mix with cubic metres of indoor air, which delays the concentration build-up. Advice such as closing windows when spikes in pollution outdoors occur therefore makes sense. However, there are also many sources indoors that emit contaminants.

3.2.5 Process-related contaminants

3.2.5.1 Dust, vapour, smoke, mist and gaseous clouds

Industries typically emit process-related contaminants. Mechanical material processing, such as stone sawing on building sites, produces dust. Vapours and gaseous clouds are the result of thermal treatment and chemical processes. Evaporating liquids cause mist, while smoke comes from fires.

In non-industrial environments, traffic and wood stove burning are the dust-producing culprits. Health effects mainly depend on particle size, nature and toxicity of the forming agents, their solubility in biological liquids and inhaled doses. Particles with sizes below 2.5 μm, also known as pm 2.5, invade the lung alveoli, whereas the mucous membranes retain the 2.5–10 μm ones. The non-industrial less than 2.5 μm particles, emitted by car traffic, mainly consist of polycyclic aromatic hydrocarbons (PAHs).An investigation in four Chinese hospitals situated along busy roads confirmed that these particles enter buildings. PAH-loaded indoor and outdoor 2.5 μm dust concentration was consistently related. Deposits that moving people whirled up even resulted in higher concentrations indoors. PAHs are carcinogenic and increase mortality, though some people still call the proof flimsy, while years of exposure to sandstone dust of less than 2.5 μm provokes brown lung disease. For a day-long exposure, the limits are:

Dust	Limit (mg/m^3)
Cadmium	0.05
Manganese	1.0
Plaster	10.0
Corn	10.0
Silica fume	0.1
All dust together	15.0
Breathable dust	5.0

There is huge diversity in vapours, smoke, mist and gaseous clouds. Inhaling is hard to avoid, which is why what is allowable must be respected, be it by source control, ventilation or wearing masks.

3.2.5.2 Fibres

Fibres are mineral filaments with a length-to-diameter ratio above 3, constant section and a diameter less than 3 mm. Asbestos, a silicate fibre that splits into ever thinner filaments, is to be feared. The filaments don't dissolve in biologic liquids. When inhaled, they finally enter the lung alveoli where the immune system becomes activated and triggers inflammation. Long-term exposure to high concentrations induces asbestosis or pleural asbestos calcification, which in turn initiates heart insufficiency. Occasional exposure to low concentrations of blue asbestos can cause mesotheliom, a deadly lung cancer. Today, strict concentration limits indoors are compulsory:

	AIC, fibres per cm^3
Blue asbestos	0.000 2
Other types of asbestos	0.001

By contrast, glass wool, mineral wool and other silicate-based man-made fibres are not carcinogenic, owing to their large fibre diameter (3–5 μm), their length (some centimetres), the fact they don't split into ever thinner filaments and their solubility in biologic liquids. However, they irritate the skin and the mucous membranes, can cause itchy dermatitis, upset the eyes and give temporary inflammation of the mucous membranes at high concentration, so protective clothing and a mask have to be worn when working with materials composed of such fibres.

3.2.5.3 Ozone

Excessive ozone concentrations outdoors arise when, in warm sunny weather, solar UV forces hydrocarbons in the air to react with the nitrous oxides emitted by diesel cars. Indoors, air purifiers and ionization apparatus are the suspects. Ozone hinders breathing to such extent that already weakened people have to be discouraged from performing heavy physical activity under such circumstances.

3.2.6 Building, insulation and finishing material related contaminants

3.2.6.1 (Semi) volatile organic compounds ((S)VOCs)

VOCs and SVOCs have boiling points of 50–250 °C. Building and insulation materials, finishes, furniture, ventilation systems, consumer articles, polluted soils and micro-organisms act as emitters. They are also present in tobacco smoke. More than 1000 compounds, including benzene, toluene, alkenes, n-alkenes, cycloalkenes, aromatics, pentane, heptanes, halogens, alcohols, ethers, aldehydes, ketenes and amides are known. The amounts emitted depend on the source and ambient conditions (temperature, relative humidity, air velocity). Some dry materials, such as the binder in mineral fibre, polystyrene and polyurethane foam, glued timber based boards, polymer floor covers, carpets and textiles, show a limited, slowly decreasing emission, mostly of VOCs with low boiling point. Wet finishes such as paints, varnish and lacquers instead show an initial boost in output, followed by a fast drop, see Figure 3.11.

House dust and inside surfaces adsorb a proportion of the (S)VOCs to re-emit them afterwards when the concentration differential with the air allows. This secondary emission increases with surface temperature and for certain (S)VOCs and materials with relative humidity. (S)VOCs cause olfactory nuisances. They irritate the eyes, the mucous membrane and the bronchi. Some are believed to be carcinogenic for people exposed to high concentrations over long periods of time – see painter's disease. The

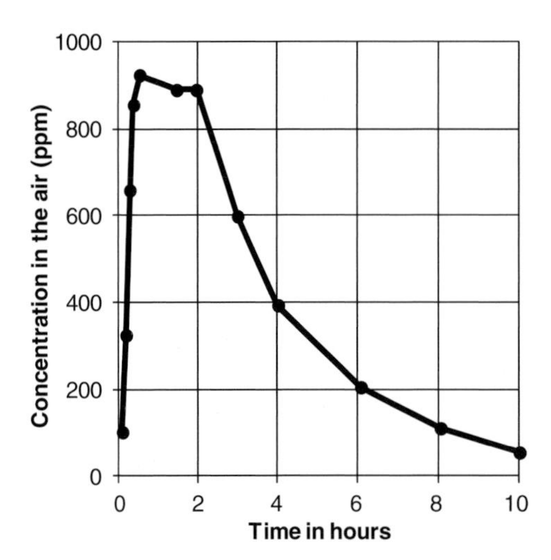

Figure 3.11 VOC-concentration in the air after application of a wet finish.

effects of many seem additive, which is why for safety reasons concentrations must obey:

$$\sum C_i / C_{\text{accept},i} \leq 1$$

with C_i the concentration present and $C_{\text{accept},i}$ the acceptable concentration per VOC_i. The following array helps in judging possible harmful effects:

Total VOC concentration (μg/m^3)	Effects
<200	No irritation expected
200–3000	Possible irritation in combination with smoking
3000–25 000	Irritation or nuisance expected
≥25 000	Neurotoxin effects possible

(S)VOCs are a problem worldwide but, due to different building traditions and materials used, the cocktail emitted differs between regions. In most buildings, the concentration per species lies far below the olfactory threshold and the value causing irritation. But this does not preclude complaints. A summing effect may exist, though there is no objective data to support this hypothesis, albeit that it may not only hold for (S)VOCs but also for any combination with other pollutants.

3.2.6.2 Formaldehyde (HCHO)

This is a colourless gas, soluble in water, that materials bonded with formalin or a phenol resin, such as chipboard, plywood, oriented strand board and certain textiles emit.

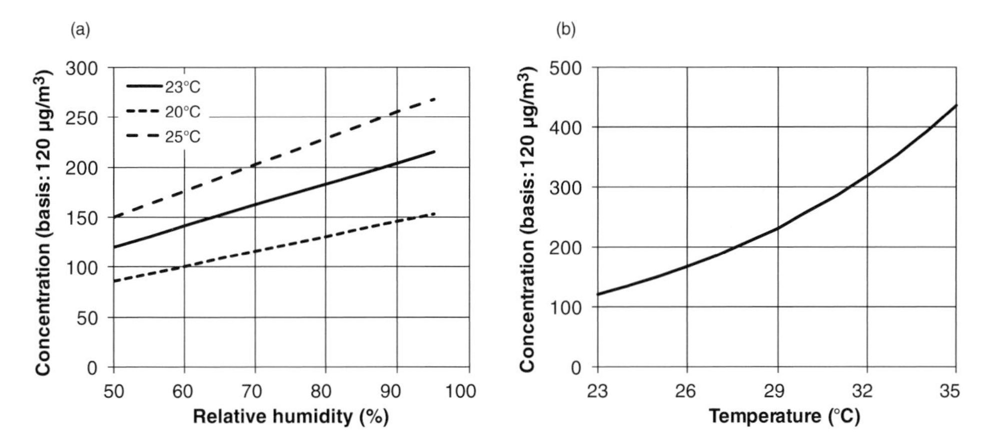

Figure 3.12 Oriented strand board. (a) the effect of relative humidity, (b) of temperature at 50% relative humidity on the formaldehyde concentration in the test chamber.

Emission increases with temperature and relative humidity, as shown by the changes measured in a test chamber on 1 m^2 oriented strand board per m^3 air:

$$C = C_o\left[1 + 0.0175(\phi - \phi_o)\right]\exp\left[-9799\left(T^{-1} - T_o^{-1}\right)\right]$$

with C, ϕ (%) and T (K) the actual concentration, relative humidity and temperature and C_o, ϕ_o, T_o the concentration, relative humidity and temperature at standard conditions (45%, 296.15 K). Figure 3.12 shows data for oriented strand board, emitting 120 μg/m^3 at standard conditions, when relative humidity increases to 95% and temperature to 35 °C.

With sensitive people formaldehyde irritates the mucous membranes of the upper and lower bronchi and provokes allergic reactions.

3.2.6.3 Phthalates

These are organic compounds used to keep synthetic materials pliable and deformable. Their low boiling point favours emission, which house dust absorbs. Loaded dust may irritate the bronchi and could be co-responsible for the increasing number of people suffering from asthma.

3.2.6.4 Pentachlorinephenols

These were the active substance used in wood preservatives. Investigations have shown that they are highly carcinogenic. Their usage is now forbidden.

3.2.7 Soil-related radon as contaminant

Radon is a radioactive inert gas with a half life of 3.7 days, belonging to the decay sequence of Uranium 238:

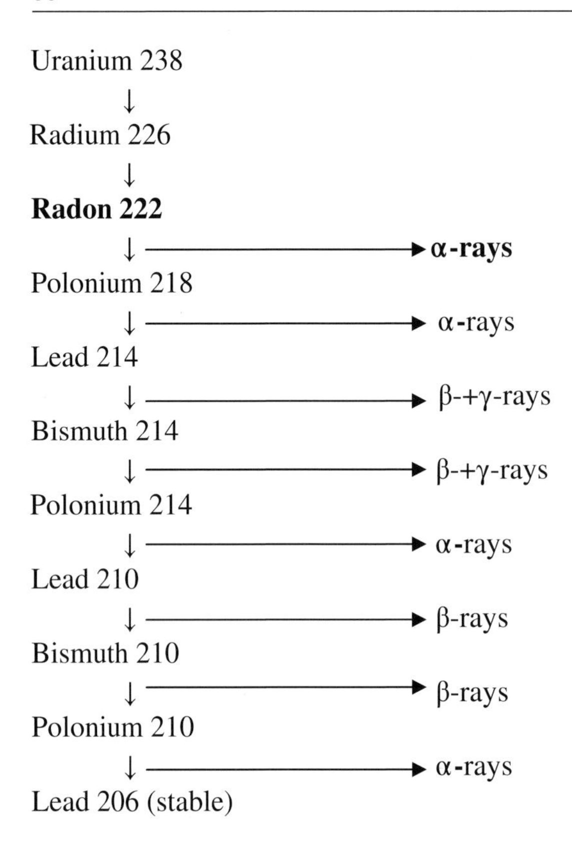

With cosmic and soil-linked γ-rays, it contributes to the background radiation outdoors. Uranium 238 is found in small amounts in primary rock formations, turning the soil into the most important radon source, with phosphoric gypsum, groundwater and the outside air as distant seconds. As a gas, it effuses out of the soil and spreads in the air. The emission rate depends on the concentration in the soil, the diffusion resistance of the soil and the presence of soil crevices purged by air. Wind lowers the concentrations outdoors to 5–10 Bq/m^3. Indoors, however, due to the limited ventilation rates, concentration can reach high values. Some measurements in buildings have given:

Geographic region	Radon concentrations, Bq/m^3		
	Median	Mean	90%-value
Flanders, northern Wallonia, Southern Luxemburg	32	39	66
Rest of Wallonia	52	77	135

In problem premises, concentrations above 1000 Bq/m^3 have been logged. Building users inhale the radon. While decaying in their lungs it emits α-rays, which increases lung cancer risk after a long enough exposure time with following percentages:

Mean concentration Bq/m^3	Lung cancer risk %
100	1.5
200	3.0
400	6.0
800	12.0

Allowed concentrations, according to the WHO and the Environmental Protection Agency (EPA) in the USA are:

Organization	Buildings, Bq/m^3	
	Existing	New
WHO	100	100
EPA	150	150

In combination with smoking, overall risk may increase more rapidly than the sum of the two separate risks suggests, though some doubt exists about this model.

3.2.8 Combustion related contaminants

Fossil fuel combustion gives water vapour, carbon dioxide (CO_2), carbon monoxide (CO), nitrous oxides (NO_X) and sulphur dioxide (SO_2), while the dust formed is loaded with VOCs and PAHs. Water vapour and CO_2 don't directly do any harm, but CO and NO_X do.

3.2.8.1 Carbon monoxide

This is an extremely toxic, colourless, non-smelling gas, which is produced by incomplete combustion. The word 'incomplete' points to a lack of ventilation or bad chimney draw. When CO is inhaled, the gas attaches to the red blood cells' haemoglobin, preventing them from adsorbing oxygen, thus disturbing its transport to the cells. A person affected loses consciousness and eventually dies. Even at very small concentrations, carbon monoxide can be lethal.

3.2.8.2 Nitrous dioxide (NO_2)

This is produced by combustion at such high temperatures that oxygen reacts with the nitrogen in the air. Long-lasting inhalation of high concentrations can inflame the lung tissue. Low concentrations aggravate breathing.

3.2.9 Bio-germs

3.2.9.1 Viruses

Both a too low and too high relative humidity can have adverse effects on the airborne transmission of viruses. A study by the US National Institute for Occupational Safety and Health showed that the transfer of the influenza virus between coughing and non-coughing people reduced significantly if relative humidity was not permanently below 40%, as is often the case indoors in winter, but rather is kept between 40% and 73%.

3.2.9.2 Bacteria

Legionella pneumophila, a feared bio-germ, can develop in water at 20–40 °C, a temperature range present in air humidifiers, cooling towers, hot water pipes and spray installations. Propagation occurs via water mist in the air. When inhaled by persons with weakened immune system, the germ can cause a deadly lung inflammation known as the legionnaire's disease. For combating the bacterium regularly boosting the domestic hot water temperature to just above 60 °C, cleaning air handling units, disinfecting cooling towers regularly, avoiding stagnant water in long hot water pipes and mixing hot and cold water at the taps suffices.

3.2.9.3 Mould

The micro-fungi called moulds have no leaf green activity, digest their food externally and reproduce by releasing spores. When these deposit on a suitable substrate, their life cycle with germination, mycelium growth, spore formation and spore release starts. The following species typically colonize indoor surfaces (all deuteromycetes):

Species	Percentage of cases (14 studies)
Aspergillus	93
Penicillium	85
Cladosporium	71
Aureobasidium	64
Alternaria	57

The spores float in the air. Surface infestation requires the right conditions: some oxygen, a preferred temperature, carbon, nitrogen and salts in the substrate, and enough humidity. Oxygen below 0.14% m^3/m^3 suffices for germination and growth. The right temperature ranges from 5 to 40 °C. More than minimal amounts of carbon, nitrogen and salts are not needed, but relative humidity is critical. The right value differs between mould families and depends on temperature. Mould-specific isopleths reflect the impact of temperature and relative humidity on growth rate, see Figure 3.13 with two simplified

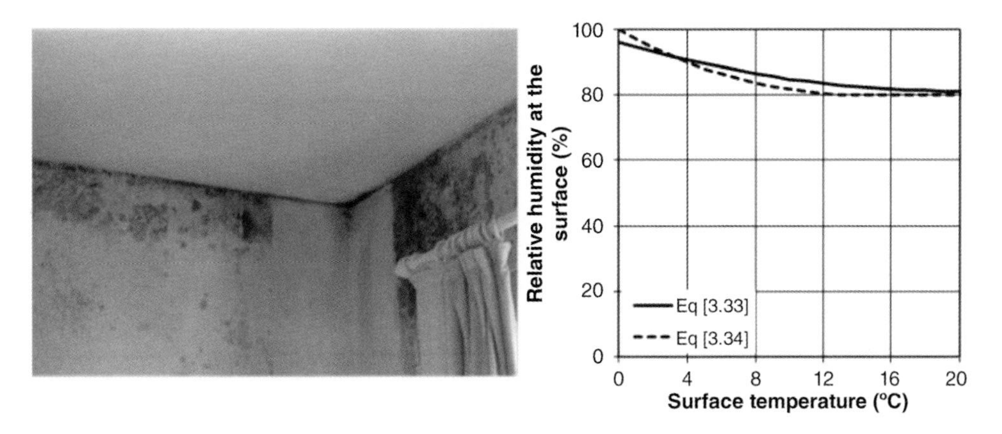

Figure 3.13 Isopleths fixing the lowest mould growth rate.

isopleths drawn that fix the lowest rate:

$$1)\quad \phi_{crit} = 0.033\theta_s^2 - 1.5\theta_s + 96 \tag{3.33}$$

$$2)\quad \begin{array}{l} \theta_s < 20\,°C\ \phi_{crit} = \max\left(80, -0.00297\theta_s^3 + 0.16\theta_s^2 - 3.13\theta_s + 100\right)\ (\%) \\ \theta_s \geq 20\,°C\ \phi_{crit} = 80\% \end{array} \tag{3.34}$$

For isopleths nearing the optimal temperature and relative humidity combinations, growth rate increases and the period before germination shortens. Beyond this, growth rate drops again. Globally, at surface temperatures of 17–27 °C, a four-week mean surface relative humidity of 80% (0.8 on a scale from 0 to 1) suffices to see aspergillus, cladosporium and penicillium colonizing substrates with limited nutritive value. This 80% figure has therefore been chosen as the design value for controlling mould risk on inside surfaces. Risk is assumed 1 when the four-week mean vapour pressure on a surface (p_s) exceeds 0.8 times the four-week mean saturation value ($p_{sat,s}$):

$$p_s \geq 0.8 p_{sat,s} \tag{3.35}$$

Mean relative humidity values beyond 80% shorten the period for germination (t in days):

$$\phi_{crit} \geq \min\left\{100, \left(0.033\theta_s^2 - 1.5\theta_s + 96\right)[1.25 - 0.072\ln(t)]\right\} \tag{3.36}$$

Of course, moulds other than the three mentioned, such as stachybotrys, may demand a higher surface relative humidity to germinate. In recent decades, investigations have refined the tools available for evaluating mould on substrates. Starting from a mixed species population, Finnish researchers have introduced a growth index to evaluate mould density on pine and spruce:

Index	Growth rate	Description
0	None	Spores not active
1	Small spots of mould on the surface	Initial growth stages
2	Less than 10% of the surface covered	
3	10–30% of the surface covered	New spores produced
4	30–70% of the surface covered	Moderate growth
5	>70% of the surface covered	Considerable growth
6	Very heavy and tight growth	100% coverage

German researchers classified substrates into four classes according to mould sensitivity:

Class	Type
0	Optimal substrate (agar)
1	Biodegradable materials
2	Typical building materials
K	Critical for mould on an optimal substrate

They dealt with mould as an additional layer at the inside, characterized by a mould-specific vapour diffusion resistance, sorption isotherm and growth rate related thickness.

Under the right conditions, moulds grow on any surface, even on the filters in HVAC and ventilation systems that, once infected, contaminate the indoor air with spores more than moulded surfaces do. But they also create other annoyances. Aesthetics are spoiled and economic loss can be caused. In fact, before selling their property, people tend to clean infected surfaces, otherwise the market value of the building would drop. Moulds also upset dwellers and, though not convincingly proven, harm health. Some assume that the mycotoxines emitted aggravate house dust allergy, while exposure to very high concentrations could impair the immune system. However, according to the John Hopkins School of Medicine, the relation between mould on inside surfaces and allergies of the upper bronchi is not substantiated, while the literature on health effects is more agitational than scientific. What is proven is that, at high concentrations, spores of aspergillus fumigatus can colonize the lungs, causing an illness called aspergillosis. Tests on laboratory animals showed that spores of stachybotrys are so toxic that inhaling high concentrations can cause a deadly lung disease called IDH. At low concentrations, however, the white blood cells eliminated the spores. And, the likelihood of encountering spore concentrations in buildings as high as those during the tests is close to zero.

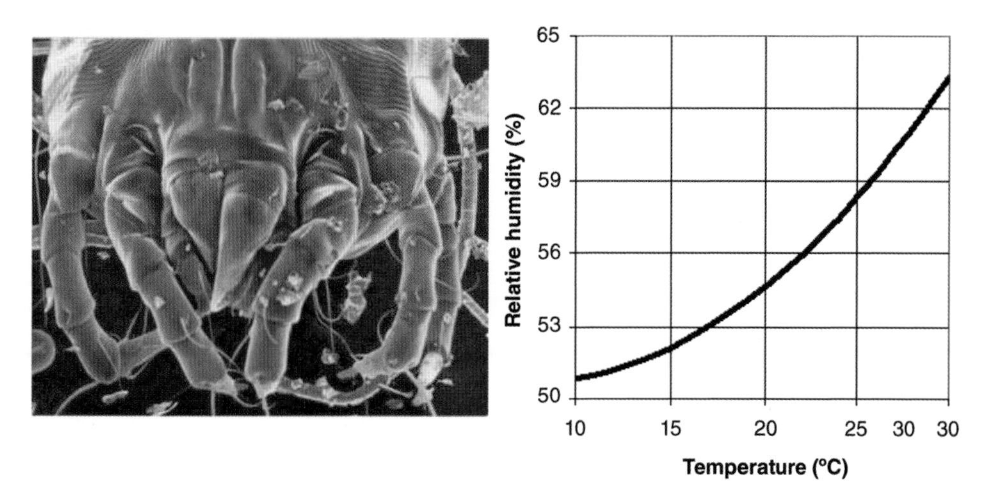

Figure 3.14 Critical relative humidity for the dust mite dermatophagoides farinae.

3.2.9.4 Dust mites

Dust mites are colourless, 0.5 mm long octopods that colonize carpets, textiles, cushions and beds and that feed on human and animal hair and scales. There are 46 different known species with 13 of them found in buildings, and three being found worldwide. Dust mites don't drink. Respiration and perspiration happens through the skin of the back, abdomen and legs. In this way, their moisture balance depends on temperature and relative humidity in their direct environment. The range of 10–30 °C is preferred, though relative humidity for dehydration differs between species. For dermatophagoides farinae the value is:

$$\phi_{\text{crit}} = 52.1 - 0.375\theta + 0.025\theta^2 \tag{3.37}$$

Above the curve shown in Figure 3.14, though depending on temperature, the growth from egg to mature mite accelerates substantially, from 122.8 ± 14.5 days at 16 °C to 15 ± 2 days at 35 °C. For other species the critical relative humidity scatters around 65%.

Dust mite density relates directly to house dust allergy that affects some 20% of the population. Risk analysis in Denmark showed that exposure to the digestive enzymes in mite excrements, suspended in the air, explained up to 60% of the asthma cases. The risk of developing hypersensitivity starts at a concentration of 2 μg per gram of house dust. In order for a sensitive person to get allergic attacks, it has to exceed 10 μg per gram.

3.2.9.5 Insects

The presence of insects indicates spots in buildings showing elevated humidity. Mosquitoes breed and hibernate in moist crawl spaces. Moving into the building

Table 3.9 Mean skull width of rodents and insectivores.

Species	Skull width mm
Common pipistrelle (protected)	7–8
Shrew mouse	9–10
House mouse	10–12
Black rat	16–22
Brown rat	17–27
Hedgehog (protected)	28–38

from there is no problem given the many leaks around pipes in the ground floor. At night the bloodthirsty females come out and attack the sleeping human victims. Humid cavities behind baseboards and rotting timber sills shelter cockroaches, fleas and other insects. In buildings with air heating, cockroaches sometimes become a real nuisance as they nestle in the air ducts. Atopic people react allergically to certain enzymes in their excrement, which the warm supply air ejects. Extermination of insects requires pesticides. Often inhabitants use spray cans, thus allowing the pesticide to pollute the indoor air.

3.2.9.6 Rodents

Due to the damage they cause, rodents are not welcome in buildings. Blood-diluting products are popular in combating them. A better method is preventing entrance by sealing all openings wider than their skull, see Table 3.9.

3.2.9.7 Pets

Cats release substances that provoke allergic and asthmatic reactions in atopic people. An investigation in 93 office buildings in the USA showed that nearly all dust on the floor contained cat allergens although hardly any cats ever entered the buildings. The average concentration sampled was $0.3\,\mu g/g$, the maximum $19\,\mu g/g$. Developing hypersensitivity starts at $8\,\mu g/g$, a value exceeded in 7 of the 93 buildings. The source proved to be the dust coming from the clothes that the employees wore from home, where they had a cat.

3.2.10 Human related contaminants

Humans also contaminate their environment. They pick up oxygen when inhaling and emit carbon dioxide and water vapour when exhaling. Perspiration and transpiration increases water vapour release, while bio-odours out-gas through the skin and are freed by respiration and flatulence. After the (re)discovery of America and the tobacco plant, people started smoking, first in Europe, later worldwide.

3.2.10.1 Carbon dioxide (CO_2)

Healthy humans inhale 4 M litres of air per hour with M their metabolism in watts. The lungs pick up some 27% of the oxygen in the inhaled air and add 4% m^3/m^3 of CO_2. This way, inhaled and exhaled air consists of the following:

	Inhaled air, % m^3/m^3	Exhaled air, %m^3/m^3
Oxygen	20.9	15.3
Nitrogen	79.6	74.5
Carbon dioxide	0.04	4.0
Water vapour	?	6.2

Individuals therefore emit 0.16 M norm-litres of carbon dioxide per hour (M in W). Until a concentration of 50 000 ppm, CO_2 is not considered poisonous, although recent research, albeit disputed, has shown that indoor air concentrations above 1000 ppm could have some effect on intellectual activity. Experiments with 22 subjects subjected to low (600 ppm), medium (1000 ppm) and high (2500 ppm) CO_2-concentrations for 2½ hours in fact showed some reductions in decision-making performance, see Table 3.10. Concentrations nearing 5000 ppm lessen attention even more and initiate drowsiness. Above 10 000 ppm drowsiness reigns, and beyond 40 000 ppm people start complaining of headaches.

3.2.10.2 Water vapour

Water vapour is not a contaminant as such. Its presence, however, activates (S)VOC outgassing and favours bio-germs. At normal activity, humans release less than 40–60 grams of water vapour per hour, though emissions increase proportionally to metabolism (see Figure 3.15).

3.2.10.3 Bio-odours

Bio-odours such as isoprene and 2-propanone are VOCs. Their release is most easily sensed when entering a badly ventilated room having been occupied shortly before by a group of people. It smells 'stuffy'.

Table 3.10 CO_2 impact on human decision-making performance.

Classes	Basic activity	Applied activity	Focused activity	Task orientation	Initiative	Information orientation	Information utilization	Breadth of approach	Basic strategy
Superior	O								
Very good	X	O,X	O,X,Y	O, X		O,X,Y			
Average		Y		Y	O,X		O,X	O,X	O,X
Marginal	Y						Y	Y	
Dysfunctional					Y				Y

O 600 ppm; X 1000 ppm; Y 2500 ppm.

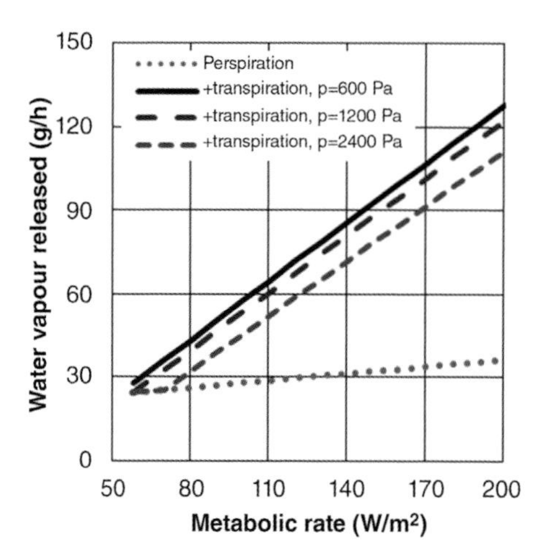

Figure 3.15 Water vapour released by people.

3.2.10.4 Tobacco smoke

Slowly burning hydrocarbons is very efficient. The reaction with oxygen produces carbon dioxide and water vapour. Tobacco, however, contains such a complex set of compounds that even slow burning emits an amalgam of airborne contaminants with nicotine as the habituating substance, see Table 3.11. At the same time, smoking figures as a source of obvious olfactory irritation.

The primary contaminant flow consists of the smoke inhaled by the smoker, the secondary one being the smoke released by the burning cigarette. An active smoker is exposed to a mixture of both, whereas the passive smoker inhales a mixture of secondary and exhaled primary flow. A considerable amount of the primary contaminants remains in the respiratory tract of the active smoker. They are therefore most exposed to the health risks that smoking induces. Although the mixture of exhaled primary and secondary flow is different, in the long run passive smokers are also exposed to analogous health risks. Researchers at the US Department of Energy's Lawrence Berkeley National Laboratory proved that third-hand smoke formed by the reaction of nicotine with indoor nitrous acid and ozone and giving potentially harmful ultrafine particles continues to have impacts for many hours after cigarettes are extinguished. In the tests, more than 50 VOCs and airborne particles were still present 18 hours after smoking took place. Up to 60% of the potential noxious effects remained up to two hours after smoking had stopped.

For chain-smokers, health risks include heart and vascular disease, pulmonary complaints, stomach complaints and lung cancer. Smoking more than one packet of cigarettes a day increases premature death occurrence with 15%, meaning that 1 in 6 chain smokers die because of their habit. For passive smokers it's 1 in 60. To avoid health problems for their child, pregnant women are firmly discouraged from smoking.

Table 3.11 Airborne contaminants in tobacco smoke.

Contaminant	Concentration mg per cigarette	MAC ppm
NO_X	1.801	
NO	1.647	
NO_2	0.198	
CO	55.10	50
Ammoniac	4.148	25
Acetaldehyde	2.500	100
Formaldehyde	1.330	
Acetone	1.229	1000
Acetonitrile	1.145	
Benzene	0.280	
Toluene	0.498	
Xylene	0.297	
Styrene	0.094	
Isoprene	6.158	
1.3 Butadiene	0.372	5000
Limonene	1.585	
Nicotine	0.218	
Pyridine	0.569	
Dust	13.67	
Hydrogen cyanide	1600 ppm	10
Methyl chloride	1200 ppm	100

3.2.11 Perceived indoor air quality

3.2.11.1 Odour

People typically judge the indoor air quality by smelling. Yet, evaluating malodour objectively is a difficult task, which is why at the end of the 1980s P.O. Fanger proposed a perception-based methodology, called the olf/decipol rationale, a source and field model with a human panel as reference instrument. The olf figures as emission unit with a value 1 representing the bio-odours emitted by an adult male with a body area of 1.8 m^2, who is lightly active, takes a shower five times a week, puts on fresh underwear every day and uses deodorant moderately. The decipol (dP) in turn represents odour intensity, where 1 decipol characterizes the olfactory pollution a 1 olf male causes in a room aired with 36 m^3 of zero olf fresh air per hour.

A human panel of ten individuals is trained to quantify decipol values. They first learn to estimate the scale using reference sources. Once trained, they can estimate malodours in spaces during a short visit. The average of the decipols noted by the members represents the perceived value. With the ventilation flow known, this value allows the olfs present to be calculated. In fact, if dP_e is the decipol value outdoors, dP_i the value indoors, n the ventilation rate in ach and V the air volume in the space in m^3, the number of olfs equals:

$$P_{olf} = (dP_i - dP_e)nV/36 \tag{3.38}$$

In most cases, more olfs are noted than people present. These come from out-gassing materials and poorly maintained ventilation systems. The investigations helped weighting sources:

Source	Olf
Reference person	1
Active person, 4 met	5
Active person, 6 met	11
Smoker with cigarette	25
Smoker without cigarette	6
Materials and ventilation systems	0–0.4 olf/m^2

The many panel visits resulted in a statistical relation between perceived malodour in decipol (dP_i) and the number of dissatisfied (PD, %, Figure 3.16):

$$dP_i = 112[\ln(PD) - 5.98]^{-4} \tag{3.39}$$

ASHRAE quotes 20% dissatisfied as acceptable, fixing 1.4 dP as the limit for perceived malodour. In 'sick buildings' the value may be more than 10 dP.

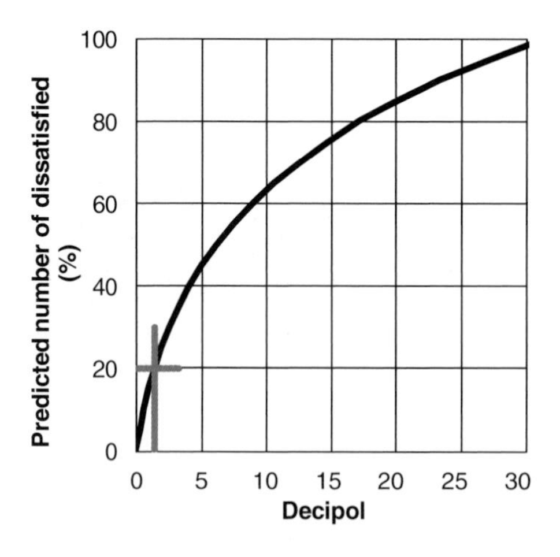

Figure 3.16 The predicted number of dissatisfied (PD) versus decipol relation.

The olf/decipol model is criticized. Using a human odour panel as a measuring instrument induces random doubts about the decipol values perceived, while questions such as 'how do you calibrate the decipol sources used for training?' cast a shadow over the method.

3.2.11.2 Indoor air enthalpy

Complaints about bad indoor air quality seem to multiply with increasing air enthalpy (h), per kg of dry air given by:

$$h = 1008\,\theta + x_v(2\,500\,000 + 1840\,\theta) \tag{3.40}$$

with θ the air temperature in °C and x_v the vapour ratio in kg/kg with as link to relative humidity:

$$\phi = \frac{x_v P_{atm}}{p_{sat}(0.621 + x_v)} \quad (0 \leq \phi \leq 1)$$

where P_{atm} is the atmospheric pressure ($\approx$100 000 Pa) and p_{sat} the vapour saturation pressure at the air temperature in Pa. The random relation between enthalpy and PD as deduced from experiments is:

$$PD = 100\frac{\exp[-0.18 - 5.28(-0.033h + 1.662)]}{1 + \exp[-0.18 - 5.28(-0.033h + 1.662)]} \tag{3.41}$$

Also here, 20% dissatisfied is taken as acceptable. Investigations in classrooms showed that values logged during teaching hours largely exceed that limit, see Figure 3.17. Ventilation with operable windows opened between teaching hours clearly fails in guaranteeing acceptable indoor air quality.

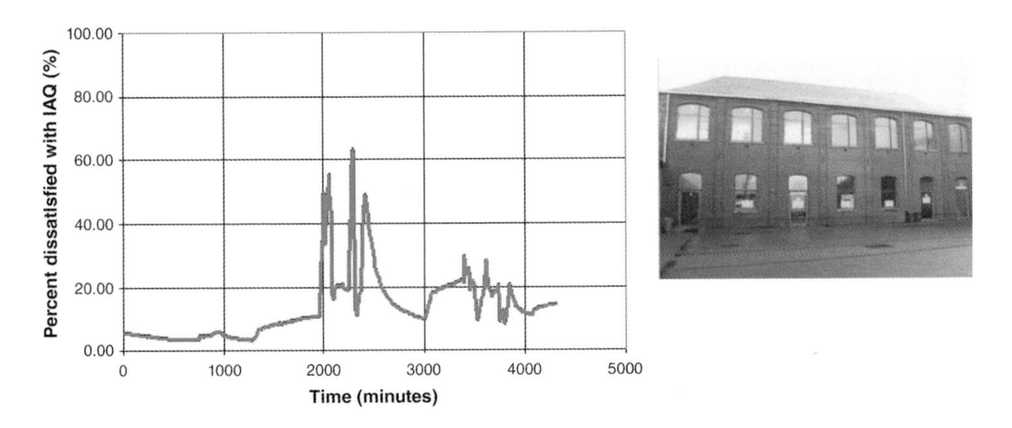

Figure 3.17 Classroom in a secondary school, percentage of dissatisfied with perceived indoor air quality based on measured air enthalpy.

Again, not everyone agrees on the role air enthalpy plays. Some presume it only gains importance when people feel thermally uncomfortable. Individuals then perceive the combination of high air temperature and high relative humidity as oppressive, though lower air temperatures may not stop the complaints. By contrast, a low relative humidity always seems more pleasing.

3.2.12 Sick building syndrome (SBS)

A much-discussed issue linked to bad environmental quality in offices is SBS. In literature, an office building is called sick when contact with outdoors is lacking, space, privacy and quietness fails, visual, acoustical and thermal comfort is questionable and unwanted contaminants pollute the air. Complaints disappear during weekends and holidays but return when work is resumed. Symptoms are eye, nose, sinus and throat irritation, difficulty in breathing, heavy chest, coughing, hacking, headaches and dry skin.

The impact of air quality has been researched intensively, among others by looking to the ventilation system and the magnitude of the ventilation flows, see Figure 3.18. Symptom prevalence at first diminishes but then at higher ventilation rates uncertainty increasingly dominates.

Many studies did not consider other probable causes. Better analyses concluded that fully air conditioned office buildings show 30–200% more SBS complaints than naturally ventilated ones. This is mainly due to dirty ductwork and contaminated filters. Air humidification looked especially guilty. Also the link with the CO_2-concentration has been investigated. Each extra 100 ppm apparently resulted in a 10–30% increase in dry mucous membrane, irritated throat, stuffy nose and coughing complaints.

Another cause of SBS complaints commented on in recent years is work stress. In fact, employees who perform mandated duties complain more than executives. In the cases investigated, the complainants called their job stressful and lacking encouragement. No relation came out with ventilation effectiveness, VOC-concentration or noise level

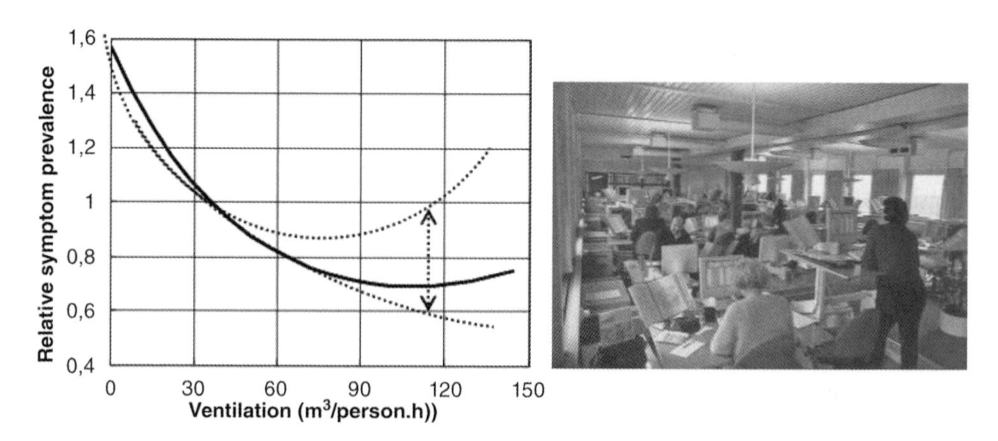

Figure 3.18 Relative symptom prevalence versus better ventilation, mean curve, 95% confidence band (dotted lines).

indoors, albeit that employees who could adapt their working environment by adjusting the heating or operating windows complained less. Clearly, SBS is not only the result of failing building and HVAC performance but is also a consequence of psychosocial stress. The author's own experiences have confirmed that conclusion, though the kind of office building and its HVAC-system also plays a part.

3.2.13 Contaminant control

As mentioned, contaminant control includes emission minimization, diluting the concentrations to acceptable levels by ventilation, disinfecting the air or taking personal protective measures.

3.2.13.1 Minimizing emission

Source control forms a first line of defence. To give examples, the use of asbestos and pentachlorinephenols is no longer allowed. Asbestos cement based corrugated plates and slates are not manufactured anymore. Tight foundation floors minimize radon immission in newly built homes. That measure of course is hardly applicable in existing premises, where footing-founded partition and outer walls divide up the floors on grade. Many countries forbid smoking in public buildings, schools, restaurants and pubs. In temperate and cold climates, well insulated, thermal bridge free, airtight envelopes figure as a source control because they minimize mould risk.

3.2.13.2 Ventilation

Dilution by ventilation offers a second line of defence. For human-related pollution it's the main way to get better indoor air quality and to keep carbon dioxide to acceptable levels. Against bio-germs, it is one of the measures.

Ventilation effectiveness concerns the way the supply air replaces the room air. A goal could be a room-wide perfect blend, though short circuits between supply and return hamper perfect mixing. Better is to displace the contaminated air by creating a laminar progressing fresh air front, but turbulences caused by furniture, apparatus and upward stack around people will likely break that front.

To evaluate the ventilation needs, assume that a pollutant in the supply air with concentration x_{oj} (kg/kg) is emitted at rate $G_{P,j}$ (kg/s). Allowed in the room air is a concentration x_{aj} (kg/kg). In case of mixing or displacement ventilation, the steady state ventilation flow ($G_{a,j}$ in kg/s) required equals:

$$G_{a,j} = \left(\frac{G_{P,j}}{x_{aj} - x_{oj}}\right)\frac{1}{\varepsilon} \tag{3.42}$$

with ε ventilation effectiveness. Ideal displacement ventilation has effectiveness 2. Real displacement scores between 1 and 2. Ideal mixing gives a value 1, real mixing a value below 1. The formula applies per pollutant. The flow needed then is the largest calculated:

$$G_a = \max\left(\in G_{a,j}\right) \tag{3.43}$$

Table 3.12 MAC- and AIC-values for carbon dioxide ($c_{CO_2,i}$).

Country	MAC (ppm)	AIC (ppm)
Canada, USA	5000	1000
Germany, the Netherlands, Switzerland	5000	1000–1500
Italy	5000	1500

Under transient conditions and perfect mixing, concentration x_{aj} of contaminant j with source strength $G_{P,j}$ totals:

$$\rho_a V_a \frac{dx_{aj}}{dt} = G_{a,j}\left(x_{oj} - x_{aj}\right) + G_{P,j} \tag{3.44}$$

with V_a room air volume. Solving this first-order differential equation can be done analytically or numerically, depending on the initial and boundary conditions for the concentration outside, the ventilation flow and the source strength (x_{oj}, $G_{a,j}$, $G_{P,j}$) as time functions.

Humans being the main pollution source means that carbon dioxide is the perfect tracer for quantifying ventilation. Although Table 3.12 shows differences in AIC values between countries, the Pettenkofer number of 1500 ppm seems an acceptable highest threshold ($c_{CO_2,i}$).

With 390 ppm ($c_{CO_2,e}$) outdoors and 0.16 M litres per hour per individual (G_{CO_2}), the ventilation flow per person per hour needed to keep the concentration indoors below 1500 ppm is:

$$\dot{V}_{a,pers} \geq \frac{1000 G_{CO_2}}{c_{CO_2,i} - c_{CO_2,e}} = \frac{160\,M}{1110} = 0.153\,M \qquad (m^3/(h.pers)) \tag{3.45}$$

with M the metabolism in watts. Also see Table 3.13. With an increasing carbon dioxide concentration outdoors, more air per person will be needed. Anyhow, the average at low activity seems to be some 20 m^3 per hour. For non-residential buildings the standard EN 13779 works with four classes, see Table 3.14.

Carbon dioxide as a tracer also helps quantifying the ventilation flow. Figure 3.19 summarizes data measured in 19 classrooms with different occupation at the end of a teaching hour. The mean largely exceeds 1500 ppm, with the MAC value of 5000 ppm as the worst result. The reason for these alarming values is the lack of classroom ventilation.

Whether a building will suffer from mould depends on several factors with temperature and relative humidity outdoors, thermal insulation of the envelope, presence of thermal bridges, water vapour released indoors, heating habits and airing intensity as the most important ones. The linked simple design rule for 'no visible mould on inside surfaces' looks like: the four-week mean relative humidity at any surface not beyond 80%.

Table 3.13 Ventilation flow as a function of activity.

Activity	Metabolic rate W/m^2	Ventilation flow, AIC 1500 ppm m^3/(h.pers)
Sleeping	41	10.0
Lying	46	11.5
Seating	58	14.6
Standing	70	17.6
Laboratory work	93	23.4
Teaching	93	23.4
Typing	62–68	15.6–17.1
Studying	78	19.7
Dancing	139–255	35.0–64.3
Playing tennis	240–267	60.5–67.2
Playing basketball	397–441	100.0–111.1

Without condensation on other surfaces, that mean is:

$$\phi_s = 100\left(\frac{p_i}{p_{sat,s}}\right) = \frac{100}{p_{sat,s}}\left(p_e + \frac{462 T_i G_{v,P}}{\dot{V}_a}\right) \tag{3.46}$$

with p_i vapour pressure indoors, $p_{sat,s}$ saturation pressure at the surface, T_i temperature indoors in K, $G_{v,P}$ vapour released indoors and $\dot{V}_a$ the ventilation flow (all four-week averages). Saturation pressure and surface temperature (θ_s) are linked:

$$p_{sat,s} = 611 \exp\left(\frac{17.08\theta_s}{234.18 + \theta_s}\right) \tag{3.47}$$

while the monthly mean inside surface temperature somewhere on the envelope equals:

$$\theta_s = \theta_e^* + f_{h_i}\left(\theta_o - \theta_e^*\right) \tag{3.48}$$

Table 3.14 The four indoor air quality classes.

Class	Indoor air quality	CO_2 excess indoors (ppm)	
		Interval	Mean
IDA 1	High	≤400	350
IDA 2	Mean	400–600	500
IDA 3	Moderate	600–1000	800
IDA 4	Low	≥1000	1200

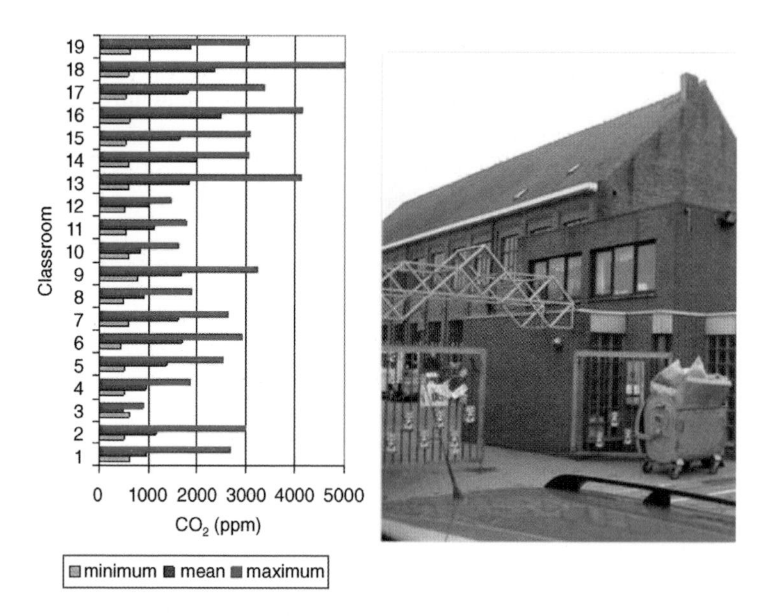

Figure 3.19 Carbon dioxide concentrations measured in classrooms.

with f_{hi} the temperature factor:

$$f_{h_i} = (\theta_s - \theta_e^*)/(\theta_i - \theta_e^*) \tag{3.49}$$

θ_e^* is the four-week mean sol-air outside temperature and θ_o the four-week mean operative temperature indoors, both in °C. The four equations fix the ventilation needed, if all other parameters are known. Figure 3.20 summarizes results for January at Uccle.

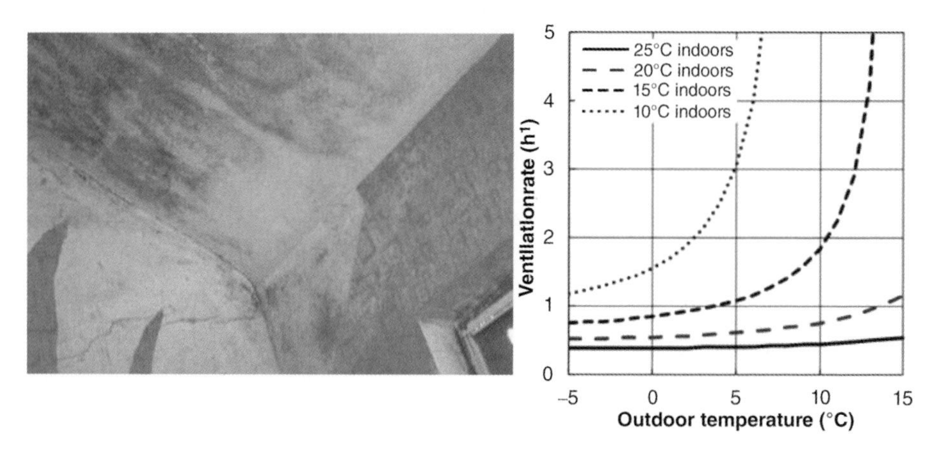

Figure 3.20 January at Uccle. Ventilation needed to avoid mould in a 200 m^3 large dwelling with lowest envelope temperature ratio 0.7, for 14 kg/day water vapour release.

Less ventilation is needed at higher temperatures indoors. Conversely, more ventilation is required when it is warmer outdoors, when more vapour is released indoors and when the temperature ratios at some locations on the opaque parts of the envelope are lower. Thus, excluding mould in temperate and cold climates is not just a matter of ventilation. It is better to ask what temperature ratio reduces mould risk to 0.05 in dwellings, where vapour release is average and ventilation plus heating guarantee a comfortable, healthy indoors. Calculations for temperate regions gave a value 0.7. In hot and humid climates, only drying the ventilation air, for example by a dedicated outdoor air system (DOAS), helps.

Another issue is avoiding dust mite overpopulation. The best here is keeping hygroscopic moisture in carpets, mattresses, sheets, blankets and wall textiles below the threshold of Figure 3.14. During winter, enough thermal insulation, hardly any thermal bridging, good ventilation and correct heating help. Mattress finishes, sheets and blankets with low specific surface, sudden chilling by airing during cold weather and washing bedding at high temperature ($\approx$60 °C) are also beneficial. In summer and in warm humid climates only air-drying gives relief.

What does good perceived indoor air quality require? As mentioned, 20% dissatisfied demands a decipol value of 1.4. If the load indoors (P_{olf}) and the decipol value outdoors (dP_{e}) are known, less than 1.4 indoors requires a ventilation flow of:

$$\dot{V}_{\text{a}} = 36P_{\text{olf}}/(1.4 - dP_{\text{e}}) \qquad (\text{m}^3/\text{h}) \qquad (3.50)$$

For the decipol values outdoors, the following numbers apply:

	dP_{e}
Mist	>1
Urban environment	0.05–0.35
At the coast, in the mountains	0.01

In urban environments, the ventilation flow needed per person thus becomes:

$$36P_{\text{olf}}/(1.35m) \leq \dot{V}_{\text{a,pers}} \leq 36P_{\text{olf}}/(1.05m)$$

with m the number of individuals present. In a typical office, building materials and HVAC represent some 0.4 olf per square metre of floor area (A_{fl}) and each employee equals some 1 olf, giving as lower and upper ventilation threshold per individual:

$$27 + 10.7A_{\text{fl}}/m \leq \dot{V}_{\text{a,pers}} \leq 34 + 13.7A_{\text{fl}}/m \qquad (3.51)$$

For 10 m^2 per employee, the result is $134 < \dot{V}_{\text{a,pers}} < 171$ m^3/(h.pers), a multiple of the value guaranteeing allowable carbon dioxide concentrations. Use of low out-gassing materials and good maintenance of the HVAC system can lower their impact to less than

Table 3.15 The four indoor air quality classes: decipol value.

Class	Perceived indoor air quality (*dP*-value)		
	Interval	Mean	% dissatisfied
IDA 1	≤1.0	0.8	13
IDA 2	1–1.4	1.2	18
IDA 3	1.4–2.5	2.0	26
IDA 4	≥2.5	3.0	33

0.1 olf/m^2, reducing the ventilation need to 54–66 m^3/(h.pers). Again, EN 13779 differentiates between four IDA-classes, see Table 3.15.

Turning to smoking, assuming that smoking indoors is allowed, ventilation then depends on the number of cigarettes smoked per hour, with 120 m^3/h per cigarette needed to ensure a *PD* with perceived indoor air quality below 20%. Smoking one cigarette takes about 7.5 minutes. A chain smoker consumes 3–6 cigarettes an hour, which requires 360–720 m^3 of fresh air per hour. In other words, a room filled with chain smokers requires vast amounts of ventilation. An alternative consists of applying the olf/decipol approach by adding 0.05 times the percentage of smokers in the room as extra olfs. The additional ventilation flow then equals:

$$\frac{36 \times 0.05 \times \%\text{smokers}}{1.4 - 0.05} \leq \Delta\dot{V}_{\text{a,pers}} \leq \frac{36 \times 0.05 \times \%\text{smokers}}{1.4 - 0.35} \qquad (\text{m}^3/(\text{h.pers}))$$

or:

$$1.33 \times \%\text{smokers} \leq \Delta\dot{V}_{\text{a,pers}} \leq 1.71 \times \%\text{smokers} \qquad (\text{m}^3/(\text{h.pers}))$$

For one chain smokers per 10 m^2, an expensive high amount of 268–328 m^3 fresh air an hour are still needed. To keep things affordable, the practice therefore is to double the ventilation flow compared to a non-smoking environment. This could result in a few complaints about perceived air quality, though investigations have shown that smokers are more tolerant than non-smokers. Happily, in many countries, smoking indoors in public and non-residential buildings is now forbidden.

All the basics just discussed still lack clarity, which is why standards took over. The four IDA classes, for example, require the ventilation flows listed in Table 3.16.

Table 3.17 gives design values for residential buildings. Read design as 'inlets and outlets to be dimensioned for these flows'.

ASHRAE standard 62.2-2013 offers the following formula for the total ventilation flow in residences:

$$\dot{V}_{\text{a}} = 0.54A_{\text{fl}} + 12.6(n_{\text{br}} + 1) \qquad (\text{m}^3/\text{h}) \qquad (3.52)$$

Table 3.16 Ventilation in non-residential buildings according to EN 13779.

Class	Ventilation flow per person present in m^3/h			
	Smoking not allowed		Smoking allowed	
	Interval	Mean	Interval	Mean
IDA 1	>54	72	>108	144
IDA 2	36–54	45	72–108	90
IDA 3	22–36	29	44–72	58
IDA 4	<22	18	<44	36

with A_{fl} the floor area in m^2 and n_{br} the number of bedrooms. The standard assumes two persons at minimum in a studio or a one-bedroom home and an additional person per extra bedroom. With higher occupancies for a given number of bedrooms, 12.6 m^3/h per person must be added. The formula results in lower flows than the EN standards. The main reason is that the latter considers entering a room that has housed people as critical, while the ASHRAE standard is based on the lasting odour perception after entering.

3.2.13.3 Air cleaning and personal protective measures

If neither source control nor ventilation is possible or feasible, air cleaning or personal protective measures have to take over. Cleaning replaces ventilation, and research has shown that it can be quite effective and energy efficient. As an example of personal

Table 3.17 Design ventilation flows in residential buildings, Belgian standard NBN D50-001.

Room	Design ventilation flow
Inlets	
Living room (inlet)	3.6 m^3/h per m^2 of floor area, with a minimum of 75 m^3/h and a maximum of 150 m^3/h
Sleeping room, playroom	3.6 m^3/h per m^2 of floor area with a minimum of 25 m^3/h and not exceeding 36 m^3/h per person
Outlets	
Kitchen, bathroom, washing room, etc	3.6 m^3/h per m^2 of floor area, with a minimum of 50 m^3/h and a maximum of 75 m^3/h
Toilet	25 m^3/h
Flow through spaces	
Corridors, staircases, day and night hall and analogous spaces	3.6 m^3/h per m^2 of floor area
Garage	Should be ventilated separate from the dwelling

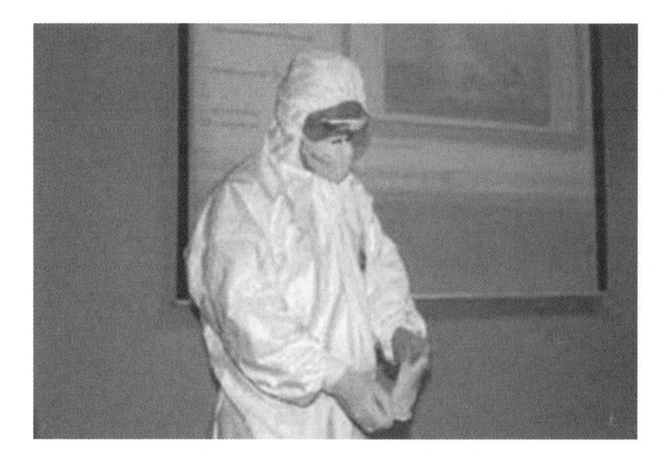

Figure 3.21 Whole body garment with dust mask.

protective measures, when installing glass fibre or mineral fibre insulation, workers must wear whole-body garments with a dust mask, see Figure 3.21.

Clearing asbestos used as insulation or fire protection in existing buildings requires workers to wear a complete protective suit included oxygen mask, while the spaces where asbestos is removed must stay under-pressurized and shielded from adjacent rooms. That is why only specialized contractors are allowed to do this job. Less severe but still quite strict precautions are imposed when stripping asbestos cement products.

3.3 Energy efficiency

3.3.1 In general

Energy keeps the world moving. Industry, traffic, residential plus equivalents and energy transformation determine the source consumption. While industry consists of big users that need energy for production, traffic and residential plus equivalents instead represent huge numbers of small consumers that appreciate the services provided without much concern about the energy required. Only when rising prices make the services more expensive, do people protest. Yet, from a societal point of view, energy is a constant challenge with availability, affordability and sustainability the main factors. Many industrialized countries have to import coal, oil and gas, which not only impacts their balance of payments but also creates political issues because large amounts of fossil fuels come from less stable countries.

Environmentally alarming, when looking at energy consumption worldwide, is the global warming issue, the chief culprits for which are the huge amounts of CO_2 that fossil fuel combustion emits. The Kyoto protocol of 1997, ratified in 2004, demanded the developed world a 5% cut in emissions between 2008 and 2012 compared to the 1990 level. That objective wasn't reached. Instead, emissions worldwide still continue to rise. The post Kyoto era therefore looks troubling. To stabilize global warming, cuts between 50 and 80% are needed in 2050 compared to 1990. This will change the economy, demand another energy production and impact transportation, town planning, building design and retrofit. In the interrim, the EU has adopted a 20/20/20 target for 2020: 20% end energy covered by renewables, 20% less global warming gas emissions than in 1990 and 20% less energy consumed compared to business as usual. For 2030, 35% less GWG is the objective.

Progress in 'residential plus equivalents' requires a consequent application of the 'trias energetica': first minimize end energy demand, then use renewables as far as economically feasible and finally, use for what's left fossil fuels as efficiently as possible. A problem with the renewables of wind and sun is their variability. In today's society power guarantee at any given moment is a necessity. Smart grids with demand-side management and energy storage may help, but back-ups will still be necessary to align production with demand.

In Europe, governments promote rational use of energy. In the building sector, with its millions of non-knowledgeable decision-makers, legal obligations are most effective. There are many arguments in favour of limiting energy use. If correctly designed and constructed, energy-efficient buildings offer better thermal comfort, a healthier indoor environment and longer service life than energy-consuming premises, while each fossil megajoule avoided helps in combating global warming.

3.3.2 Some statistics

According to the International Energy Agency (IEA-OECD) residential plus equivalent represents some 40% of the annual end use in developed countries, see Figure 3.22 for the Flanders region in Belgium.

Despite all mandatory efficiency measures since 1992, annual end use there, corrected on degree days, is still increasing:

$$E_{\text{end,ann,resid+eq.}} = -3789 + 0.086 D_{15}^{15} + 1.997\,\text{Year} \qquad (\text{PJ/annum})$$
$$r^2 = 0.646 \quad \sigma_{0.085} = 0.018 \quad \sigma_{1.997} = 0.591$$

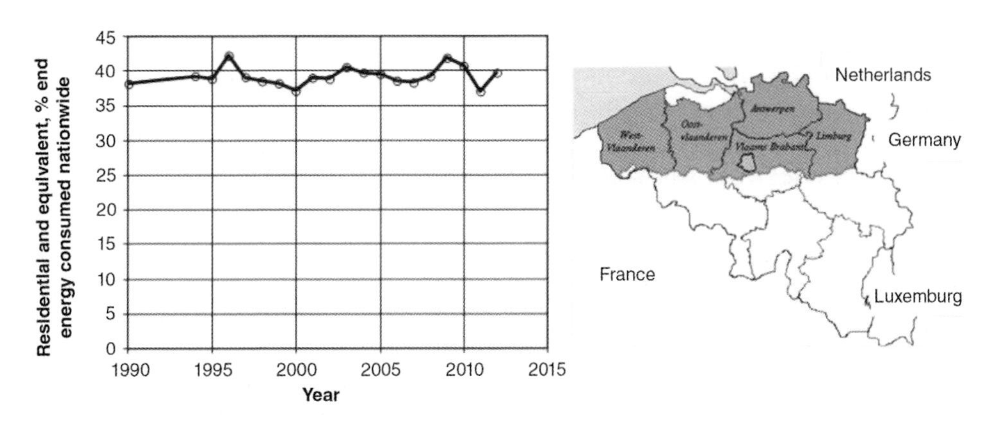

Figure 3.22 Flanders, Belgium, residential and equivalents: share in the annual end use.

The larger number of homes needed due to smaller households, a slow population growth and the prevalence of detached housing in new construction has the effect of neutralizing the ever more severe mandatory metrics. If the energy that is used for manufacturing materials and components, transporting them to building sites, constructing, retrofitting and demolishing buildings and conditioning the manufacturing plants needed is added, the buildings' share in the overall total consumed is definitely the largest. As most energy uses still involve fossil fuels, the built environment's share in the total amount of CO_2 emitted is very important.

3.3.3 End energy use in buildings

Over the past decade, buildings seem to have been constructed and retrofitted to conserve energy. This of course is untrue. First, buildings must offer good comfort, a healthy environment, functionality, adaptability and global sustainability, but this should, of course, be done as energy efficiently as possible.

End energy includes lighting and appliances (LA), domestic hot water (DHW) and heating, ventilating and air conditioning (HVAC) – but mainly heating in temperate and cold climates. As an example, Table 3.18 summarizes data for 201 dwellings built between 1910 and 1995 in a country with temperate climate. On average, lighting and appliances took 16% of the total. Domestic hot water had an equal share, while heating consumed the remaining 68–75%. The three together gave an average of 180 kWh per square metre floor per year. Of course, quite a lot of energy has to be added to this for traffic, surely when ribbon development and scattered habitation prevail as settlement patterns.

3.3.3.1 Lighting and appliances

Household, office work, teaching and nursing, all demand appliances that require energy, in most cases electricity. Everyone also wants task-appropriate comfortable lighting, again consuming electricity.

In residential buildings, the annual electricity so consumed can statistically be written as:

$$Q_{\mathrm{LA,a}} = \sum Q_{\mathrm{basis}} + P_{\mathrm{eff}} \sum Q_{\mathrm{extra,P_{eff}}} + \max[0, 0.8(V - 540)] \sum Q_{\mathrm{extra,A}}/3 \qquad (3.53)$$

with Q_{basic} the usage per appliance as listed in Table 3.19, $Q_{\mathrm{extra,Peff}}$ a correction factor per effective household member, V the protected volume in m^3, equal to the total volume

Table 3.18 Annual end energy use in 201 dwellings built between 1910 and 1995.

		Mean	St.dev.	Minimum	Maximum
Gross floor area	m^2	242	89	98	485
Total end use (H+DHW+LA)	$kWh/(m^2.a)$	180	68	66	488
End use for H and DHW	$kWh/(m^2.a)$	149	71	66	428
End use for LA	$kWh/(m^2.a)$	28	14	8	74

Table 3.19 Annual end energy for appliances and lighting.

Appliance	Energy label	Q_{basis} kWh/a	$Q_{extra,Peff}$ kWh/(a.P_{eff})	$Q_{extra,A}$ kWh/(m^2a)
Stove	A++	244	57	0,03
	A+	284	66	0,03
	A	382	89	0,05
	B	512	120	0,06
	C	650	152	0.08
Dishwasher	A	263	63	0,02
	C	360	86	0,03
Washing machine	A	228	65	0,02
	C	300	86	0,03
Dryer	A	356	46	0,02
	C	440	57	0,03
Lighting, mainly incandescent lamps		700	190	0.69
Lighting, low-energy light bulbs, LEDs		233	53	0.23
Fridge	A+	138	44	0,03
	C	270	85	0,06
Freezer	A+	205	68	0,03
	C	400	133	0,06
Fridge with freezing part	A+	274	86	0,06
	C	959	302	0,21
PC, cathode screen		150	0	0
PC, LCD screen		75	0	0
Pump central heating, not controlled		300	0	0.97
Pump central heating, controlled		150	0	0.49
Others		400	161	2.08

minus the volume taken by the never-heated spaces (garage, attics), all measured using the outside dimensions, and $Q_{extra,A}$ a correction factor per appliance for each m^2 of gross area.

The number of effective household members (P_{eff}) equals:

$$P_{eff} = \sum_{j=1}^{m} \left(n_j/168\right)$$

with n_j the hours per week each is at home.

Both equations were evaluated using the field data of Table 3.20, which applies for the worst labelled appliances and incandescent lighting. The difference between the

Table 3.20 Electricity use for appliances and lighting measured in 26 homes.

		Mean	Stand.dev.	Minimum	Maximum
End energy	kWh/a	5804	1995	2025	9925
Primary energy[a)]	kWh_{th}/a	14 510	4990	5060	24 800

a) The end to primary conversion factor for electricity in the country considered is 2.5

calculated and measured mean was −4.5%, the difference between their standard deviations −21.7%, the difference with the measured maximum −16.6% and the difference with the measured minimum −44.5%.

With one effective member per 150 m^3 of protected volume permanently present and the worst labelled light bulbs and appliances installed, the relation between annual electricity use and protected volume (*V*) becomes:

$$Q_{LA} = 3742 + 6.79V$$

With the best labelled installed, the result is:

$$Q_{LA} = 2152 + 4.12V \quad (\text{kWh/a})$$

Figure 3.23 compares both with measured annual electricity uses in 125 residences. The least square line through the cloud of dots gives:

$$Q_{LA} = 3772 + 2.84V \qquad r^2 = 0.16 \qquad (\text{kWh/a})$$

For a protected volume less than 1250 m^3, the line passes the one for the best labelled light bulbs and appliances.

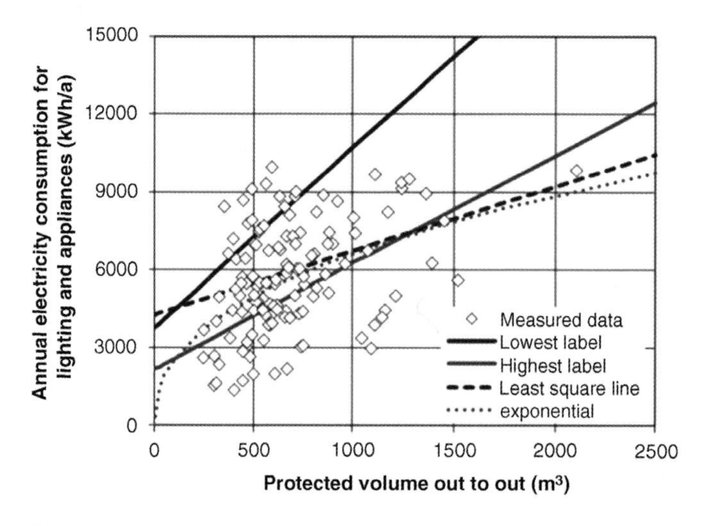

Figure 3.23 Logged annual electricity use for lighting and appliances in 125 residences.

Of course, as Table 3.18 confirms, end energy for lighting and appliances does not seem to be a prime concern in poorly insulated residences. Heating dominates, but with the move to low and nearly net zero energy homes, primary energy use for lighting and appliances will end up ahead, with related annual electricity cost largely exceeding fossil fuel costs. Or, the quest for energy-efficient types becomes pressing, though getting rid of all energy devouring choices will of course lower the internal heat gain.

In non-residential buildings energy use by appliances and lighting has to be evaluated in full detail. Which appliances? What power is required? What is the average use in terms of hours per week? How many lamps? What luminous efficiency and power? What switching systems are needed? Is daylight integrated? Table 3.21 collects some information for office buildings.

Table 3.21 Appliances and lighting in office buildings.

Appliance	Peak power, W	Mean, %peak	Convective, %	Radiant, %
Fluorescent lamp			50	50
Incandescent lamp			20	80
15″ monitor	220	35	63	37
Desktop computer	575	23	78	22
Laser printer	836	30	89	11
Copier	1320	14	86	14
Fax			65	35
Internal gains by appliances per m^2				
		Mean, W/m^2	Convection, %	Radiant, %
Low occupation	6.5 workstations per 100 m^2, printer, fax	5.4 W/m^2	70	30
Moderately occupied	8.5 workstations per 100 m^2, printer, fax	10.8 W/m^2	70	30
Quite well occupied	11 workstations per 100 m^2, printer, fax	16.1 W/m^2	70	30
Densely occupied	13 workstations per 100 m^2, printer, fax	21.5 W/m^2	70	30

3.3.3.2 Domestic hot water

In residential buildings annual end energy use for domestic hot water depends on the volume of water used, the temperature and the system efficiency. The volume changes with the number of household members, their age and activities, the standard of living and the hygienic habits. The water temperature required differs according to usage: hygiene, recreation, and so on. To avoid legionella, a daily burst to 60 °C is recommended. Heating, storage and distribution finally fix the system efficiency. The following formula estimates the related monthly net energy demand:

$$Q_{\text{dhv,net}} = \max[80, 80 + 0.275(V - 196)]\Delta t/1000000 \qquad \text{(MJ)} \qquad (3.54)$$

with Δt the month in seconds. The assumptions behind this are that a statistical relation exists between the protected volume and the number of effective household members, that the water is heated from 10 to 60 °C and that each member consumes on average 33 litres a day.

In many non-residential buildings, apart from those involving habitation (e.g. hotels and hospitals), the domestic hot water volumes needed are too low to have more than a marginal impact on the overall annual end energy used.

3.3.3.3 Space heating, cooling and air conditioning

In temperate climates passive measures such as restricted glass area, solar shading, massive construction and night ventilation easily render active cooling in domestic premises redundant, while a relative humidity control is mostly not needed. So, heating is the main end energy user there. In hot climates, however, active cooling and humidity control are a necessity, as are heating and humidity control in really cold climates.

3.3.4 Space heating

3.3.4.1 Terminology

Net energy demand for heating	Energy needed to keep a building on set-point temperature, assuming infinite system capacity and 100% efficiency
Gross energy demand for heating	Net energy demand for heating over a given period of time, divided by the corresponding building related system efficiency
End energy use for heating	Energy consumed by the boiler or heat pump to keep the building on set point temperature over a given period of time. Encompasses the energy used by the heating system to function properly (pumps, controls, fans)
Primary or source energy use for heating	End energy for heating, added the energy needed to produce and transport the energy carriers involved to the building site
Total net energy demand	Energy needed for heating, cooling, (de)humidification and domestic hot water assuming infinite system capacity and 100% efficiency

Total gross energy demand	Total net energy demand for heating, cooling, (de)humidification and domestic hot water, each divided by the corresponding building related system efficiency
Total end energy use	End energy consumed by the system to keep buildings on set point temperature, added the end energy for the system to function, for producing domestic hot water and for lighting and appliances, all over the same time span.
Total primary energy use	Total end energy over a given time span, added the energy needed to produce and transport the energy carriers involved to the building site
Exergy	Indicates which part of the energy is transformable into labour; the higher the exergy content, the better the quality of the energy
Steady state	No inertia effects included. Done by looking to time-averaged boundary conditions. Periods go from one day to a heating season. For calculations in moderate climates one month is typically used
Transient	Applies to calculations using instantaneous boundary conditions, stretching from minutes for systems and controls to hourly means for the building fabric
Protected volume	The sum of all spaces in a building where thermal comfort is required. In residential premises this volume includes the day- and night-time zone with exception of spaces left unheated (garages, attics, etc.). Quantification is done by measuring on the inside or on the outside, called in-to-in or out-to-out
Zone	Each part of a protected volume at different operative temperature. In residential buildings, zones may correspond to rooms and corridors. Often a split in day- and night-time suffices. Quite common is considering the protected volume as one zone
Envelope or enclosure, also called loss surface	The surface enclosing the protected volume. Includes roofs, facade walls, lowest floors and all partitions between the protected volume and adjacent unheated spaces, except party walls with neighbouring buildings at same operative temperature. The loss surface is measured in the same way as the protected volume: in-to-in or out-to-out
Loss surface	Each envelope face with different orientation and/or slope
Partition	Each wall or floor between adjacent zones
Mid plane	The middle of a vertical (wall) and the upper side of a horizontal partition (floor) between zones
Envelope part	Each fraction of the envelope with different physical properties (thermal transmittance, transient characteristics, longwave emissivity, shortwave absorptivity, solar transmittance)
Partition part	Each fraction of a partition with different physical properties (thermal transmittance, transient characteristics, longwave emissivity, shortwave absorptivity)

Inside temperature	The operative temperature averaged over the time span considered in each zone. Where the protected volume figures as one zone, the inside temperature equals the operative temperature in the protected volume that gives the same net energy demand for heating as all rooms on separate operative temperatures do
Ventilation rate	Represents the average ratio between the outside air flow in m^3 per hour and the air volume in a zone over the time span considered. Where the protected volume acts as one zone, the ventilation rate equals the mean outside air flow entering that volume over the time span considered, divided by the air volume, typically set equal to 0.8 times the out-to-out measured protected volume

3.3.4.2 Steady state heat balance at zone level

Assume that a building has j zones, each with volume Vj. The protected volume is:

$$V = \Sigma V_j \qquad (\mathrm{m}^3)$$

A monthly mean (operative) temperature (θ_j) and a monthly net heating demand ($Q_{\mathrm{heat,net},j,\mathrm{mo}}$) characterize each zone. When the temperatures are known, the net demands figure as unknowns. Otherwise it's vice versa. Zones are seen as nodes that exchange heat and enthalpy with adjacent nodes (Figure 3.24).

Transmission links all neighbouring nodes and the outdoors (together called the surroundings) to the node considered, while enthalpy exchange between these nodes flows in the direction that the air moves. Solar gains, internal gains and enthalpy flows from a mechanical ventilation system are injected in, or for the extract air, removed from the nodes.

The monthly energy balance for node j thus becomes (algebraic sum, energy flows can be negative or positive):

$$Q_{\mathrm{heat,net},j,\mathrm{mo}} + \left[\begin{array}{l} \left(\Phi_{\mathrm{T},j,\mathrm{mo}} + \Phi_{\mathrm{V},j,\mathrm{mo}}\right)\Delta t_{\mathrm{mo}}/10^6 \\ \qquad + \eta_{\mathrm{use},j,\mathrm{mo}}\left(Q_{\mathrm{SLW},j,\mathrm{mo}} + \Phi_{\mathrm{I},j,\mathrm{mo}}\Delta t_{\mathrm{mo}}/10^6\right) \end{array} \right] = 0 \quad (\mathrm{MJ}) \qquad (3.55)$$

with $\Phi_{\mathrm{T},j,\mathrm{mo}}$ the transmission flow from the surroundings, $\Phi_{\mathrm{V},j,\mathrm{mo}}$ the total enthalpy flow by air exchanges with the surroundings, included the flow injected mechanically, $Q_{\mathrm{SLW},j,\mathrm{mo}}$ the monthly solar gains corrected for longwave losses and $\Phi_{\mathrm{I},j,\mathrm{mo}}$ the internal gain, all injected in node j. The gains are multiplied by a utilization efficiency $\eta_{\mathrm{use},j,\mathrm{mo}}$, whose value without heating equals 1. $Q_{\mathrm{heat,net},j,\mathrm{mo}}$, the net heating demand in node j, positive when the losses win and zero otherwise, closes the equation. Δt_{mo} is the month in s. All flows are monthly means with watts as unit, except $Q_{\mathrm{SLW},j,\mathrm{mo}}$ and $Q_{\mathrm{heat,net},j,\mathrm{mo}}$ which are both totals in MJ. A lower net heating demand means fewer losses or more gains. In heating climates with hardly any sun in winter, the losses dominate, so the best is to optimize their algebraic sum.

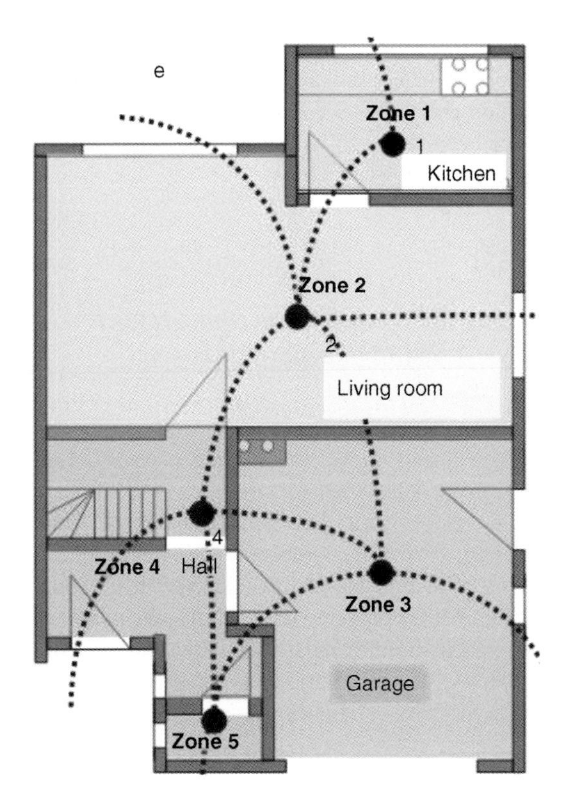

Figure 3.24 Zones as nodes.

Transmission includes flows from outdoors, from all neighbouring nodes and from adjacent protected volumes at different temperature:

$$\Phi_{\mathrm{T},j,\mathrm{mo}} = H_{\mathrm{T},j\mathrm{e},\mathrm{mo}}\left(\theta_{\mathrm{e},\mathrm{mo}} - \theta_{j,\mathrm{mo}}\right) + \sum_{k=1}^{m} H_{\mathrm{T},jk,\mathrm{mo}}\left(\theta_{k,\mathrm{mo}} - \theta_{j,\mathrm{mo}}\right) \tag{3.56}$$

with $H_{\mathrm{T},j\mathrm{e},\mathrm{mo}}$ the monthly mean transmission heat transfer coefficient between node j and outdoors (air, the soil and sometimes water) and $H_{\mathrm{T},jk,\mathrm{mo}}$ the monthly mean transmission heat transfer coefficient between nodes j and k:

$$H_{\mathrm{T},j\mathrm{e},\mathrm{mo}} = \sum_{\mathrm{opaque}} U_{\mathrm{e,op}} A_{\mathrm{op}} + \sum_{\mathrm{windows}} U_{\mathrm{e,w}} A_{\mathrm{w}} + \sum_{\mathrm{ground}} a_{\mathrm{fl}} U_{\mathrm{e,fl}} A_{\mathrm{f}l} + \sum_{\mathrm{linTB}} \psi_{\mathrm{e}} L + \sum_{\mathrm{localTB}} \chi_{\mathrm{e}} \tag{3.57}$$

$$H_{\mathrm{T},jk,\mathrm{mo}} = \sum_{\mathrm{opaque}} U_{\mathrm{i,op}} A_{\mathrm{op}} + \sum_{\mathrm{windows}} U_{\mathrm{i,w}} A_{\mathrm{w}} + \sum_{\mathrm{linTB}} \psi_i L + \sum_{\mathrm{localTB}} \chi_{\mathrm{i}} \tag{3.58}$$

The Us in both formulas stand for the thermal transmittances of all envelope and partition parts. A is their area, ψ the linear thermal transmittances of all linear thermal bridges contained with L their length and χ the local thermal transmittances of all local thermal bridges present. The suffix e refers to the envelope, suffix i to the partitions. The summations remind us that all parts interact.

Ventilation and infiltration cover a complex pattern of air flows G_{ajk} with the suffix indicating flow direction ($_{jk}$ node j to node k, $_{je}$ node j to outdoors, $_{ej}$ outdoors to node j, $_{in,j}$ supplied to and $_{out,j}$ extracted from node j by a mechanical system). Related sensible heat balance in node j is:

$$\Phi_{V,j,mo} = \begin{aligned}&\Big[-H_{V,je,mo}\theta_{j,mo} + H_{V,ej,mo}\theta_{e,mo} - \sum_{k=1}^{m} H_{V,jk,mo}\theta_j + \sum_{k=1}^{m} H_{V,kj,mo}\theta_{k,mo}\\ &+H_{V,in,j,mo}\theta_{in,j,mo} - H_{V,out,j,mo}\theta_{j,mo}\Big]\end{aligned} \quad (3.59)$$

where the $H_{V,xy,mo}$s are the ventilation heat transfer coefficients, to outdoors for $xy = je$, from outdoors for $xy = ej$, to node k for $xy = jk$ and from node k for $xy = kj$:

$$H_{V,xy,mo} = \left(1 - \eta_{rec,mo}\right) c_a G_{a,xy,mo} \quad (3.60)$$

with $\eta_{rec,mo}$ the heat recovery efficiency (zero without), c_a the specific heat capacity of air (1000–1030 J/(kg.K)) and $G_{a,xy,mo}$ the air flow in kg/s. Consider heat recovery. When infiltrating flows pass through leaks in the envelope, some heat recovery occurs, while exfiltrating flows temper the transmission losses. However, combined, the effects are too small to deserve consideration. Of course, flows passing through unheated spaces will pick up lost heat, whereas a ducted ventilation system with supply and extract fans may include a heat recovery unit. All quantities listed again relate to monthly means. The flows follow from solving the system of nodal air balances. However, information on air permeances of partitions and envelope is mostly missing, which is why default values are commonly used.

The solar gains and longwave losses are given by:

$$Q_{S/LW,j,mo} = \sum Q_{S/LW,j,op,mo} + \sum Q_{S/LW,j,w,mo} \quad (3.61)$$

with $Q_{S/LW,j,op,mo}$ the monthly gains through each opaque envelope part and $Q_{S/LW,j,w,mo}$ the resulting gains through each transparent part linked to node j. The first are given by:

$$Q_{S/LW,j,op,mo} = \frac{U_{e,op}A_{e,op}J}{h_e}\left[\alpha_K r_{sT} Q_{sT,mo} - 100 r_{sd} e_L F_{op/sky} t_{mo}/10^6\right] \quad (3.62)$$

with $\Phi_{sT,mo}$ total monthly solar radiation in MJ under clear sky conditions impinging on the unshaded surface, J the ratio between radiation under real and clear sky conditions, r_{sT} the shadow factor for total solar and r_{sd} the shadow factor for diffuse longwave radiation, both default equal to 0.9, $F_{op/sky}$ the view factor between the surface considered and the sky and t_{mo} the month in seconds. Solar gains across opaque parts clearly increase when less insulated (higher U-value), better wind protected (surface film coefficient h_e lower) and having an outside surface with high shortwave absorptivity and low longwave emissivity.

Those across the transparent parts depend on the glass and solar shading used:

$$Q_{S/LW,j,w,mo} = A_{e,w}J\left[0.95 \cdot 0.9 g_n f_{wgl} r_{sT} Q_{sT,mo} - 100\frac{U_{e,w}}{h_e} r_{sd} e_L F_{w/sky} t_{mo}/10^6\right] \quad (3.63)$$

with f_{wgl} the ratio between glass and window area, g_n the solar transmittance for beam radiation normal to the glass, corrected for the presence of a shading device, 0.9 a conversion factor to account for the beam angle and the diffuse radiation effect on the solar transmittance and 0.95 a multiplier correcting for dust deposit or a transparent drape shielding the glass indoors. The corrected solar transmittance is often simplified to:

$$g_n = [aF_{\text{shad}} + (1 - a)]g_{\text{glas,n}}$$

with a the average time fraction for which the shading is down per month and F_{shad} the reduction the glass solar transmittance experiences with the shading down. F_{shad} is set to 0.5 for outside shading, 0.6 for vented shading between two glass panes and 0.9 for inside shading.

Transparent insulation (TIM) transmits shortwave and stops longwave radiation, while its structure impedes conduction and convection. Solar gains and longwave losses by an assembly finished outside with TIM become:

$$Q_{\text{S/LW},j,\text{TIM,mo}} = A_{\text{e,TIM}} J \left[\tau_{\text{gl}} \tau_{\text{TIM}} \alpha_{\text{K}} (1/h_{\text{e}} + R_{\text{TIM}}) f_{\text{r}} r_{\text{sT}} Q_{\text{sT,mo}} - 100 \frac{U_{\text{e,TIM}}}{h_{\text{e}}} r_{\text{sd}} e_{\text{L}} F_{\text{TIM/sky}} t_{\text{mo}} / 10^6 \right] \tag{3.64}$$

with R_{TIM} the thermal resistance of the TIM layer, $U_{\text{e,TIM}}$ the thermal transmittance of the whole assembly, τ_{gl} the shortwave transmissivity of the glass covering the TIM, τ_{TIM} shortwave transmissivity of the TIM, α_{K} the shortwave absorptivity of the surface behind the TIM and f_{r} the ratio between the surface of the assembly covered with TIM and its total surface.

When quantifying solar gains and longwave losses, simplifications apply. The ratio J between solar radiation under real and under clear sky conditions is assumed constant, independent of slope and azimuth, while the ratio between the solar radiation on an arbitrary and a horizontal surface under clear sky conditions at the month's central day is considered representative for the whole month.

The shadow factor r_{sT} for transparent parts combines shielding with shadowing by overhangs and protrusions. The vertical angle α_{e} between the horizontal plane across the part's centre and the mean horizon height defines the shielding. The vertical angle α_{v} between the same horizontal plane and the line representing the average overhang border in turn quantifies its shadowing, while the horizontal angles $\alpha_{\text{h,l}}$ and $\alpha_{\text{h,r}}$ between the horizontal projection of the normal across the part's centre and the verticals coinciding with the average depth of the protrusions left and right fixes their shadowing (Figure 3.25). Beam radiation becomes zero for a solar height lower than the horizon angle α_{e}, a solar height above the vertical angle α_{v}, a time angle larger than $a_{\text{s}} + \alpha_{\text{h,l}}$ and a time angle smaller than $a_{\text{s}} - \alpha_{\text{h,r}}$, with a_{s} the azimuth of the surface containing the transparent part. In all other cases, beam radiation reaches the window unhindered.

The shadow factor r_{sd} for longwave sky radiation is:

$$r_{\text{sd}} = \frac{\cos(\pi - \alpha_{\text{v}}) - \cos(\pi - \alpha_{\text{e}})}{1 + \cos(s_{\text{s}})} \left(\frac{\alpha_{\text{h,l}} + \alpha_{\text{h,r}}}{\pi} \right)$$

with s_{s} the slope of the surface containing the transparent part.

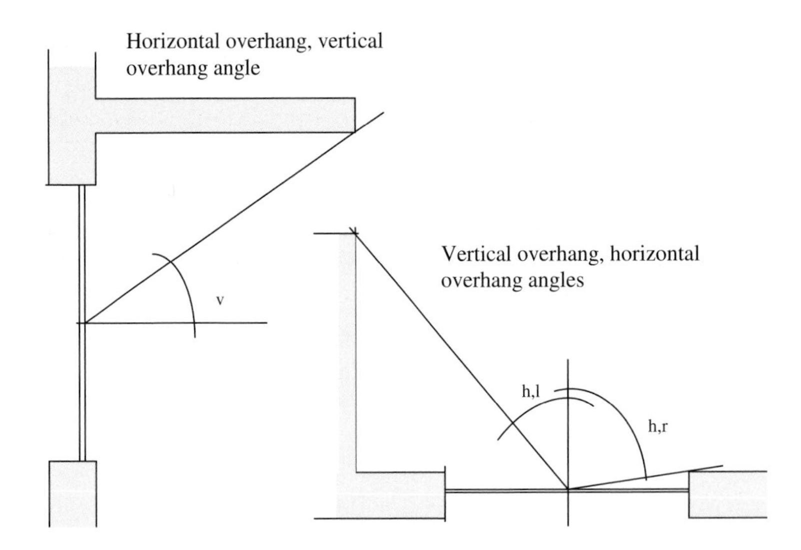

Figure 3.25 Horizontal protrusions (α_{hl}, α_{hr}) and the one vertical overhang angle (α_v).

The internal gains are stochastic in nature. They include the sensible heat released by the occupants, the heat that artificial lighting emits and the heat produced by using appliances. In residential buildings, following least square line gives the electricity on average consumed annually by lighting units and appliances:

$$Q_{LA} = 3772 + 2.84V \qquad \text{(kWh/a)}$$

Yet, not all appliances stay in the protected volume (V) and not all electricity converts into sensible heat. Cooking for example produces up to 50% latent heat. In addition there is the sensible metabolic heat, lost by the effective number of household members. If 47% of the electricity used converts to latent heat or to sensible heat outside the protected volume and if the one effective member per 150 m^3 protected volume sleeps 8 hours a day, then the gains become:

$$\Phi_I = \left(\frac{3772 \times 0.53 \times 3600000}{3600 \times 24 \times 365.25}\right) + \left[\left(\frac{2.84 \times 0.53 \times 3600000}{3600 \times 24 \times 365.25}\right) + \frac{79}{150}\right]$$

$$V = 228 + 0.70V(\text{W})$$

which is close to the commonly advanced formula:

$$Q_I = (220 + 0.67V)t_{mo}/10^6 \qquad \text{(W)}$$

For multi-zone calculations, the monthly gains are often distributed over the zones:

$$Q_{I,j} = \left[(220 + 0.67V)t_{mo}/10^6\right]V_j/V \qquad \text{(W)} \qquad (3.65)$$

where V_j is the volume per zone. All these formulas reflect statistical means. Looking at individual dwellings, the internal gains may vary as follows:

$$\left[(130 + 0.78V)t_{mo}/10^6\right]V_j/V \leq Q_{I,j} \leq \left[(226 + 0.94V)t_{mo}/10^6\right]V_j/V$$

In non-residential buildings, gains from lighting must be calculated based on lighting design, its projected use and the controls chosen. Human heat released follows from usage patterns and the mean occupancy (70–90 W per adult for office work). Guessing the heat produced by appliances is more difficult. Data from buildings with similar function may help. Otherwise, data from literature is used, see Table 3.21.

In terms of how the solar and internal heat gain are used, clearly the 'shoulder' months give warmer days where the gains push the zone temperature beyond set point, even when a monthly calculation gives positive demands. This will be considered by multiplying all gains with a utilization efficiency ($\eta_{\text{use},j,\text{mo}}$)

3.3.4.3 Whole building steady state heat balance

The balance per zone can be rewritten as:

$$-\left(H_{\text{T},je,\text{mo}} + H_{\text{V},je,\text{mo}} + \sum_{k=1}^{m}\left(H_{\text{T},jk,\text{mo}} + H_{\text{V},jk,\text{mo}}\right)\right)\theta_{\text{i},j,\text{mo}}$$
$$+\sum_{k=1}^{m}\left(H_{\text{T},jk,\text{mo}} + H_{\text{V},kj,\text{mo}}\right)\theta_{\text{i},k,\text{mo}} + \frac{Q_{\text{heat,net},j,\text{mo}}}{t_{\text{mo}}}$$
$$= -\left(H_{\text{T},je,\text{mo}} + H_{\text{V},ej,\text{mo}}\right)\theta_{\text{e,mo}} - \eta_{\text{use},j,\text{mo}}\left(\Phi_{\text{S/LW},j,\text{mo}} + \Phi_{\text{I},j,\text{mo}}\right)$$

Combining the transmission and ventilation heat transfer coefficients into transfer coefficients named $H_{j,\text{mo}}$, $H_{jk,\text{mo}}$ and $H_{ej,\text{mo}}$ gives:

$$-H_{j,\text{mo}}\theta_{\text{i},j,\text{mo}} + \sum_{k=1}^{m} H_{kj,\text{mo}}\theta_{\text{i},k,\text{mo}} + \frac{Q_{\text{heat,net},j,\text{mo}}}{\Delta t_{\text{mo}}}$$
$$= -H_{ej,\text{mo}}\theta_{\text{e,mo}} - \eta_{\text{use},j,\text{mo}}\left(\Phi_{\text{S/LW},j,\text{mo}} + \Phi_{\text{I},j,\text{mo}}\right)$$

The result for a building with n zones is a system of n equations with $2n$ unknowns, n temperatures $\theta_{\text{i},j,\text{mo}}$ and n net energy demands for heating $Q_{\text{heat,net},j,\text{mo}}$:

$$\left|H_{j,\text{mo}}\right|\left|\theta_{\text{i},j,\text{mo}}\right| + |1|\left|Q_{\text{heat,net},j,\text{mo}}/\Delta t_{\text{mo}}\right| = -\left|H_{je,\text{mo}}\theta_{\text{e,mo}} - \eta_{\text{use},j,\text{mo}}\left(\Phi_{\text{S/LW},j,\text{mo}} + \Phi_{\text{I},j,\text{mo}}\right)\right| \tag{3.66}$$

with $|H_{j,\text{mo}}|$ an $[n \times n]$ heat transfer matrix, $|\theta_{\text{i},j,\text{mo}}|$ a column matrix of the n zone temperatures, $|1|$ an $[n \times n]$ unit matrix, $|Q_{\text{heat,net},j,\text{mo}}/\Delta t_{\text{mo}}|$ a column matrix of the n net heating demands and $|H_{je,\text{mo}}\theta_{\text{e,mo}} - (\Phi_{\text{SLW},j,\text{mo}} + \Phi_{\text{I},j,\text{mo}})|$ a column matrix with the n known terms. With the temperatures in $n - m$ zones and the net heating demand, usually zero, in the m other zones known, the system is solved for each month. When some months give negative net demands in zones with known temperature, a second step is added, solving the system for no heating there.

3.3.4.4 Heat gain utilization efficiency

As stated, the utilization efficiency accounts for the days in a month that the gains in the heated zones cause overheating. This adds a third calculation step: its quantification per zone. First, per heated zone j, the ratio between monthly gains and losses, the second step gives, is calculated as

$$\gamma_{j,\mathrm{mo}} = \frac{\Phi_{\mathrm{SLW},j,\mathrm{mo}} + \Phi_{\mathrm{I},j,\mathrm{mo}}}{\Phi_{\mathrm{T},j,\mathrm{mo}} + \Phi_{\mathrm{V},j,\mathrm{mo}}} \tag{3.67}$$

Next, the time constant per zone j equals:

$$\tau_{j,\mathrm{mo}} = C_j / \left(H_{\mathrm{T},j,\mathrm{mo}} + H_{\mathrm{V},j,\mathrm{mo}}\right)$$

where C_j is the total heat storage capacity in the zone considered. For envelope parts with a layer thicker than 10 cm at the inside of the insulation, only the first 10 cm from inside are assumed acting as storage, whereas for thinner layers the whole does (Figure 3.26). For partitions the rule is 10 cm or, if thinner than 20 cm, half the thickness

$$C_j = \sum_{\text{all parts}} \left[A \sum_{i,d \leq 0.1\ \mathrm{m}} (\rho c d) \right] \tag{3.68}$$

The monthly utilization efficiency per zone is then set:

$$\gamma_{j,\mathrm{mo}} \neq 1 : \eta_{\mathrm{use},j,\mathrm{mo}} = \frac{1 - \gamma_{j,\mathrm{mo}}^{a_{\mathrm{ma},j}}}{1 - \gamma_{j,\mathrm{mo}}^{a_{\mathrm{ma},j}+1}} \qquad \gamma_{j,\mathrm{mo}} = 1 \quad : \eta_{\mathrm{use,ma},j} = \frac{a_{j,\mathrm{mo}}}{1 + a_{j,\mathrm{mo}}} \tag{3.69}$$

with

$$a_{j,\mathrm{mo}} = 1 + \tau_{j,\mathrm{mo}} / 54\,000$$

So, the monthly gains are multiplied per zone by the corresponding efficiencies and the net heating demands corrected by subtracting these recalculated gains from the second step losses. In all non-heated zones, the utilization efficiency is by definition 1.

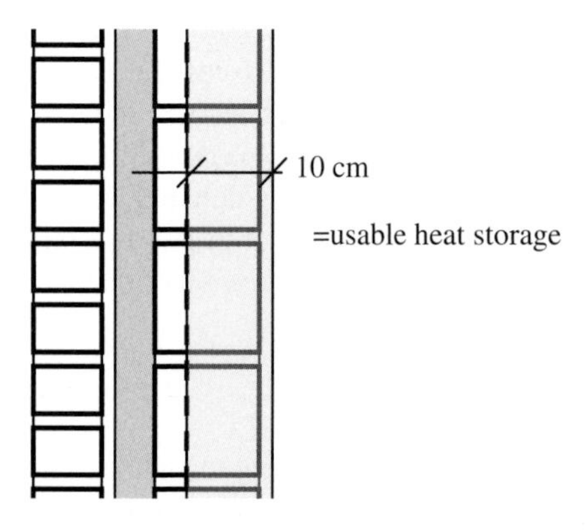

Figure 3.26 Envelope part, the first 10 cm acting as heat storage (left is outside)

3.3.4.5 Annual end use for heating

Summing up the corrected monthly demands for all zones gives the net heating demand per month:

$$Q_{\text{heat,net,mo}} = \sum_{j=1}^{n} Q_{\text{heat,net},j,\text{mo}}$$

Where one heating system warms the building, dividing each monthly net demand by the building related monthly system and production efficiencies ($\eta_{\text{heat,sys,mo}}$, $\eta_{\text{heat,prod,mo}}$) gives the monthly end energy used for heating:

$$Q_{\text{heat,use,mo}} = Q_{\text{heat,net,mo}} / \left(\eta_{\text{heat,sys,mo}} \eta_{\text{heat,prod,mo}}\right) \tag{3.70}$$

Where several heating systems serve the building, Eq. (3.70) applies per group of zones heated by the same system. Annual end use finally equals:

$$Q_{\text{heat,use.a}} = \sum_{mo=1}^{12} \left(Q_{\text{heat,use},mo}\right) \tag{3.71}$$

Including a solar heating system with collectors and storage, the end use drops to:

$$Q_{\text{heat,use,a}} = \sum_{mo=1}^{12} \left(\frac{Q_{\text{heat,gross},mo} - \eta_{\text{S},mo} Q_{\text{S},mo}}{\eta_{\text{heat,prod},mo}} \right) \tag{3.72}$$

with $Q_{\text{S},mo}$ the monthly solar radiation on the collectors and $\eta_{\text{S},mo}$ the monthly efficiency of the solar heating system:

$$\eta_{\text{S},mo} = \min \left[\max \left(0, 0.16 + \frac{0.015}{Q_{\text{heat,net},mo} / \left(\eta_{\text{heat,sys},mo} Q_{\text{S},mo}\right)} \right), 0.8 \right] \tag{3.73}$$

3.3.4.6 Protected volume as one zone

The one zone at a given temperature approach simplifies the methodology substantially. Per month, losses are calculated. Then the solar and internal gains are quantified, after which the monthly utilization efficiency is estimated for the whole protected volume, with the envelope and all partitions acting as storage as explained. Multiplying the gains by these efficiencies gives the useful gains that, when subtracted from the losses, fix the monthly net heating demand. Stepping to the monthly gross demand requires knowledge of the monthly mean system efficiency, while quantifying the monthly end use requires knowledge of the monthly mean production efficiency. Contribution of a solar heating system is evaluated in the way just explained. Annual end energy use then follows from summing the monthly usage.

3.3.5 Residential buildings, parameters shaping the annual net heating demand

3.3.5.1 Overview

Table 3.22 lists all the outdoor climate, building use and building design parameters involved.

Table 3.22 Parameters impacting the annual net heating demand.

Set	Parameter	Impact on the annual net heating demand, the other parameters remaining unchanged
Outdoor climate (heating season mean)	Temperature	Increases when lower
	Solar radiation	Decreases when more
	Wind velocity	Increases when higher
	Precipitation	Increases when more
Building use (heating season mean)	Inside temperature	Increases when higher
	Temperature control	Decreases when the principle of 'heating a zone only when used' is better applied
	Ventilation	Increases when more
	Internal gains	Decreases when higher
Building design (and construction)	Protected volume	Increases when larger
	Compactness	Decreases when higher
	Floor organization	Decreases when the floor organization results in a better use of solar gains and when the north side utility rooms shield living zones
	Surface area, type, orientation and slope of transparent surfaces (windows)	Increases when the envelope contains larger transparent parts Decreases when these have a higher solar transmittance, a lower thermal transmittance, are more facing the sunny sides and are vertical
	Thermal insulation	Decreases with lower mean thermal transmittance of the envelope
	Thermal inertia	Hardly impacts net demand in temperate to cold climates
	Envelope airtightness	Decreases when tighter

3.3.5.2 Outdoor climate

The net heating demand increases at lower heating season mean outdoor temperature, all other parameters being equal. This temperature drops the nearer to the poles, the farther inland and the higher above sea level the location. Although situated farther north than the USA, north-west Europe has a more temperate climate thanks to the warm Gulf Stream.

The opposite effect gives a sunnier heating season, all other parameters being equal. Solar radiation during the heating season is quite low in countries at higher latitudes with a maritime climate. Regions inland and countries closer to the tropics have less cloudy skies and more sun, albeit that less cloudy means clearer nights and more longwave losses.

A higher wind speed during the heating season in turn increases the net heating demand, all other parameters being equal. More wind in fact heightens infiltration and often the airflows that purpose-designed ventilation systems deliver, while it also enlarges the outside surface film coefficients. All three cause extra losses. Higher wind speeds are typical for locations close to the sea, on hilltops and in valleys parallel to the prevailing wind direction. Also the absence of nearby buildings and no screening by trees favours wind.

More precipitation may also give an increase. The impact anyhow depends on the way buildings are constructed. Unlike in North America where timber-framed buildings with a sheathed finish are preferred, north-west Europe loves brick cavity walls. These ensure rain-tightness, among others due to a water buffering veneer. Its drying, however, requires heat, for a capillary saturated veneer up to 45 MJ/m^2. Part comes from outdoors and part from indoors. The last and the lower thermal resistance of the wet veneer increases heat loss somewhat, although a filled cavity minimizes this drawback.

So, although climate looks given, mankind has an impact, as global warming proves. The built environment itself alters local microclimates, with cities being warmer, less sunny and less windy than rural regions. Spread settlements benefit from planting trees on the prevailing wind side, while broadleaf trees let the sun pass in winter and give shade in summer.

3.3.5.3 Building use

Indoor temperature, its control, ventilation and the internal gains are the four parameters impacting the net heating demand. All link to user habits, the big unknown in predicting end use. Statements such as 'net heating demand increases when a higher indoor temperature is maintained, net heating demand drops when the rule 'only heat rooms when used' is better applied, 'more ventilation augments demand', 'more internal gains reduces it' sound logical. The main question however is: how do users act?

What heating secures is thermal comfort. However, when related energy costs are perceived as too high, people tend to use less. The best way to prove whether habits govern reality is by logging end energy use in as many homes as possible, and comparing the data with a reference, such as the annual end use calculated as mandated. That reference includes assumptions and simplifications. The protected volume, for example, forms one zone, at a temperature of 18 °C. In Belgium, the year of Table 3.23 describes the outdoors. The purpose-designed ventilation flow equals:

$$\dot{V}_{\text{a,dedic}} = mV\left[0.2 + 0.5\exp(-V/500)\right] \qquad (\text{m}^3/\text{h})$$

with m a factor reflecting the quality of the system used. Infiltration is linked to the n_{50} air change rate, with as default value:

$$n_{50} = 12A_{\text{T}}/V \qquad (\text{h}^{-1})$$

Table 3.23 Reference year for Belgium.

Month	θ_e (°C)	$Q_{ST,hor}$, MJ/m^2	$Q_{sd,hor}$ MJ/m^2	[θ_e,18] degree days
January	3.2	71.4	51.3	459
February	3.9	127.0	82.7	409
March	5.9	245.5	155.1	375
April	9.2	371.5	219.2	264
May	13.3	510.0	293.5	146
June	16.3	532.4	298.1	0
July	17.6	517.8	305.8	0
August	17.6	456.4	266.7	0
September	15.2	326.2	183.6	84
October	11.2	194.2	118.3	211
November	6.3	89.6	60.5	351
December	3.5	54.7	40.2	450
Total				**2748**

A_T being the envelope area in m^2. The internal gains are:

$$Q_{I,m} = (220 + 0.67V)t_{mo}/10^6 \qquad \text{(MJ/month)}$$

with t_{mo} the number of seconds per month.

A system ($\eta_{sys,heat}$) and production efficiency ($\eta_{gen,heat}$) transpose the net demand ($Q_{heat,a,net}$) into the annual end energy use for heating ($Q_{heat,a,ref}$):

$$Q_{heat,a,ref} = Q_{heat,a,net}/\left(\eta_{sys,heat}\eta_{gen,heat}\right) \qquad \text{(MJ/a)}$$

For the system efficiency, a combination of the distribution, control and emission efficiency, default values are used, see Table 3.24, while production efficiency equals the boiler efficiency at 30% part load. For heat pumps the seasonal performance factor (SPF) applies.

Table 3.24 Central heating, system efficiency.

Control inside temperature	Supply temperature control			
	All pipes or ducts within the protected volume		Part of the pipes or ducts outside the protected volume	
	Constant	$F(\theta_e)$	Constant	$F(\theta_e)$
Per room	0.870	0.890	0.827	0.846
Other	0.850	0.870	0.808	0.827

Before comparing measured data with that reference the first has to be multiplied by the ratio between the 2748 [θ_e,18] degree days in the reference year of Table 3.23 and the [θ_e,18] degree days in the weeks logging run ($D_{\theta_e}^{18}$). The result is given the name 'normalized':

$$Q_{\text{heat,a,meas,normalized}} = 2748 Q_{\text{heat,meas}} / D_{\theta_e}^{18}$$

The idea that user habits may fix annual end energy for heating in residential buildings was first looked for in 17 homes by guessing the temperature indoors that made the calculated reference equal to the logged data after normalization. Following least square line came out:

$$\theta_i = 15.7 - 0.00425 H_T \qquad (°\text{C})$$

with H_T the specific transmission losses, the product of envelope area (A_T in m^2) and its mean thermal transmittance (U_m in W/(m^2.K)). These least square values are lower than 18 °C and drop with higher specific transmission losses. Users clearly don't heat the whole protected volume, a fact pointing to habits, though the effect decreases with either a better insulation for a same enclosing surface or the same insulation but a smaller enclosing surface. In a second step, measured annual end uses in homes with direct electric heating were compared with fuel and gas heated homes. After normalization following least square lines came out:

Direct electricity	16 homes	$Q_{\text{heat,a,meas,norm}} = 69 H_T + 18520$	(MJ/a)
Fuel or gas	211 homes	$Q_{\text{heat,a,meas,norm}} = 109 H_T + 55900$	(MJ/a)

Clearly, direct electric heated ones consume less. The reason fits into a rebound hypothesis. Electricity is more expensive per MJ than oil and gas, enough to temper comfort aspirations. In a third step, the owners of a non-insulated villa, built in the 1930s with compactness 1.6 m and a mean thermal transmittance 1.94 W/(m^2.K), and a detached low energy house, built in the mid 1980s with compactness 1.35 m and a mean thermal transmittance 0.40 W/(m^2.K), were asked to keep their whole dwelling day and night at comfort temperature for a period of 6 weeks, before returning to their usual heating habits. Figure 3.27 gives the results after normalization. While the villa benefitted from the habits, the low-energy house didn't, underlining again that higher energy efficiency lessens user impact.

Final step consisted of quantifying a most likely factor, for simplicity reasons called the reboundfactor, applicable to residential as a whole, applying as definition:

$$a_{\text{rebound}} = 1 - Q_{\text{heat,a,meas,normalized}} / Q_{\text{heat,a,ref}} \tag{3.74}$$

First a best fit for the reference was calculated, based on audit results in 95 residences. Compiling the data in a specific transmission loss versus annual end use for heating diagram, both per m^3 of protected volume [H_T/V, $Q_{\text{heat,a,ref}}/V$] gave (also see Figure 3.28):

$$\frac{Q_{\text{heat,a,ref}}}{V} = 386\left(\frac{H_T}{V}\right) - 4.4 \qquad r^2 = 0.79 \qquad (\text{MJ}/(\text{a.m}^3)) \tag{3.75}$$

Then, normalized data for 1050 homes, ranging from low energy to non-insulated, from detached over end of the row and terraced to apartments, one to four stories high, formed

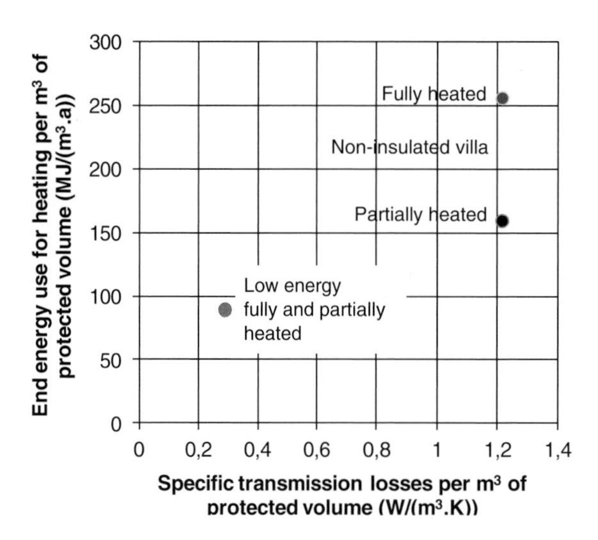

Figure 3.27 Non-insulated villa and low energy detached house, end energy for heating.

the ordinate in a figure with on the abscissa the specific transmission losses, again both per unit of protected volume, see Figure 3.29. The trend is a power function with exponent less than 1:

$$\frac{Q_{\text{heat,a,real}}}{V} = 255\left(\frac{H_{\text{T}}}{V}\right)^{0.81} \qquad r^2 = 0.82 \qquad (\text{MJ}/(\text{a.m}^3)) \tag{3.76}$$

At very low specific transmission losses per unit of protected volume this curve passes the reference but drops below higher up, underlining the fact that user habits have quite

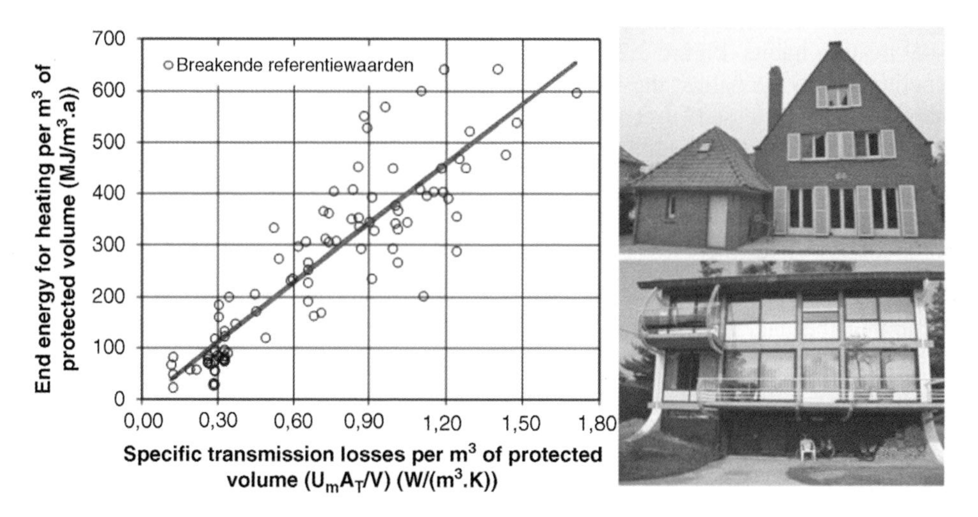

Figure 3.28 End energy for heating in 95 homes, reference.

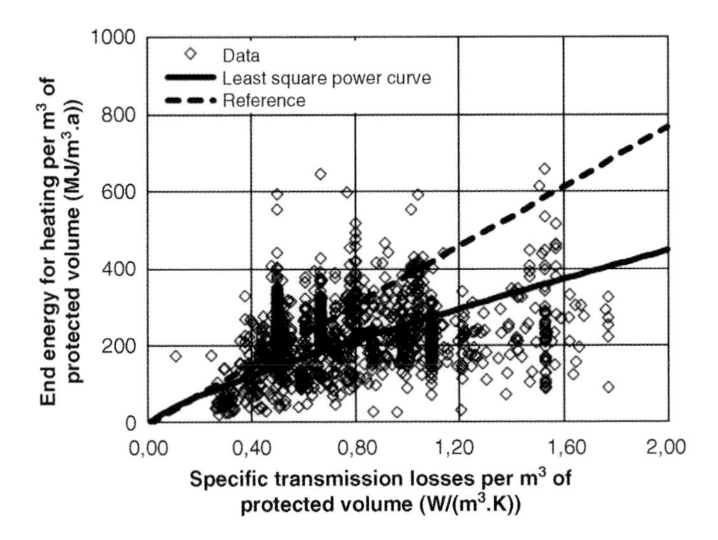

Figure 3.29 Measured annual end energy for heating after normalization in 1050 homes as a function of the specific transmission losses, both per unit of protected volume.

an impact on end use for heating. Of course, the outliers show that some heat enthusiastically, whereas other do it very economically.

Transposing both curves into the rebound factor formula gives as most likely value for that factor (Figure 3.30):

$$a_{\text{rebound}} = 1 - 0.66(H_T/V)^{-0.19} = 1 - 0.66(U_m/C)^{-0.19} \quad (3.77)$$

where C stands for the compactness, units m, the ratio between the protected volume (V) in m^3 and the enclosing surface (A_T) in m^2.

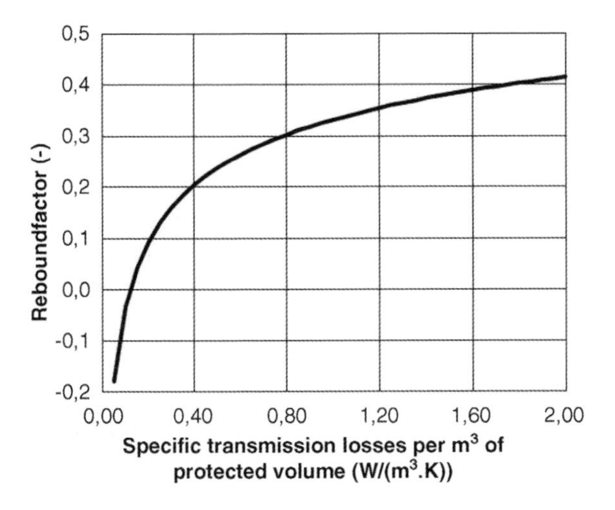

Figure 3.30 Statistically best fitting rebound factor.

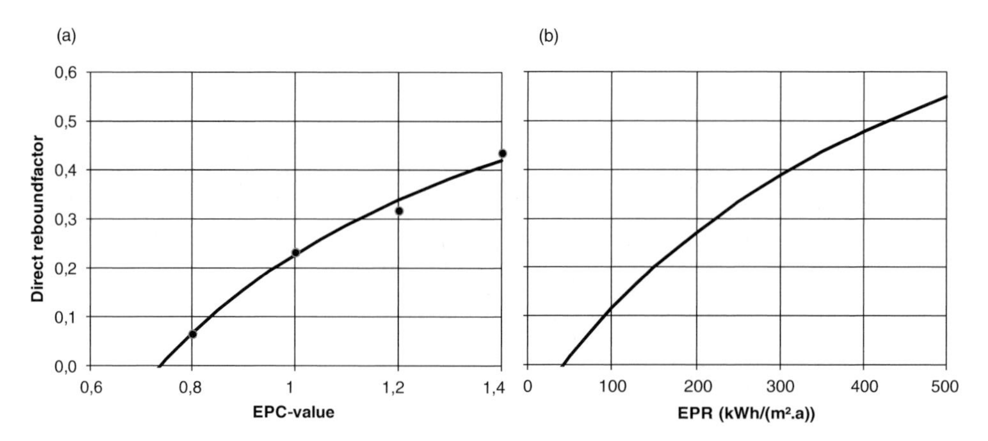

Figure 3.31 Rebound factor, (a) the Netherlands, (b) Germany.

When known, annual end energy for heating in a large enough sample of residential buildings ($Q_{\text{heat,a,total}}$) will approach:

$$Q_{\text{heat,a,total}} = \sum_{j=1}^{n} \left[\left(1 - a_{\text{rebound},j}\right) Q_{\text{heat,a,ref},j} \right] \tag{3.78}$$

Similar exercises in the Netherlands and Germany gave as rebound factor the functions shown in Figure 3.31.

As an equation:

$$\text{Netherlands}: a_{\text{rebound}} = 1 - 0.77(EPC)^{-0.85} \quad \text{Germany}: a_{\text{rebound}} = 1.2 - \frac{1.3}{1 + EPR/500}$$

with for the Netherlands as abscissa the EPC values, for Germany the primary energy per m^2 of floor area, called the EPR-value:

$$EPC = \frac{1}{1.12}\left(\frac{Q_{\text{prim,a,calc}}}{330A_{\text{fl}} + 65A_{\text{T}}}\right) \qquad EPR = \frac{Q_{\text{prim,a,calc}}}{A_{\text{fl}}} \quad (\text{kWh}/(\text{m}^2.\text{a})$$

Habits have important effects. More energy efficient designs save less energy compared to less efficient ones than the reference calculations suggest. This worsens the economic feasibility of efficiency measures and could set limits on how severe future legal requirements for new construction and retrofit should be. At the same time, energy consumption for lighting and appliances has a larger share in the primary energy used annually than perhaps realised, as a measuring campaign on an estate, built in the 1950s, underlined, see Figure 3.32: 38% on average! Actual legislation covers neither of the two.

Even the net zero concept often only looks to the annual energy for heating and domestic hot water and its compensation by renewables. Favoured is PV. The electricity produced partly serves for lighting and appliances but surely in summer a major fraction goes to the local distribution grid. That it lowers fossil fuel use for heating and domestic hot

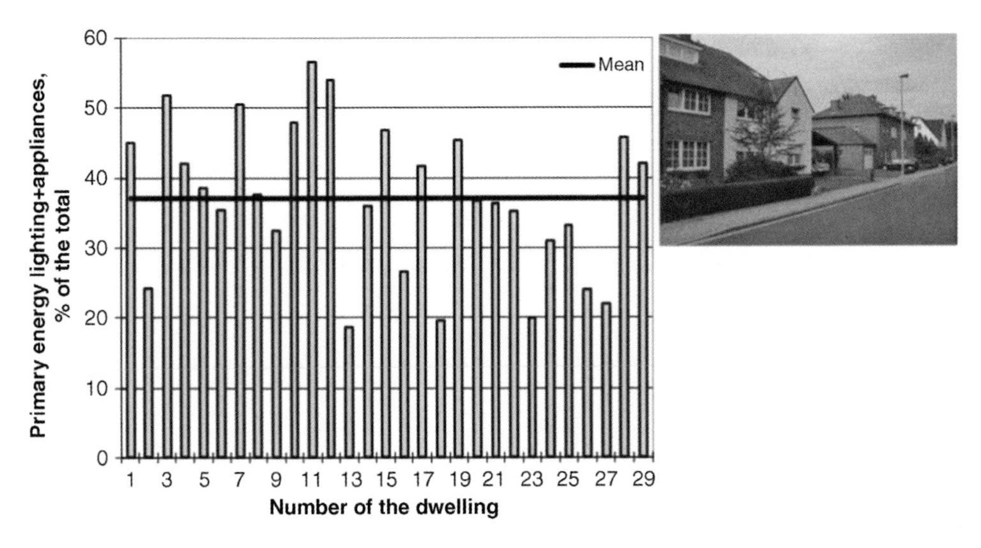

Figure 3.32 Estate built in the 1950s, primary energy for lighting and appliances, percentage of the annual total.

water is a fiction in north-western Europe. In winter, roof-based PV installations hardly produce any electricity (see Figure 3.33), while in summer too much injected into the local grid may overload it to a level where the inverters have to gradually shut off PV roofs. Part of the solution consists of compelling owners to opt for smart inverters and to

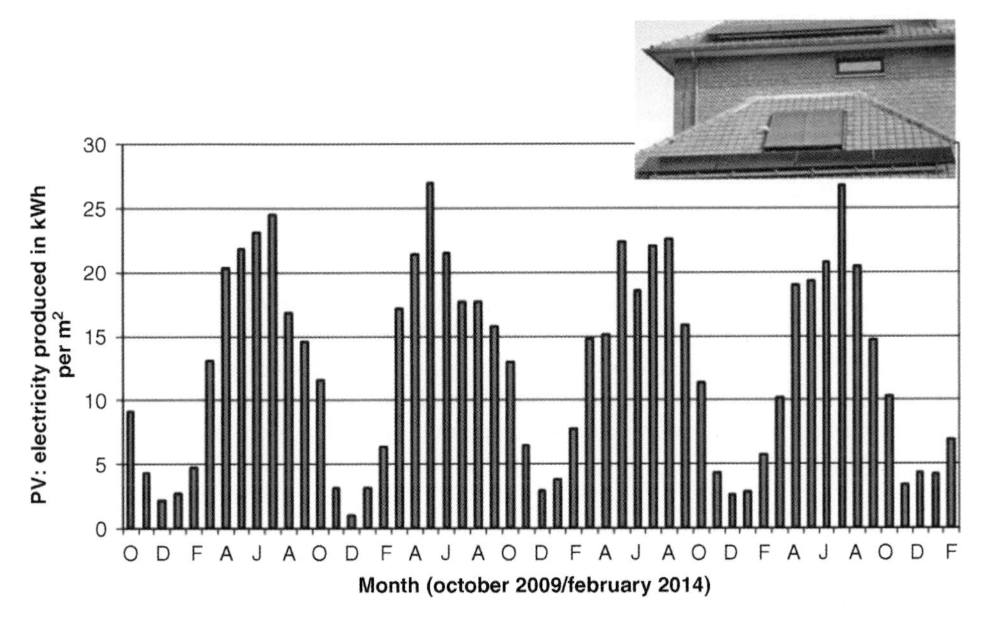

Figure 3.33 North-western Europe, roof mounted PV, electricity produced.

Table 3.25 Adventitious ventilation rates in ach by window airing.

Apartment buildings						
Shielding	Two facade apartments **Airtightness envelope**			One facade apartment **Airtightness envelope**		
	Low	Average	High	Low	Average	High
None	1.2	0.7	0.5	1.0	0.5	0.5
Moderate	0.9	0.6	0.5	0.7	0.5	0.5
Strong	0.6	0.5	0.5	0.5	0.5	0.5
Dwellings	Low	Average	High			
None	1.2	0.7	0.5			
Moderate	0.9	0.6	0.5			
Strong	0.6	0.5	0.5			

install battery storage, which of course makes PV more expensive and reduces the likelihood of the investment seeing a payback over the service life of the installation.

Which habits are quite common? In temperate climates, few households heat bedrooms, unless instantaneously when used in the daytime. During the week, most heat in the morning and the evening, most apply a night-time setback and a daytime setback during the hours that the parents are out at work and the children are at school.

For ventilation, adventitious airing is often translated into the rates listed in Table 3.25. However, habits again determine the actual rates. Indeed, during the colder months, the outside air requires heat to reach the comfort temperature, so costs make less ventilation attractive. But, less is not a free lunch. Enough fresh air is necessary to keep the indoor environment healthy. But, what is enough? Adventitious ventilation relies on infiltration and operating windows. Airtight envelopes reduce infiltration as a mean of guaranteeing enough with the windows closed. Only a few people seemed to care. Observation in fact showed that operating windows was mainly a warm season activity. In winter, households kept them closed or open only in the morning for a very short period of time. For that reason, purpose-designed ventilation systems became mandatory. Choices are: trickle vent, supply, extract and balanced ventilation, all but trickle vents having demand control as an option. Where extract ventilation allows the use of exhaust air as a heat source for a heat pump, balanced ventilation permits direct heat recovery. This requires a really airtight building ($n_{50} < 1$ ach) and correctly designed airtight ducting that is insulated where necessary. Otherwise, balanced ventilation can become an energy-devouring device. Some households having balanced ventilation with heat recovery also go on loving open windows even in winter, a nice habit but one that jeopardizes controllability and heat recovery efficiency. Temperature differences between heated and unheated spaces also matter. When a dwelling is not really airtight, recovery

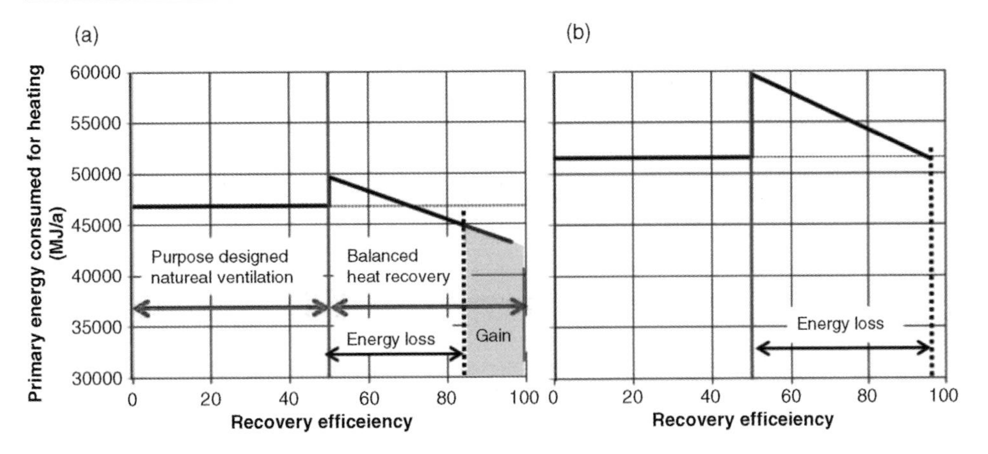

Figure 3.34 Dwelling, 799 m^3, ground floor 153 m^2, unheated first floor 117 m^2, $C = 1.35$ m, $U_m = 0.47$ W/(m^2.K). Balanced ventilation with heat recovery, perfect workmanship. (a) for $n_{50} = 1$ ach a recovery >72% gives energy benefit, (b) for $n_{50} = 3.1$ ach, >96% is needed.

efficiency must be extremely high before the balance between fan electricity used and heat recovered gives an energy benefit, see Figure 3.34.

However, regardless of what's decided upon, peak ventilation must remain possible, which is why each room should have operable windows with a surface equal to x times the floor area in m^2, where x is often set 0.064. During the warmer months, window airing remains the most energy efficient option in temperate climates.

As regards the internal gains, these include the occupants and the heat released by lighting and appliances. The drop in net heating demand that these two give should not be read as 'more lighting and appliances save energy'. In fact, if produced in thermal power plants, electricity requires much more primary energy than the gas or fuel saved.

3.3.5.4 Building design and construction

The seven building related parameters involved last as long as the building exists. Finding a creative combination that realizes a net heating demand as low as economically viable is the design team's job, while good workmanship by the contractors must ensure the result as predicted

The importance of the protected volume (V in m^3) can be seen when writing the losses for a single zone building as:

$$\Phi_T + \Phi_V = V\left[U_m/C + 0.34\left(n_V + n_{inf}\right)\right](\theta_i - \theta_e) \quad \text{(W)} \tag{3.79}$$

with n_V the ventilation and n_{inf} the infiltration rate, both in ach. Heat loss and related net heating demand clearly increase proportionally, all other parameters being equal. Energy efficient buildings should thus have a volume dictated by the functions housed, not by the ambition to show prosperity and related spacious living spaces dreamed of.

As regards compactness, the ratio between protected volume and enclosing surface in m, Eq. (3.80) underlines the fact that the net heating demand changes inversely proportional

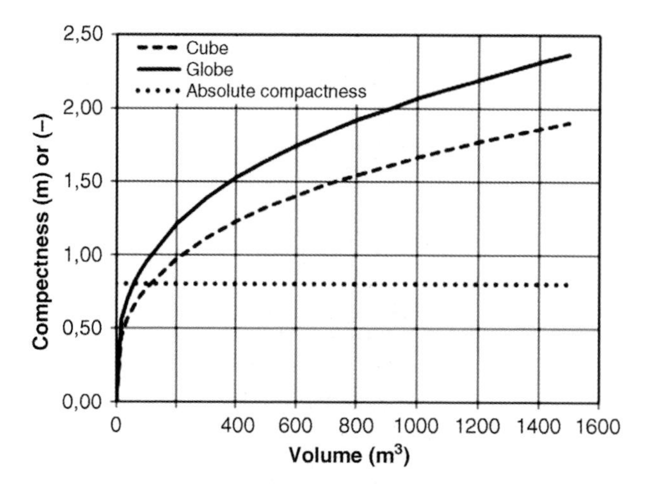

Figure 3.35 Compactness and absolute compactness as function of volume.

to its value. Globes are the densest volumes but lack functionality: lots of space is lost, and placing furniture is a problem. Conversely, a cube with compactness one sixth of the side is very usable. Compared to a globe with same volume, even a cube is 24% less compact. Compactness also increases with the volume, though absolute compactness, the ratio between the compactness of a given volume (V) and a globe with same volume, does not (see Figure 3.35):

$$C_{\text{abs}} = 4.836V^{2/3}/A_{\text{T}} \qquad (-) \tag{3.80}$$

Terraced dwellings with simple volumes are more compact than detached ones. A detached cube of 500 m^3, for example, has compactness of 1.32 m (absolute 0.806), while terracing five dwellings increases the compactness of the middle three to 1.98 m (absolute 1.21).

Compact buildings are investment-friendly. In fact, it's not the volume but the envelope and all the partitions that determine the cost of the structural work. Consider a building with a volume of 600 m^3. Compactness 1.405 means an envelope of 427 m^2. Fragmenting that volume to a compactness of 1 m demands 600 m^2 of envelope and requires more partitions, extra foundations, extra roof edges and so on, which is pricier. Poorly insulated compact buildings also offer more thermal comfort than poorly insulated fragmented ones. High compactness nonetheless is not a paradigm. Tall buildings are very compact. Those with rectangular foot print have a value of (Figure 3.36):

$$C = 0.5/\left(\frac{L+B}{LB} + \frac{1}{H}\right) \tag{3.81}$$

with H height, L depth and B width, all in m. The more storeys, the more compact, although the curve flattens once above 30. On the other hand, more storeys demands extra space for lifts and shafts, thus diminishing the ratio between gross and usable floor area.

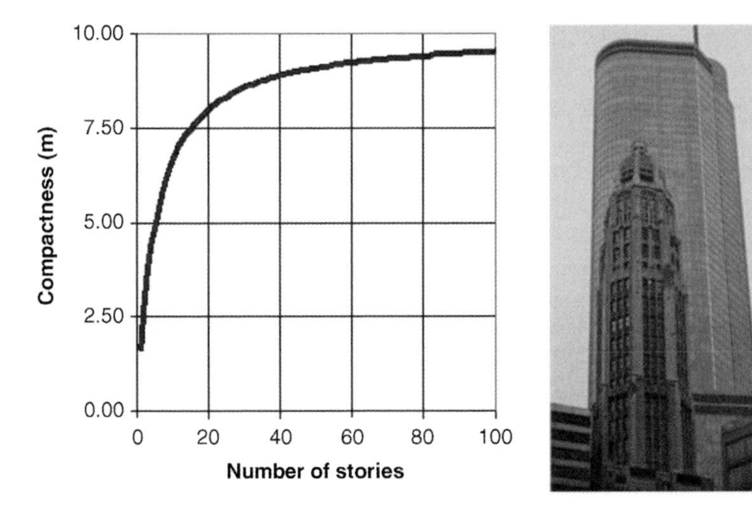

Figure 3.36 Tall buildings, compactness.

An energy-efficient floor organization that lowers net heating demand, all other parameters being equal, requires closable rooms. Otherwise, forget any differentiation in temperature and ventilation. A hall with a door to the living room limits infiltration. To allow optimal use of solar gains, windows in the daytime rooms should, if possible, look south-east through south to south-west (north-east through north to north-west in the southern hemisphere). But remember that, without solar shading, overheating and too strong differences in luminance could be a problem.

It seems logical that the net heating demand, all other parameters being equal, increases with more glass, unless better insulating types are used. But even these still give higher transmission losses than well insulated opaque assemblies, although a beneficial orientation and slope helps in maximizing the solar gains. Table 3.26 lists most glass

Table 3.26 Glass types.

Types	U-value W/(m2.K)	g-value —	$\tau_{visible}$ —
Single glass	5.9	0.81	0.90
Double glass	3.0	0.72	0.80
Low-e double glass	1.8	0.63	0.70
Argon filled low-e double glass	1.3	0.58	0.75
Krypton filled low-e double glass	1.0–1.1	0.58	0.75
Xenon filled low-e double glass	0.9–1.0	0.58	0.75
Krypton filled low-e triple glass	0.7	0.50	0.65
Xenon filled low-e triple glass	0.6	0.50	0.65

types actually available. While the thermal transmittance drops 8.5 times between worst and best, for the solar and visible transmittance the ratio is about 1.6.

Despite the impressive decrease in thermal transmittance, large glass areas still don't necessarily advance energy efficiency in the cloudy climate of north-western Europe. Even with the best types, more still means replacing better insulating opaque by less insulating transparent assemblies, a difference not brushed away by the solar gains. It is advisable therefore to limit the glass surface per room to 1/5 of the floor area and orientate, as already said and if functionally possible, the larger surfaces south-east through south to south-west (north-east through north to north-west in the southern hemisphere). South is preferred because in winter gains are largest, while in summer the solar height makes shading easier. That rule of course has its limits. If for any reason bedrooms face south, their glazed surface should not exceed what's needed for day-lighting. A living room facing north in turn loses quality when equipped with such small windows that contact with outdoors becomes too restricted. It is also good to mount glass vertically at higher latitudes because in addition to easy maintenance, the winter solar gains are then as large as possible. One drawback is that large glazed surfaces looking east through south to west (east through north to west in the southern hemisphere) increase overheating risk in springtime, summer and autumn. Happily, providing exterior shading may prevent dwellers installing from active cooling.

Insulating the envelope remains a very effective measure for lowering net heating demand in temperate and cold climates, all other parameters being equal. The quality mark for an airtight envelope assembly is the whole wall thermal transmittance:

$$U = U_o + \left(\sum_{k=1}^{m} \psi_k L_k + \sum_{l=1}^{p} \chi_l \right) / A \qquad (\mathrm{W/(m^2.K)}) \qquad (3.82)$$

with U_o the clear wall value, A the parts surface, ψ the linear thermal transmittance and L length of the linear thermal bridges present and χ the local thermal transmittance of the local thermal bridges present. For envelopes it's their mean thermal transmittance (U_m):

$$U_m = \sum_{\text{all}} aUA/A_T, A_T = \sum_{\text{all}} A \qquad (\mathrm{W/(m^2.K)}) \qquad (3.83)$$

In this formula A_T is the envelope surface in m^2 and a a reduction factor for partitions that separate the protected volume from unheated neighbouring spaces, for parts at grade and parts contacting the soil. In theory, the annual end energy for heating should be proportional to the specific transmission losses ($H_T = U_m A_T$), which apparently fits with Figure 3.37, and the consumption data logged in 20 homes seemed to confirm:

$$Q_{\text{heat,use,a}} = (216 \text{ to } 360) H_T = (216 \text{ to } 360) A_T / \left[a + (d/\lambda)_{\text{ins}} \right] \qquad (3.84)$$

This is a simple equation usable for estimating end energy for heating at an early design stage. Practice meanwhile has shown that user behaviour negates such built environment wide proportionality between end energy use and the specific transmission losses, both per m^3 of protected volume.

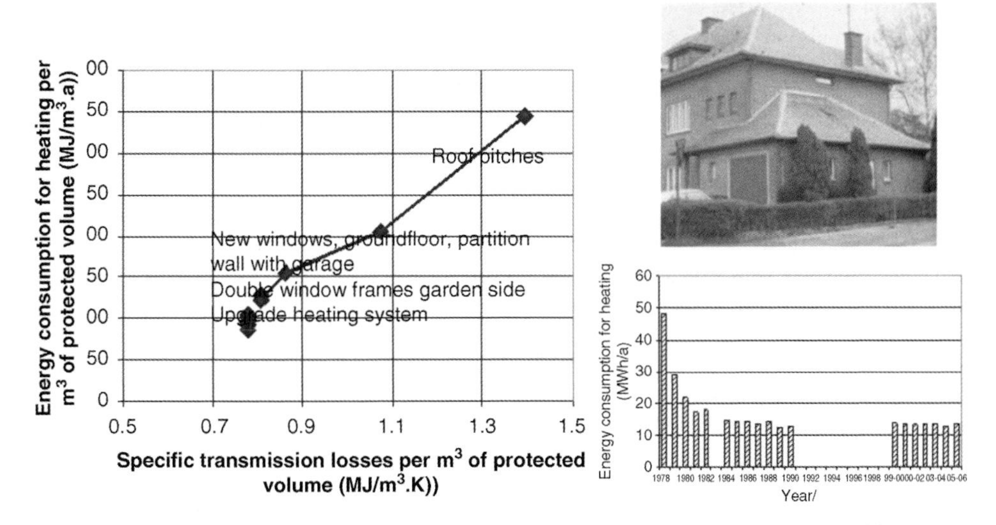

Figure 3.37 Stepwise retrofit of an end of the row house: annual end energy for heating in relation to the specific transmission losses (both per unit of protected volume).

Insulating buildings however has its limitations. The additional drop in end energy use shrinks when mean insulation thickness further increases (d_{ins}), a fact that user behaviour accentuates, as Figure 3.38 illustrates.

Halving the mean thermal transmittance doubles the insulation thicknesses needed. Stepping to very low values ($<0.1\,W/(m^2.K)$) requires thicknesses that generate important extra costs due to more complex details, higher roof edges, larger exterior surface or less living space. Of course, the lower the insulation's thermal conductivity (λ_{ins}), the less thickness is needed for the same thermal transmittance. Vacuum

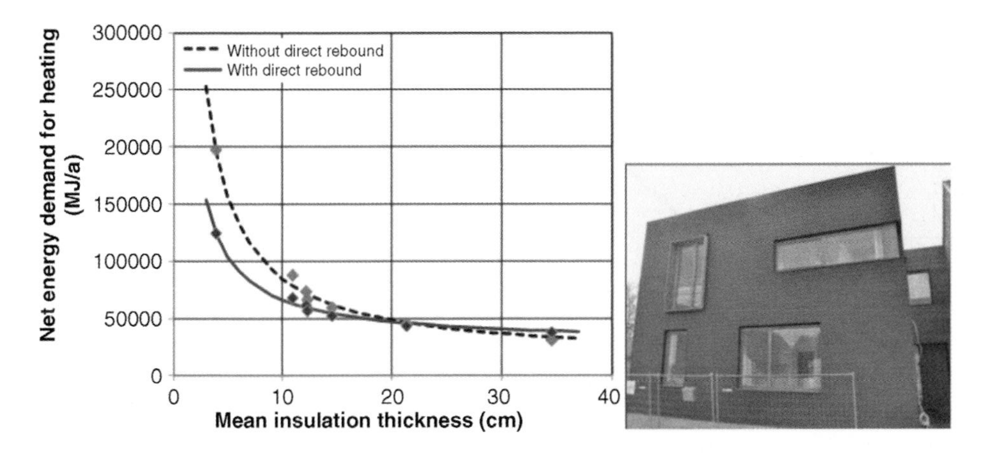

Figure 3.38 Detached dwelling, decreasing benefit of higher mean insulation thickness.

insulation panels (VIPs) would offer benefits but their vulnerability and the fact that they can't be cut makes their use in construction barely practicable.

How should thermal transmittances be chosen? At sketch design, fix the annual end energy target for heating. Then calculate related specific transmission losses $U_m A_T$ using formula (3.84) with as multiplier (216 + 360)/2. Divide by 1.05 to account for thermal bridging. Then equate this with the sum of weighted products of the thermal transmittances of roofs, opaque facade walls, lowest floors, windows, and so on and their respective surfaces:

$$U_m A_T / 1.05 = U_{roof} A_{roof} + U_{facade} A_{facade} + \sum a_{fl} U_{fl} A_{fl} + U_w A_w + a U_{h,nh} A_{h,nh}$$

Set all thermal transmittances minus one equal to what's mandatory ($U_{roof} \leq U_{roof,max}$; $U_{facade} \leq U_{facade,max}$; $U_{floor} \leq U_{floor,max}$; . . .) and calculate the one remaining. This done, check if that one is lower than mandatory. Finally, if possible, go for uniformity in thermal transmittances, which is economically the most viable and most helpful in case of partial heating.

Looking to thermal inertia, although a higher value brings the gain utilization efficiency closer to 1, the effect is quite unimportant and slows down with increasing storage capacity. Of course, in climates where short heating periods alternate with no heating, the effect gains importance compared to climates where heating seasons last for 200 consecutive days or more. Even well insulated massive buildings cannot even out the fluctuating weather over such long periods.

The simple electric circuit of Figure 3.39 helps to prove that statement. The resistance-capacitance-resistance circuit $R_{i,x1} - C_{x1} - R_{x1,e}$ represents the opaque envelope with $R_{i,x1}$ the equivalent thermal resistance between indoors and the envelope's capacitance centre of gravity (C_{x1}) at temperature θ_{x1} and $R_{x1,e}$ the equivalent thermal resistance between this centre and outdoors. All partitions, exposed surface A_i, are coupled to the indoor node by a capacitance-resistance circuit $C_{x2} - R_{i,x2}$ where $R_{i,x2}$ is the equivalent thermal resistance between indoors and the partition's capacitance centre

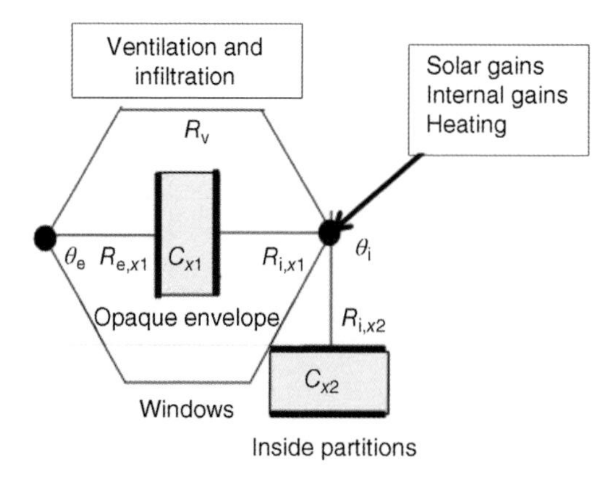

Figure 3.39 Electric analogy.

of gravity C_{x2} at temperature θ_{x2}. The ventilation and infiltration related thermal resistance in turn is given by:

$$Rv = R_{\text{v,i}} = 1/\left[0.34\left(n_{\text{V}} + n_{inf}\right)V\right]$$

That resistance links the indoor node directly to outdoors. The same applies to the heat losses through the windows. Solar gains, internal gains and heating by the system are again injected in the indoor node. The balance equations therefore become:

Indoors:

$$\frac{(\theta_{x1} - \theta_{\text{i}})\left(A_{\text{T}} - \sum A_{\text{w}}\right)}{R_{\text{i},x1}} + \sum U_{\text{w}} A_{\text{w}}(\theta_{\text{e}} - \theta_{\text{i}}) + 2\frac{(\theta_{x2} - \theta_{\text{i}})A_{x2}}{R_{\text{i},x2}} + 0.34nV(\theta_{\text{e}} - \theta_{\text{i}})$$
$$+ \Phi_{\text{SLW}} + \Phi_{\text{I}} + \Phi_{\text{heat,net}} = (\rho c)_{\text{eq,a}} V_{\text{a}} \frac{\text{d}\theta_{\text{i}}}{\text{d}t}$$

Envelope

$$\left(\frac{\theta_{\text{i}} - \theta_{x1}}{R_{\text{i},x1}} + \frac{\theta_{\text{e}} - \theta_{x1}}{R_{\text{e},x1}}\right) = C_{x1}\frac{\text{d}\theta_{x1}}{\text{d}t}$$

Partitions

$$\left(\frac{\theta_{\text{i}} - \theta_{x2}}{R_{\text{i},x2}}\right) = C_{x2}\frac{\text{d}\theta_{x2}}{\text{d}t}$$

In these equations, θ_{i} and θ_{e} are the indoor and outdoor temperatures, both in °C, Φ_{SLW} the solar gains, Φ_{I} the internal gains, $\Phi_{\text{heat,net}}$ the heating power, all in watts, A_{w} the surface and U_{w} the thermal transmittance of the windows, $(\rho c)_{\text{eq,a}}$ the equivalent volumetric specific heat capacity indoors (air, furniture and furnishings), and V_{a} the air volume. For massive buildings insulated outside or massive buildings with filled cavity walls, the difference between the temperature indoors and the temperature of all capacitances (θ_{x2} and θ_{x1}) is minimal. They and their derivatives can be set equal, melting the three equations into one with the equivalent resistance $R_{\text{e},x1}$ replaced by the mean for the whole envelope:

$$\underbrace{\left(\frac{A_{\text{T}}}{R_{\text{e}}} + 0.34nV\right)\theta_{\text{e}} + \Phi_{\text{SLW}} + \Phi_{\text{I}} + \Phi_{\text{heat,net}}}_{(1)} - \underbrace{\left(\frac{A_{\text{T}}}{R_{\text{e}}} + 0.34\,nV\right)}_{H}\theta_{\text{i}}$$
$$= \underbrace{\left(\rho_{\text{eq.a}} V + C_{x1}A_{\text{T}} + 2C_{x2}A_{x2} + (\rho c)_{\text{eq,a}}\right)}_{C_b}\frac{\text{d}\theta_{\text{i}}}{\text{d}t}$$

With inside insulation the opaque envelope's capacitance hardly has any effect:

$$\underbrace{\left(\frac{A_{\text{T}}}{R_{x1}} + 0.34nV\right)\theta_{\text{e}} + \Phi_{\text{SLW}} + \Phi_{\text{I}} + \Phi_{\text{heat,net}}}_{(1)} - H\theta_{\text{i}}$$
$$= \underbrace{\left(\rho_{\text{eq.a}} V + 2C_{x2}A_{x2} + (\rho c)_{\text{eq,a}}\right)}_{C_b}\frac{\text{d}\theta_{\text{i}}}{\text{d}t}$$

Solving both differential equations for a step change of term (1) at time zero gives:

$$\theta_i = \theta_{i\infty} + (\theta_{i\infty} - \theta_{\text{io}})\exp(-Ht/C_b) \qquad (3.85)$$

where θ_{io} is the indoor temperature at time zero and $\theta_{i\infty}$ its final value. The inverse of C_b/H is the time constant τ_b of the building in seconds. For a harmonic change with period T of the outdoor temperature and the solar gains with amplitude $\hat{\theta}_e$ and $\hat{\Phi}_{SLW}$, the complex indoor temperature without heating ($\alpha_{i,WH}$) equals (internal gains constant):

$$\alpha_{i,WH} = \frac{\alpha_e}{\left[1 + i\frac{2\pi\tau_b}{T}\right]} + \frac{\hat{\Phi}_{SLW}}{H\left[1 + i\frac{2\pi\tau_b}{T}\right]} \qquad (3.86)$$

with $\left[1 + i\,2\pi\tau_b/T\right]$ the complex building impedance. Annual net heating demand then becomes:

$$\text{Average}: \quad \overline{Q}_{\text{heat,net}} = 31.5576H\left[\overline{\theta}_i - \left(\overline{\theta}_e + \frac{\Phi_{SLW} + \Phi_I}{H}\right)\right] \quad (\text{MJ/a}) \qquad (3.87)$$

$$\text{First harmonic}: \quad \hat{Q}_{\text{heat,net}} = H\int_0^T \left(\alpha_i - \alpha_{i,WH}\right)\text{d}t \quad (\text{MJ/a}) \qquad (3.88)$$

In the case of permanent heating, thermal inertia only dampens the indoor temperature without heating and the net heating demand when the one-year complex impedance exceeds 1. Otherwise thermal inertia doesn't come into it. Moderately insulated, heavy buildings have time constants around 3–5 days. An excellent thermal insulation could give some 10 days. The time constant of a well-insulated lightweight building can drop to less than a day. Poor insulation reduces this to a number of hours. Or, in all four cases, complex impedances remain close to 1, see Table 3.27.

Or, in temperate and cold climates thermal inertia hardly impacts the net heating demand. Intermittent heating may even be negative. Warming up the assessable capacity

Table 3.27 Impact of thermal inertia on net heating demand.

Building	Capacitive		Well insulated		$\hat{Z}$
	Yes	No	Yes	No	
Moderately insulated, heavy structure	✓			✓	1.0012/1.0037
Excellently insulated, heavy structure	✓		✓		1.015
Well insulated, timber frame		✓		✓	1.0
Poorly insulated, timber frame		✓	✓		1.0

after each setback in fact demands extra energy, the more compared to a non-inert building the larger the inertia. Of course, simultaneous heat gain utilization ameliorates this. For night setback, the two effects neutralize one another. However, longer setbacks see extra energy prevailing.

In well-insulated dwellings with optimal gains, capacity nonetheless offers benefit in spring and autumn, when periods of full and no heating alternate. Its effect on the net demand over the whole heating season, however, is marginal. All this does not mean that building inertia doesn't matter in temperate climates. It helps moderating overheating during warm spells; in other words it supports passive cooling.

Although a tighter envelope lowers the net heating demand, all other parameters being equal, some nevertheless become suspicious when the word 'airtightness' slips out. What does ventilation mean in such a case? The answer is: for ventilation, don't rely on random infiltration across leaks and air permeable building parts, but use a purpose-designed system, see Table 3.28.

Envelope airtightness is measured using a blower door with calibrated fan, mounted in the entrance door opening (Figure 3.40). Testing is done as follows: All other outer doors, windows and ventilation inlets and outlets are closed or taped, and all inner doors opened. For under-pressure and overpressure, the airflow (G_a in kg/s) and pressure differential (ΔP_a in Pa) with outdoors is noted, giving as equation:

$$G_a = C\Delta P_a^n$$

Table 3.28 Airtightness and ventilation.

n_{50},ach	Climate	Way to ventilate
>13	Temperate	Too air permeable
	Cold	Too air permeable
8–13	Temperate	Adventitious natural ventilation by infiltration
	Cold	Too air permeable
5–8	Temperate	Too airtight for adventitious ventilation by infiltration, too air permeable for purpose-designed natural or extract ventilation
	Cold	Too air permeable
3–5	Temperate	Purpose designed natural or extract ventilation
	Cold	Extract ventilation
1–3	Temperate	Balanced ventilation
	Cold	Balanced ventilation
<1	Temperate	Balanced ventilation with heat recovery
	Cold	Balanced ventilation with heat recovery

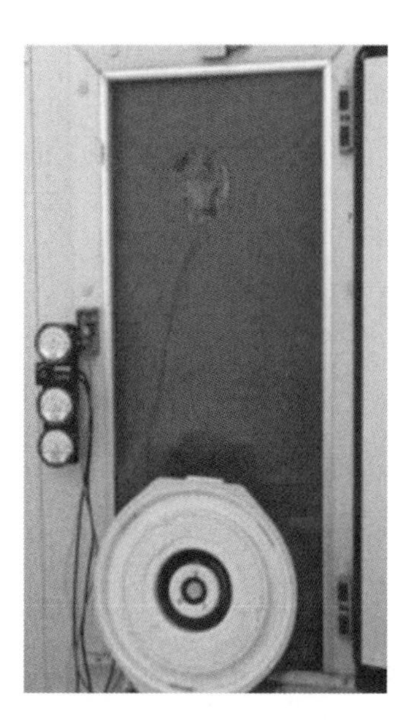

Figure 3.40 Blower door.

with C the air leakage coefficient and n the air leakage exponent. Both values follow from a least square analysis on the measured data. Airtightness then is typically expressed in terms of a ventilation rate (ach) at 50 Pa air pressure difference (n_{50}):

$$n_{50} = C\,50^n / V_{in} \tag{3.89}$$

with V_{in} the volume of air in the protected volume in m^3. The lower n_{50}, the more airtight the envelope. For a 3–4 Pa mean pressure difference with outdoors, a leakage exponent 0.6–0.8 and the leakage area equally distributed over the envelope, the sum of the mean infiltration and adventitious ventilation rate by window airing and door use (n_U) obeys following rule of the thumb:

$$n_{50}/20 + n_U \leq n \leq n_{50}/17 + n_U$$

3.3.6 Residential buildings, parameters fixing net cooling demand

The same set of parameters as for heating governs cooling but the effects differ, see Table 3.29. In temperate climates, active cooling of residential buildings is perfectly avoidable. Limit the glass area, a method that competes with more solar gains during the heating season, shade all east through south to west looking glazed surfaces (east through north to west in the southern hemisphere), go for massive construction and apply night ventilation. If active cooling is nonetheless necessary, thermal insulation will have little impact on the net demand but, as heating in such climates prevails, insulation still pays off.

Table 3.29 Parameters influencing the annual net cooling demand.

Set	Parameter	Impact on the annual net cooling demand, all other parameters equal
Outside climate (cooling season means)	Temperature	Decreases when lower
	Solar radiation	Increases when more
	Wind speed	Impact depends on the difference between outdoor and indoor temperature. If negative, a decrease, if positive, an increase
Building use (cooling season means)	Inside temperature	Decreases when higher
	Temperature control	Decreases when the principle 'cooling a zone only when used' is better applied
	Ventilation	May increase or decrease when higher. Depends on the difference between outdoor and indoor temperature. If negative, decreases, if positive, increases. More night ventilation gives a drop in regions with cool nights on condition the building is capacitive
	Internal gains	Decreases when lower. An incentive for energy-efficient lighting, daylighting and presence control
Building design	Protected volume	Increases when larger
	Compactness	Impact depends on the difference between outdoor and indoor temperature. If negative, a decrease, if positive, an increase
	Floor organization	Increases when floor organization augments solar gains in daytime rooms
	Transparent surfaces (typically windows)	Increases when the envelope contains larger transparent parts that face the sun and have high solar and low thermal transmittance
	Solar shading	Decreases when transparent parts are better shaded
	Thermal insulation	Impact depends on the difference between the mean sol-air and indoor temperature. Increases if negative, decreases if positive
	Thermal inertia	Decreases when higher. The effect is more pronounced when high ventilation rates are maintained during cool nights
	Envelope airtightness	Impact depends on the difference between the outdoor and indoor temperature. If negative, a decrease, if positive, an increase

3.3.7 Residential buildings, gross energy demand, end energy use

Gross equals net plus all the unrecovered distribution, control and emission losses. The ratio between net and gross defines the system's efficiency. End use in turn counts the energy consumed by boilers, heat pumps and chillers to deliver the gross demand. Production efficiency or, for heat pumps and chillers, the seasonal performance factor (SPF) then equals the ratio between gross demand and end use, see Table 3.30.

3.3.8 Residential buildings ranked in terms of energy efficiency

3.3.8.1 Insulated

As insulation has the largest impact in cold and temperate climates, as U-values are calculable and as insulating doesn't hinder architectural creativity, the focus is on well-insulated envelopes. Requirements include a maximum mean envelope thermal transmittance and limit values per part.

3.3.8.2 Energy efficient

The objective now is a low net energy demand, where thermal insulation, correct ventilation and optimal use of solar and internal gains cooperate. Evaluation is based on a standardized annual net demand calculation. Requirement is a maximum per m^2 of floor area, pre m^3 of protected volume or per m^2 of envelope area.

Table 3.30 Parameters of influence on the annual end energy use.

Set	Parameter	Impact, all other parameters being equal
Heating and cooling	System efficiency	Decreases when closer to 1
	Boilers, production efficiency	Decreases when closer to 1
	Heat pumps, SPF	Decreases when higher
	Chillers, SPF	Decreases when higher
Ventilation	Extract	Decreases when the extract air is used as enthalpy source for a domestic hot water heat pump
		Decreases when demand control is better applied
	Balanced	Decreases with heat recovery, providing the building is very airtight, the system perfectly installed and the inhabitants instructed on how to use it
		Decreases when demand control is better applied

3.3.8.3 Low energy

Primary energy use for heating and domestic hot water is the target here. A uniform calculation method is imposed by legislation. The epithet 'low energy' is granted for residences that consume less than 60 MJ/(m^3.a) primary energy for heating, have a good indoor climate in summer without active cooling and show a primary energy use for domestic hot water and household substantially lower than the country's average. In temperate climates, such performances are achievable with a compactness-related mean envelope thermal transmittance of 0.3–0.6 W/(m^2.K), an envelope air leakage not exceeding 3 ach at 50 Pa, a well-designed natural or extract ventilation and an energy-efficient heating plus domestic hot water system.

3.3.8.4 Passive

Passive buildings are comparable with low energy buildings. The epithet passive is granted to residential buildings that have a net heating demand below 18 MJ/(m^3.a), a good indoor summer climate without active cooling and a total primary energy consumption not exceeding 144 MJ/(m^3.a). To achieve that, air leakage is limited to 0.6 ach at 50 Pa, with a compactness-related envelope mean thermal transmittance of around 0.15–0.3 W/(m^2.K). Correctly installed and used balanced ventilation with 75% heat recovery is a must. A question mark is whether these requirements are really economically optimal, while measured data has shown that, including lighting and appliances, owing to dweller habits the overall primary energy consumed is not by definition lower than in low-energy residences.

3.3.8.5 Near zero energy

See low energy, be it that the primary energy used for heating and domestic hot water is further lowered, while a fraction of what's consumed must come from renewables (solar boilers, PV, heat pumps).

3.3.8.6 Net zero energy

Such buildings produce as much (primary) energy annually, mostly by PV, as used for heating, domestic hot water and, sometimes, cooling, lighting and appliances. With the last three included, really large PV surfaces are needed. Without battery storage, smart inverters and distribution grid upgrades, zero energy estates may in summer create such an imbalance between electricity produced and consumed that inverter shut-offs are needed. Gone then is the net zero objective.

3.3.8.7 Net plus energy

The next step is net plus energy buildings, which produce more (primary) energy than used for heating, domestic hot water and, sometimes, cooling, lighting and appliances. At an estate scale, even larger imbalances between electricity produced and consumed loom.

3.3.8.8 Energy autarkic

Solar collectors, PV, micro wind turbines, bio-fuels and heat and electric storage all offer the end energy needed for heating, ventilating and lighting the building and for using all appliances.

3.3.9 Non-residential buildings, net and gross demand, end and primary energy use

3.3.9.1 In general

What was said for residential buildings is still valid, even if the balance between gains and losses is different. Internal gains in open-plan offices, for example, are so high that even in temperate climates they need active cooling. If so, thermal insulation loses part of its importance as fewer losses are a disadvantage in cooling mode.

3.3.9.2 School retrofits as an exemplary case

Table 3.31 characterizes the eight schools that were audited.

Regarding the performances evaluated, first came the compactness and the envelope thermal transmittance. Calculation demands knowledge of layer thickness and sequence in all enclosure assemblies. When design documents fail, gathering this data remains uncertain. So the values of Table 3.32 are approximate. However, from a thermal insulation point of view, the eight schools perform poorly.

Second was the air tightness. In each school, the the infiltration rate in ach at a pressure difference of 50 Pa (n_{50}) was measured in one, two or three classrooms. Table 3.33 summarizes the results (V air volume in m^3, n air permeance exponent). A negative exponent indicates that in some classes the leaks present saw their air permeability decreasing at higher air pressure differentials! An example is the leakage across window sashes that close as a result of the air pressure differential pulling the operable part against the fixed frame.

The table shows a large spread in leakage: 2.4–12 ach. The actual infiltration rate of course depends on facade orientation, pressure differential with the corridor, leakage across the partition and the way the corridor acts as a discharge volume. In none of the cases was infiltration enough to maintain correct ventilation during class hours, which resulted in quite high differences in indoor/outdoor vapour pressure, in unacceptable peaks in CO_2 concentration and in a potentially high predicted percentage of dissatisfied with perceived air quality, see the Figure 3.41 for class 2 in school 8.

Third came the annual end energy use and a comparison with the predicted values. In each school, consumption data was gathered by analysing the bills and through data logging. Table 3.34 summarises the results after transposition to the reference year (see Table 3.23), together with the calculated reference values.

With the exception of school 5, the calculated value overestimates the energy consumption considerably. In other words, school heating is subject to a kind of rebound. There are many reasons for this. The reference assumes that ventilation guarantees an

Table 3.31 The eight schools.

School	Type	Floors m^2	Volume m^3	Characteristics
1	Primary	1083	4614	First built in 1950. Not insulated. First extension in the 1960s (no insulation) last in the 1990s, (insulated). No ventilation system. Hydronic central heating using gas
2	Special primary + secondary	6738	23 975	Built in the 1960s. Not insulated. No ventilation system. Hydronic central heating using oil
3	Special education	3742	16 935	Oldest building from the 17 th century. Gradually extended. No insulation. No ventilation system. Hydronic central heating using gas
4	Primary	1553	5001	Built in the 1920s. Part of the building retrofitted. Container class added. Poorly insulated. No ventilation system. Hydronic central heating using fuel oil
5	Primary + secondary (part of a larger school)	7002	19 070	Main building from the 17 th century. Reconstructed after World War II. New part built in the 1980s. Very little insulation. No ventilation system. Hydronic central heating using gas
6	Secondary + technical (part of a larger school)	12 651	48 453	Built in the 1980s. Roof insulated, walls not, mixture of single and double glazing. Ventilation system not used. Hydronic central heating using gas
7	Primary	2347	14 496	Built between 1909 and 1911. Part of the building retrofitted. No insulation. No ventilation system. Hydronic central heating using gas, renewed in 2002
8	Primary	582	1963	Built in the 1950s. No insulation. No ventilation system. Hydronic central heating using fuel oil

IDA3 indoor environment during class hours. For that, 29 m^3 of fresh air per hour and per pupil are needed. In reality, in all eight schools, infiltration provided ventilation with, as a result, much lower flows. The reference also exaggerates the hours that ventilation is needed: not 2628, but only 1332. Schools are also not continuously heated at 19 °C. Corridors remain unheated, while 17.5 °C as average in the heated zones is a better guess. The reference further underestimates the internal gains, by taking 1 pupil per 4 m^2 of floor area. A class of 20 should thus have 80 m^2 of floor area, while in reality the values are closer to 55 m^2 for 20–30 pupils. And, even more important, money spent on heating cannot be used for didactics.

Table 3.32 Compactness, mean envelope thermal transmittance.

School	What volume?	Compactness, m	U_m-value, W/(m^2.K)
1	Whole	1.50	1.53 ± 0.08
2	Part	1.58	1.98 ± 0.1
3	Whole	2.33	1.24 ± 0.06
4	Main building	1.78	1.83 ± 0.09
	Second building	1.32	1.88 ± 0.09
	Container	0.69	0.98 ± 0.05
5	Old building (O)	3.90	2.20 ± 0.11
	New building (N)	3.10	1.60 ± 0.08
6	Whole	3.72	1.78 ± 0.08
7	Whole	2.73	1.59 ± 0.08
8	Whole	1.94	1.38 ± 0.07

What are the consequences for retrofitting? Proposed were roof insulation, new windows with better insulating glass and correctly dimensioned trickle vents on top, solar shading where necessary, if doable outer wall and ground floor insulation, upgraded heating systems, relighting and, mandatory, extract ventilation with demand control. Judging those measures economically is done by comparing the annuity on investment with the avoided energy costs. When these are greater than the annuity, the measures are cost effective, so more money is left for didactics. Due to the rebound observed, avoided reference end use ($\Delta Q_{\text{heat,a,ref}}$) was no basis for evaluation. Instead, reality ($\Delta Q_{\text{heat,a}}$) had to be estimated as:

$$\Delta Q_{\text{heat,a}} = \Delta Q_{\text{heat,a,ref}}(1 - a_{\text{rebound}})$$

Table 3.33 Airtightness of separate classes (exterior walls only).

School	Class 1			Class 2			Class 3		
	V m^3	n_{50} ach	Exponent b (–)	V m^3	n_{50} ach	b–	V m^3	n_{50} ach	b–
1	288	6.9	0.20	260	6.0	–0.066	330	7.1	–0.058
2	128	4.0	0.67						
3	271	3.9	0.68	124	2.4	0.62	144	3.8	0.70
4	219	4.4	–0.027	218	8.2	–0.003			
5	137	10.5	0.67	120	4.3	0.77			
6	147	10.7	–0.024	147	4.6	0.002			
7	300	5.4	0.62	300	12.0	0.54			
8	165	2.6	0.76	162	3.7	0.69			

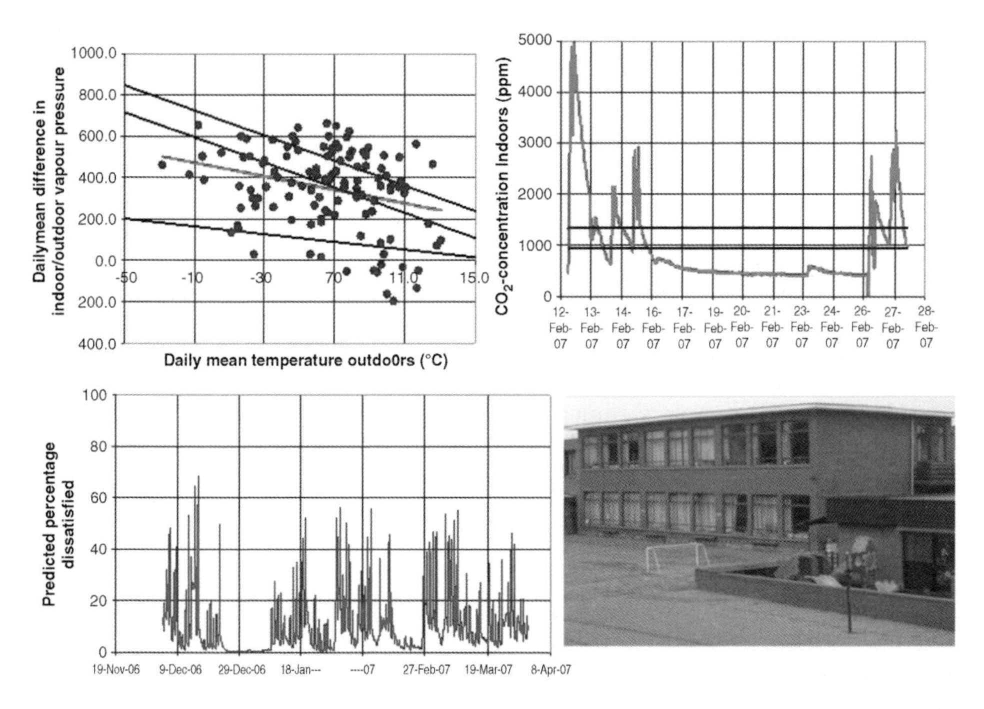

Figure 3.41 School 8, class 2, differences in indoor/outdoor vapour pressure difference, CO_2 concentration, predicted percentage of dissatisfied with the perceived air quality.

For the rebound factor ($a_{rebound}$) the values listed in Table 3.34 applied. As an IDA 3 ventilation system is mandatory, just upgrading the insulation, increasing heating system efficiency and relighting did not compensate for the extra energy such a system requires. Take school 4. Infiltration and adventitious ventilation added

Table 3.34 End energy consumption, measured and normalized versus the calculated reference, rebound factor.

School	Reference calculation, MJ/a		Measured, MJ/a		$a_{rebound}$ Heating
	Heating ±10%	Electricity (light, auxill.)	Heating	Electricity (all)	
1	2 395 000	191 800	601 000	156 100	0.74
2	344 900 (part)	No data	3 240 000 (tot)	No data	—
3	4 350 000	No data	1 517 900	202 830	0.66
4	1 080 000 (part)	88 459 (total)	308 350 (part)	67 600 (total)	0.71
5	2 818 000 (part)	No data	2 677 600 (part)	893 400 (total)	0.05
6	8 530 000 (part)	1 167 850 (part)	11 736 000 (tot)	2 990 600 (total)	—
7	2 950 700	No data	1 215 600	No data	0.58
8	737 000		247 000		0.66

41 000 MJ/a to the main building's net energy demand. An IDA3 extract ventilation, however, gave a 77 000–191 000 MJ/a increase. Compensation was sought by insulating the pitched roof, whose U-value was between 2.7 and 3.1 W/(m^2.K). Yet, even 0.2 W/(m^2.K) did not erase that extra, although a reference calculation predicted a 236 000 MJ/a drop in net demand. In reality, some 67 000 MJ/a or perhaps less could be expected. Or the need for correct air quality, which presumes extract ventilation, and rebound made it hard to create an overall gain by retrofitting.

3.4 Durability

3.4.1 In general

Durability is a building-fabric-related requirement. Absolute durability does not exist; even granite stones weather. Defining relative durability as correct functioning during a preset period of time makes more sense. What constitutes an acceptable service life depends on cultural and historical backgrounds, economics, building tradition, building usage and which enclosure part is being considered. Roofing membranes, for example, are expected to function properly for 10–15 years. Public buildings should last longer than industrial premises. A timber-framed construction has a shorter life expectancy than brick. Finishes are more subject to fashion, so have relatively short service lives.

Necessary conditions for a durable fabric include an adequate design from a structural and building physics point of view, materials withstanding the loads typical for the building part they belong to and their position in it, good workmanship, correct maintenance included timely replacement of layers and parts with shorter service life and timely cleaning, repairing and refinishing the inside and outside surfaces.

When analysing durability, terms like behaviour, load, limit state, ageing, damage and risk pertain. Behaviour relates to the evolution in time of the properties of a material or assembly. Loads combine all external conditions that impact on and change properties and behaviour over time. The idea of limit state points to damages that hinder proper functioning. Ageing reflects degradation caused by inevitable time-related physical, chemical and biological processes. Damage relates to too fast degradation due to design flaws, wrong material choices, poor workmanship or inadequate maintenance, and all these induce damage risks:

$$Risk = p \times Size \tag{3.90}$$

with p the probability that a failure will happen and 'Size' the gravity of the negative consequences. So high risk includes rare damages with severe consequences as well as common damages with moderate consequences. While structural misfits can cause collapse and take human lives, building-physics-related failures degrade usability, comfort, indoor air quality, energy efficiency and environmental impact but rarely cause collapse, though the risk is not zero, as Figure 3.42 proves. Individuals and society often judge consequences differently. Dwellers hate bad odour, draughts, rain penetration and mould, while the energy used or the CO_2 emitted are of less concern. Not so for society as a whole.

Figure 3.42 Natatorium. Left collapsed roof beam, right collapsed roof deck, both due to long-lasting interstitial condensation.

3.4.2 Loads

Distinction must be made between permanent and accidental loads. Permanent are structural, thermal, hygric and electromagnetic loads plus age. Own weight, dead load, live load, wind, snow and slant cause stress, strain and displacements. Time-related consequences include creep, relaxation, fatigue and sometimes collapse. Temperature differences and changes cause stress, strain, deformation, displacement and cracking, while below-zero values combined with wetness trigger frost damage. Hygric loads are most damaging. They induce stress and strain, deformation, displacement and cracking, degrade the thermal and structural quality of materials and are a necessary condition for biological attack. Electromagnetic loads include visible light, which fades colours and encourages biological attack, and UV, which changes the chemical composition of certain materials. Finally ageing translates into a slow degradation of the properties of certain materials.

Accidental loads include earthquakes, hurricanes, explosions and fire. In quite a few countries, earthquakes are so likely that it is mandatory to construct for maximum stiffness, while being resilient enough to withstand the axial, bending, shear and torsion movements. Fire is so devastating that all developed countries mandate that buildings should be designed to be fire safe.

Degradation could be chemical, physical or biological. Chemistry includes damage by moisture and air pollution. Physical processes, such as out-gassing by diffusion of blowing agents, alter the properties of some insulation materials. Biological damage is inflicted by bacteria, algae, moss, moulds, fungi, birds and rodents.

3.4.3 Damage patterns

3.4.3.1 Decrease in thermal quality

Water conducts heat 20 times better than air. If for any reason water displaces part of the air in an open-porous material, heat flow will increase. Latent heat of evaporation and condensation in the pores reinforces the rise, while air movements can add enthalpy flow. How large the increase is, depends on the air flow pattern, the water content and which layer becomes moist. Most negative is wetness at the warm side of the insulation, a humid insulation, air looping around, wind washing behind and indoor air washing in front of the insulation. To give an example, for a 20 cm thick aerated concrete roof and a cavity wall with 6 cm wide filled cavity, the thermal transmittances value:

Aerated concrete roof	U, W/(m^2.K)	Cavity wall	U, W/(m^2.K)
Air-dry	0.8	Air-dry	0.64
With production moisture	1.5	Veneer wet by rain	0.65
Wet by interstitial condensation	1.3	Cavity fill saturated with water	1.80
Wet by rain	1.7		
Saturated with water	2.3		

Where wetting degrades the roof, the cavity wall only degrades when the fill becomes wet.

3.4.3.2 Decrease in strength and stiffness

Strength and stiffness of most open-porous building materials decreases at higher moisture content. For glued materials and materials such as particle board, plywood, fibre board and oriented strand board this process is irreversible. In fact, hydrolysis cuts the glue polymers into monomers what lets the boards swell and lose cohesion, as Figure 3.43 illustrates by showing the drop in bending strength by interstitial condensation in ureum-formaldehyde (UF), melamine (Mel) and phenol formaldehyde (PF) particle boards.

3.4.3.3 Stress, strain, deformation and cracking

The thermal expansion coefficient (α) governs reversible temperature-related deformation of a material:

$$\alpha = 1/L(\mathrm{d}L/\mathrm{d}\theta) \qquad (\mathrm{K}^{-1}) \tag{3.91}$$

The highest values are those of synthetic materials:

Material	α (×10^{-6}), K^{-1}
Wood-based	4–30
Stony	5–12
Synthetic	25–200
Metallic	12–29

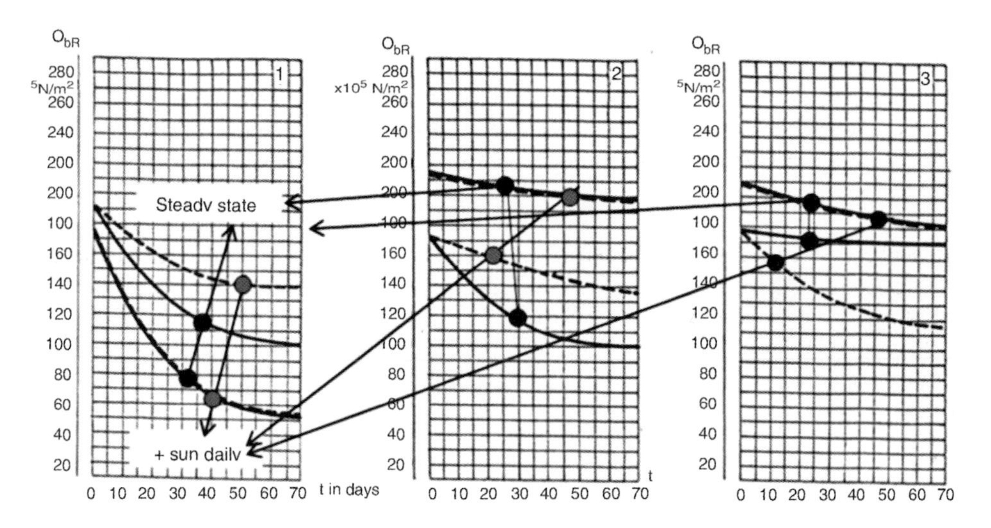

Figure 3.43 Particle boards. Bending strength under long-lasting interstitial condensation (1 = UF, 2 = Mel, 3 = PF), upper and lower thresholds.

If combined with less deforming materials, detailing must allow the synthetic parts to move independently.

The maximum temperature differences with the annual mean outdoors define the ultimate expansion and contraction. When changes in length are unconstrained:

$$\Delta L = L_{\mathrm{o}}[\exp(\alpha \Delta \theta) - 1] \approx \alpha L_{\mathrm{o}} \Delta \theta$$

with L_{o} the length at the annual mean outdoors. Usually, assemblies and their layers cannot move unhindered, so deforming causes stress (σ), for a fix-ended elastic layer:

$$\sigma = \alpha E \Delta \theta \qquad \text{(MPa)}$$

with E the modulus of elasticity. This stress increases when materials are stiffer and have larger thermal expansion coefficients. Once exceeding the tensile strength locally, micro-cracks form. Under alternating load, fatigue forces these to become macro-cracks. However, the likelihood of damage drops when the product of the thermal expansion coefficient and the modulus of elasticity (αE), divided by the tensile strength ($\sigma_{\mathrm{t.m}}$) decreases. Or, for layers requiring a high cracking resistance such as outside stuccoes, paints, roofing felts a low value of that ratio is required, unless the thermal load can be tempered.

Plastic foams suffer from irreversible thermal deformation. Successive temperature peaks induce a permanent, anisotropic volume change. This is a result of the foam's structure: an aggregate of bubbles deformed by gravity from spherical to ellipsoid during hardening. Once the material is applied, higher temperatures lower the modulus of elasticity of the bubble walls and increase gas pressure, which tends to restore the spherical form, resulting in shrinkage along the long axis and expansion across the short

Figure 3.44 Low-sloped roof above a timber kiln, irreversible deformation of PUR.

axis of the ellipsoids. Where the pores contain moisture, fluctuating vapour pressures exacerbate the deformation and possible damage, see Figure 3.44.

Hygric strain (ε). which defines the reversible deformations caused by wetness of open-porous materials, is given by:

$$\varepsilon(\phi) = \mathrm{d}L/L \quad \text{or} \quad \varepsilon(w) = \mathrm{d}L/L \tag{3.92}$$

As deformation becomes non-linear, advancing a single-valued hygric expansion coefficient is not sufficient. The derivatives $\mathrm{d}\varepsilon(\phi)/\mathrm{d}\phi$ and $\mathrm{d}\varepsilon(w)/\mathrm{d}w$ are largest at low relative humidity or low moisture content and tend to zero for a relative humidity nearing 100% or a moisture content nearing capillary, see Figure 3.45. In many materials, relative humidity after processing reaches 100%. Its drop once in use determines the initial shrinkage. The lower the mean relative humidity (ϕ_o) in use

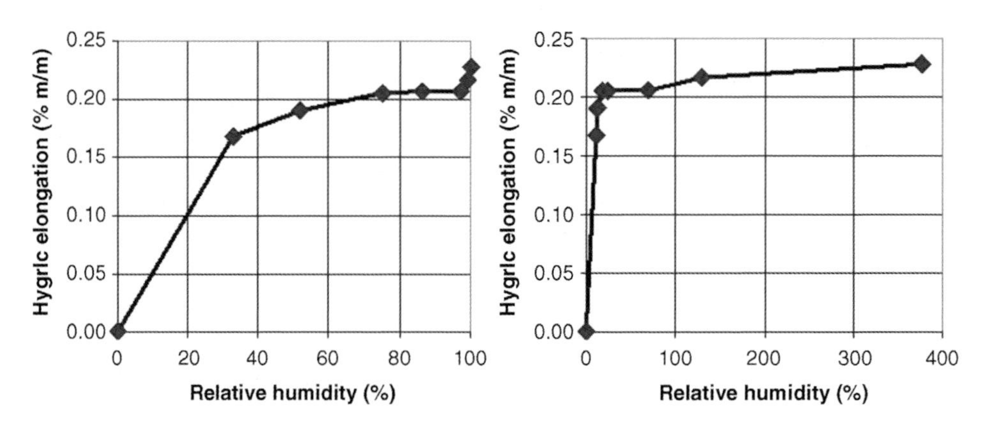

Figure 3.45 Aerated concrete, hygric expansion.

is, the more wood, lime and cement-based materials shrink. The changes in relative humidity thereafter induce expansion and contraction, which, if unhindered, are equal to:

$$\Delta L = L_o \left[\varepsilon(\phi_1) - \varepsilon(\phi_o)\right]$$

with L_o the length at the mean relative humidity (ϕ_o).

Again, the probability that movement remains unhindered is low. Part of the deformation then transposes into stress, for a fix-ended layer:

$$\sigma = E\left[\varepsilon(\phi_1) - \varepsilon(\phi_o)\right] = E(\mathrm{d}\varepsilon/\mathrm{d}\phi)_m \Delta\phi \qquad \text{(MPa)}$$

with $(\mathrm{d}\varepsilon/\mathrm{d}\phi)_m$ the average derivative of the hygric strain in the relative humidity interval $\Delta\phi$.

A higher product $E(\mathrm{d}\varepsilon/\mathrm{d}\phi)_m$ and larger relative humidity fluctuations increase stress. Once exceeding the tensile strength, micro-cracks form. A lower ratio between the product $E(\mathrm{d}\varepsilon/\mathrm{d}\phi)_m$ and tensile strength diminishes their occurrence. For layers that require a high cracking resistance, such as outside stucco, paints and roofing felts, a low product $E(\mathrm{d}\varepsilon/\mathrm{d}\phi)_m$ and high tensile strength are essential.

Irreversible hygric deformation mainly results from chemical processes or binder hydrolysis, causing materials that are glued under pressure to swell. If chemical, the irreversible part reinforces reversible initial shrinkage. If hydrolysis-related, swelling develops proportional to the drop in strength and stiffness, reflecting past moisture loading.

Whether or not hygric and thermal loads will induce cracking however is hard to predict. The two counteract each other. A temperature increase usually invokes a lower relative humidity and vice versa. Some facts are nonetheless known. Timber spalls when moisture content drops below fibre saturation. Controlled drying must exclude this, while at delivery moisture content should be close to the equilibrium in use. Combined chemical and hygric shrinkage induces cracks in fresh cement-based materials such as concrete, though when fresh, final strength and stiffness are not yet reached. Once bound, the much larger penetration depth of temperature waves compared to relative humidity waves with equal frequency makes thermal movement dominant in massive parts. However, in thin layers, hygric movement wins because temperature waves with periods shorter than a day only create small gradients, while relative humidity produces large ones. Stucco so shrinks when sun radiated and expands when cooled!

When does cracking cause damage? In general, it happens as soon as the structural integrity, rain-tightness, airtightness or sound attenuation are jeopardized. Cracks can also accelerate chemical attack, while often the appearance degrades (Figure 3.46). Formulating clear performance requirements nonetheless remains difficult. For stucco, the metric could be to limit the stress, thus making cracking unlikely.

Figure 3.46 Cracks degrade appearance.

3.4.3.4 Biological attack

Wood-based materials are attacked by bacteria, mould, fungi and insects. Aerobic and anaerobic bacteria aid the rotting process. The materials are a preferred substrate for the deuteromycetes mould family once, depending on temperature, relative humidity at their surface reaches 80–90%. The mould's low metabolism happily limits damage.

Fungi in turn are giant moulds. Those shown in Figure 3.47 belong to the basidomycetes family, which causes wood decay. Germination starts when spores settle on wood-based materials with a moisture ratio above 20% kg/kg and temperatures of 8–35 °C. The hyphae work their way into the wood while the fruiting body develops on the outside of it. Fungi have a high metabolism. Brown rot feeds on the cellulose in wood, white rot attacks the lignin skeleton. Both secrete water during digestion, creating their own moist environment. Infected wood slowly loses strength, stiffness and cohesion. Fungi colonize large surfaces. The hyphae even penetrate mortar joints and porous bricks

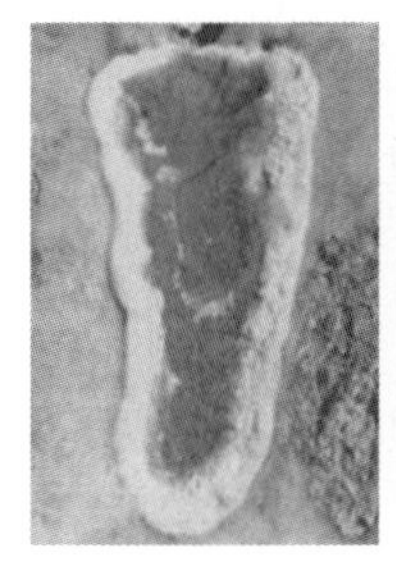

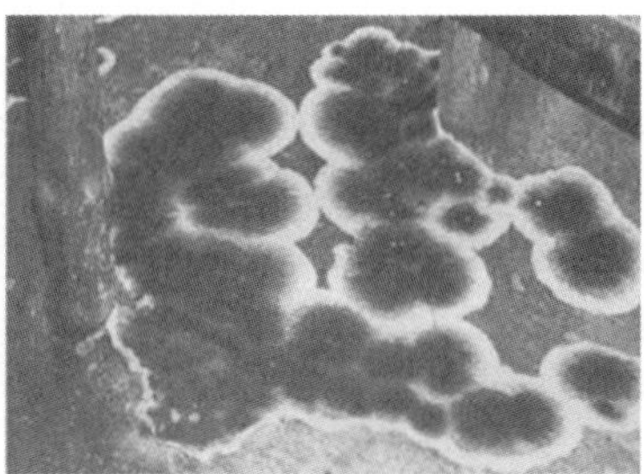

Figure 3.47 Fungi.

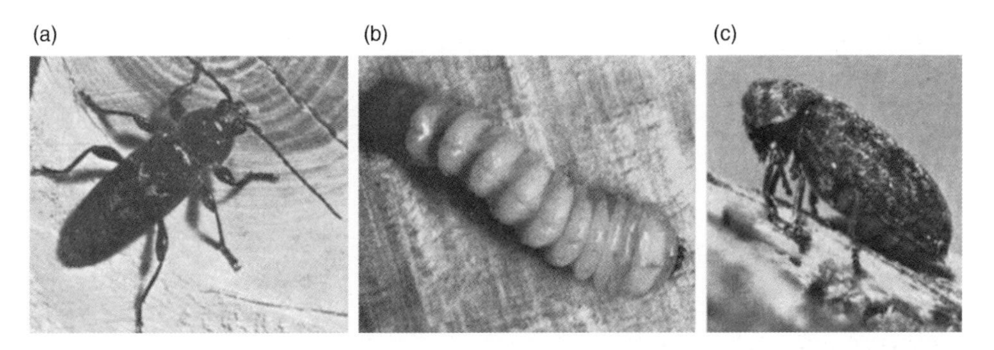

Figure 3.48 (a) A longicorn, (b) its larva, (c) anobia.

in search for food substances. The damage inflicted is enormous with timber floors and doors destroyed and wooden finishes demolished. Often all this remains concealed for a long time as paints are left intact. Avoidance demands moisture ratios to be kept below 20% kg/kg. Combating fungi when present is demanding. Affected wood must be replaced, intact parts treated with preserving liquid, brickwork injected with poisonous liquid, moisture sources removed and dismantled wood burnt.

Anobia, lycti and longicorns are common wood bugs in temperate regions (Figure 3.48), while termites are common in warm climates. Even wasps attack wood, extracting cellulose to build their nests. Anobia prefer wood infected by fungi. Some also look for dry, older wood. Lycti colonize wood having pores that contain yeast and a vitamin that their larva need to grow. Longicorns like pine and sapwood especially when stored in warm, humid environments. Termites live below grade in colonies that include workers, soldiers and females. They feed on decayed and dead organic material such as humid construction timber at grade. The damage inflicted in the USA is enormous. In Europe they are active south of the 48° latitude.

A low moisture ratio keeps insects out of wood. Because 'low' is rather vague, some countries require wood for construction to be treated. Also species-specific measures apply. To avoid damage inflicted by termites, construction timber must not contact grade. For timber frame walls, this requires an impenetrable steel barrier under the wooden bottom plates.

Rodents love nesting in fibrous and synthetic foam insulation, while birds use the mineral fibres for building nests (Figure 3.49). Avoidance requires some form of shielding. It is also possible to modify the binder in fibrous boards so that they lose their attractiveness as nesting material.

On stony materials, algae and moss that colonize surfaces, joints and overlaps need light, nutrients, moisture and stable temperatures. Moss grows best in the overlaps between slates on roof pitches that face north-west through north to north-east (northern hemisphere, higher latitudes). Relatively stable temperatures are normal there, and light is no problem. Nutrients come from the organic dust accumulating on the slates, and the capillary water filling the overlaps suffices as a moisture source. Thanks to

Figure 3.49 (a) Rodent nest in glass wool, (b) mineral fibre pecked by birds.

surface condensation by under-cooling and dust delivering the nutrients, EIFS can suffer from algae growth (Figure 3.50(a)). Whether greenery actually induces damage depends on the problems caused. Moss widens the overlaps between slates, facilitating rain penetration. Plants invading masonry walls erode the joints with their roots (Figure 3.50(b)). The organic acids that moss, algae and some plants secrete can also cause chemical attack.

3.4.3.5 Frost damage

As pore water expands by 10% when freezing, wet brittle, open-porous stony materials can crack, spall and pulverize if subjected to frost and thaw cycles. In wide pores ice forms at 0 °C. In the finer ones freezing requires temperatures below $-x$°C, where x becomes lower the smaller the pore diameter. Correct detailing and finishings that keep moisture content low can exclude frost damage. Also frost-resisting materials display zero risk. Decisive is the frost number (F), given by the difference between the critical moisture content for frost (w_{crf}) and the actual moisture content (w):

$$F = w_{cr,f} - w \tag{3.93}$$

Figure 3.50 (a) Stucco with algae above the terrace floor, (b) stone wall with greenery.

Table 3.35 Critical saturation level for frost of some materials.

Material	Critical saturation level for frost, $S_{cr,f}$ (%)
Natural stone	65–96
Normal concrete	90
Concrete with air bubbles	81
Lightweight concrete	60
Aerated concrete	40
Bricks	87
Sand-limestone	65
Cement mortar	90
Lime mortar	65

F being negative means damage. Frost resistance thus is a relative notion as materials with known critical moisture content for frost will resist in one application and deteriorate in another, depending on the actual wetness building up. The 10% expansion makes damage acute when moisture fills 90% of the pore volume, a percentage fixing the limit for the critical value:

$$w_{cr,f} \leq 0.9 w_m$$

with w_m the saturation moisture content. Materials where less than 10% expansion induces stresses beyond the tensile strength, degrade below that limit. High tensile strength or high deformability can of course buffer the 10% without damage. In micro-porous materials freezing starts so far below 0 °C that it rarely happens in temperate climates. In other words, the critical saturation degree for frost depends on pore structure, tensile strength and deformability, see Table 3.35.

Actual moisture content changes with the application. Regardless of whether the cavity is filled or not, non-painted veneers never have a moisture content passing capillary. Painted ones instead do because they absorb rain where the paint has micro-cracks, while this same paint hinders drying. Due to rain and under-cooling condensation, tiles on insulated roof pitches near saturation in winter. Also unused chimneys wet beyond the capillary moisture content. In all cases where wetness passes the critical value for frost, the frost number becomes negative. Detailing and the way frost attacks of course is also important. Double-sided freezing gives different stress distributions from single-side freezing. Stresses are larger in massive than in thin layers. In temperate climates, roof tiles must withstand 5–70 frost/thaw cycles a winter. This produces visible damage after a material-specific number of cycles.

Evaluation of the frost resistance is based on indirect and direct testing methodologies. An indirect method has been developed for natural stone, tiles and bricks. Measured are the capillary and saturation moisture content, the capillary water absorption coefficient (A) and the coefficient of secondary water uptake (A'). Analysing the correlation between pore structure, frost test results and observations on site has revealed that damage risk increased with less difference between capillary and saturation, while it decreased with a higher ratio between the capillary water absorption coefficient and the capillary moisture content. Correlating frost test result under several exposure conditions to two ratios, one between the capillary and saturation moisture content and the other between the capillary water absorption coefficient and moisture uptake at saturation ($w_m h$, with h the sample height) led to an empirical relation between the two, named the frost constant (G_c).

Bricks and stone	Tiles
Suction curve showing primary and secondary water uptake	
$G_c = -14.53 - 240\frac{A}{w_m h} + 20.3\frac{w_c}{w_m}$	$G_c = -21.04 - 504\frac{A}{w_m h} + 23913\frac{A'}{w_m h} + 22.6\frac{w_c}{w_m}$
Suction curve showing no secondary water uptake	
$G_c = -6.35 + 16676\frac{A}{w_m h}$	$G_c = -6.35 + 16676\frac{A}{w_m h}$

Its value allowed differentiating between five material classes:

	Material	Class
$G_c < -2.5$	Frost resisting even under the most severe exposure	D
$-2.5 \leq G_c < -0.95$	Usage subjected to increasingly more restrictions	C
$-0.95 \leq G_c < 0$		B
$0 \leq G_c < 4.5$		A
$G_c \geq 4.5$	Material not frost resisting	O

The approach looked statistically sound. That the class was correct had an 87% probability. The error not exceeding one class up or down was 97% probable. For bricks and stone, class and application are linked, see Table 3.36. Class D has since been renamed class F2[D] and class C class F2[C], while the classes B, A and O are grouped under F1.

Also direct tests are used to classify bricks and stones. The most extreme one consists of putting samples in a drained, sand-filled container with one side weather-exposed. If after four winters no visual damage appears, the materials belong to class D. If damaged, they fit into one of classes O to C. For C, samples are brick-laid on a horizontal concrete floor and weather-exposed. If after four winters no damage appears while they degraded in a class D test, it's C. For B, samples are stored on racks and weather-exposed. If after four winters no damage appears while they degraded in the D and C tests, it's B. For A, samples are brick-laid in a south-west facing veneer of an unfilled weather-exposed cavity wall. If after four winters no damage appears while they degraded in the D, C and B tests, it is A. O means even this test caused damage.

Classic frost/thaw testing in turn consists of one day long cycles with the wet samples freezing in the air or in a sand-filled container during $(24 - y)$ hours, followed by y hours thawing in water. Variables are the initial moisture content (saturation under vacuum,

Table 3.36 Bricks and stone, links between class and application.

Application, detailing and orientation Building part	Cavity wall		Heated inside?	Protected thanks to correct detailing?			
	Unfilled	Filled		Yes		No	
				S+W	N+E	S+W	N+E
Above grade masonry	✓		Yes	B	B	C	B
	✓		No	C	C	D	C
		✓	Yes	C	C	D	C
		✓	No	C	C	D	C
At grade masonry	✓		Yes	B	B	C	C
	✓		No	C	C	D	D
		✓	Yes	C	C	D	D
		✓	No	C	C	D	D
Masonry roof edges	✓	✓				D	C
Horizontal masonry parts	✓	✓		D	C	D	D
Chimneys	✓	✓		C	C	D	D
Masonry in contact with water						D	D

saturation at 1 atmosphere, capillary saturated), the number of cycles, the hours of freezing and the thawing temperature. Most important is the initial moisture content. Class D materials don't damage when vacuum saturated, class C materials don't damage when capillary moist, while classes B, A and O materials slightly, moderately or heavily turn damaged then. To see the effects of fatigue, at least 25 cycles are needed. The frost period (24-y) must be long enough to get the complete samples iced. Freezing at −10 to −25 °C is preferred, while thawing is best done in water at 20 °C.

Which remedies are applicable when stony constructions are frost-damaged? With limited frost damage, eliminating the cause, for example rain absorption, often suffices. If severely damaged, replacement is the only remedy.

3.4.3.6 Salt attack

Salts loading building materials are carbonates, sulphates, nitrates and chlorides of lime (Ca), sodium (Na), potassium (K) and magnesium (Mg). Phosphates and ammonium nitrates show occasional signs of being present. In many cases, the materials are the source of the salts. The clay used for brick-making may contain ferro-sulphates, which react with oxygen to form SO_3 during firing that then binds with the alkali in the clay to form alkali sulphates (Na_2SO_4, $CaSO_4$, K_2SO_4, $MgSO_4$). If fired at temperatures above 950 °C these sulphates, except $CaSO_4$, react with the silicates present. Otherwise they deposit in the pores where reactions go on after bricklaying. Sometimes bricks also contain vanadium and manganese salts.

Stone may be salt-loaded when quarried. Concrete is by definition alkali-rich ($Ca(OH)_2$, KOH, NaOH). Portland cement contains $Ca(OH)_2$, KaOH and NaOH, which all react with CO_2 in the air to form alkali-carbonates (Na_2CO_3, $CaCO_3$, K_2CO_3, $MgCO_3$). In the presence of moisture SO_2 transforms these into sulphates. At the brick/mortar interface, NaOH and KOH of the mortar reacts with the lime sulphate ($CaSO_4$) in the brick to form Na_2SO_4 and K_2SO_4 with $Ca(OH)_2$ as a by-product, which in turn forms $CaCO_3$ with CO_2 in the air. Fresh lime mortar is a mixture of $Ca(OH)_2$ and sand. Curing gives calcium carbonates ($CaCO_3$).

The environment can also be a source. Groundwater contains calcium sulphates, sodium nitrates, potassium sulphates, potassium nitrates and ammonium nitrates. Air transports sodium chlorides in coastal areas and other salts in industrial areas. Polluted air can also contain CO_2, SO_2, NO_X and compounds of sodium carbonates, potassium carbonates, calcium carbonates, magnesium carbonates and sulphates plus nitrates of the same reagents. Used as thawing salt is calcium chloride ($CaCl_2$), while cesspits are an active source of nitrates.

Moisture content and salt solubility now determine the amounts of salt that liquid water in an open-porous material can contain, see Table 3.37. As the table indicates, solubility diminishes at lower temperatures, which means that crystallization will be more intense then. Some of the salt ions bind to the pore walls. The dissolved ones move with the free pore water, while ion diffusion transfers salt from higher to lower concentration. Everywhere where capillary flow turns into vapour diffusion, the remaining liquid over-saturates and the salts it contains crystallize.

Table 3.37 Solubility of salts in water.

Salt	Solubility In g per 100 ml of water		
	0 °C	20 °C	100 °C
$CaCO_3$		0.0015	0.019
$CaSO_4.2\,H_2O$		0.24	0.22
Na_2SO_4		4.8	
$Na_2SO_4.10\,H_20$		11.0	92.0
NaCl		36.0	39.0
K_2CO_3	105.5		156.0
K_2SO_4			24.1

Each moisture source has a specific impact. Rain absorption followed by drying moves the moisture content from capillary to hygroscopic and vice versa, forcing salt dissolution to alternate with crystallization. Construction moisture acts as a temporary salt driver, rising damp as a permanent one. Salt hydration then occurs each time the relative humidity and temperature pass the threshold at which water molecules bind to the salt crystals.

What are the consequences of salt presence? It changes the sorption isotherm compared to the salt-free material. Each time relative humidity in the pore air passes the equilibrium value for a solved salt, sorption increases proportional to its concentration. For $MgCl_2$, K_2CO_3, Mg(NO3)$_2$, NaCl, KCl and K_2SO_4 that equilibrium is 33, 45, 52, 75, 86 and 97%, respectively. A salt cocktail moves the whole isotherm up. In humid climates, salt-saturated walls may therefore stay wet all year round. Keeping them dry requires a relative humidity below the lowest equilibrium value of all salts present.

Crystallized salts and the saturated solution left take up a larger volume than the over-saturated solution. In pores lacking space, pressure exerted is:

$$P_{\text{crystal}} = \mathrm{R}T/v_{\text{m}} \ln(c/c_{\text{sat}}) \tag{3.94}$$

with R the universal gas constant (831.4 J/(kg.K)), v_{m} molar salt volume and c/c_{sat} the ratio of salt concentration in the over-saturated to that in the saturated solution. Table 3.38 gives the values. NaCl ranks highest, but thanks to its high solubility, it is not most eroding. That's Na_2SO_4, a salt combining low solubility with high crystallization pressure.

Table 3.38 Crystallization pressures.

Salt	Molar volume ml	Crystallization pressure, MPa			
		$c/c_{\text{sat}} = 2$		$c/c_{\text{sat}} = 10$	
		0 °C	50 °C	0 °C	50 °C
$CaCO_4$.1/2 H_2O	46	33.5	39.8	112.0	132.5
$CaSO_4$.2 H_2O	55	28.2	33.4	93.8	111.0
$MgSO_4$.1 H_2O	57	27.2	32.4	91.0	107.9
$MgSO_4$.6 H_2O	130	11.8	14.1	39.5	49.5
$MgSO_4$.7 H_2O	147	10.5	12.5	35.0	41.5
Na_2SO_4	53	29.2	34.5	97.0	115.0
Na_2SO_4.10 H_20	220	7.2	8.3	23.4	27.7
NaCl	28	55.4	65.4	184.5	219.0
$NaCO_3$.1 H_2O	55	28.0	33.3	93.5	110.9
$NaCO_3$.7 H_2O	154	10.0	11.9	33.4	36.5
$NaCO_3$.10 H_2O	199	7.8	9.2	25.9	30.8

Table 3.39 Hydration pressures for $CaSO_4$.

RH↓	Hydration pressure, MPa		
Temp→	0 °C	20 °C	50 °C
100	219	176	93
70	160	115	25
50	107	58	0

Also hydration induces internal pressures. Depending on temperature and relative humidity, volume expansion by the changing equilibrium between salt crystals and water molecules in the pore air results in:

$$P_{\text{hydrat}} = RT/v_{\text{cw}} \ln(p/p_{\text{h}}) \tag{3.95}$$

v_{cw} being the volume of 1 mol of water, p the actual vapour pressure and p_{h} the hydration vapour pressure at the present temperature (T in K). Examples of reversible hydration are:

$$CaSO_4.2H_2O \xrightleftharpoons{120\text{–}180\,^\circ C} CaSO_4.1/2H_2O(\text{Gips})$$
$$NaSO_4.10H_2O \xrightleftharpoons{32.4\,^\circ C} NaSO_4$$
$$Ca(NO)_3.4H_2O \xrightleftharpoons{30\,^\circ C} Ca(NO)3.3H_2O \xrightleftharpoons{100\,^\circ C} Ca(NO)_3$$
$$CaCl_2.6H_2O \leftrightarrow CaCl_2.2H_2O \xrightleftharpoons{260^\circ C} CaCl_2$$
$$Na_2CO_3.10H2O \xrightleftharpoons{32\,^\circ C} Na_2CO_3.7H_2O \xrightleftharpoons{35.4\,^\circ C} Na_2CO_3.1H_2O$$

$Na_2SO_410\ H_2O$ is again the most damaging. Table 3.39 gives related hydration pressures for $CaSO_4$.

Salt deposits change the porous system by narrowing larger pores and filling finer pores. Water with dissolved salts has higher density, higher viscosity and different surface tension compared to pure water. This changes the material's moisture properties. Osmotic suction, which salt solutions initiate, impacts the chemical potential, while dissolved salts can react with the material matrix, forming for example Candlot salt with related swelling pulverizing mortars:

$$\begin{aligned} &Al_2O_3.2CaO.12H_2O + 3CaSO_4.2H_2O + 15\ H_2O \\ &\rightarrow \underbrace{Al_2O_3CaO.3CaSO_4.32H_2O}_{\text{Candlot salt}} + Ca(OH)_2 \end{aligned}$$

Looking to the type of degradations, carbonates create few problems, but chlorides, nitrates and sulphates are really damaging. There ar four types of degradation (Figure 3.51), each demanding a specific remedy. Type 1 concerns the efflorescence of sodium sulphate (Na_2SO_4) and potassium carbonate (KCO_3). It degrades the appearance, though rain washes the efflorescence away. A remedy when painting is

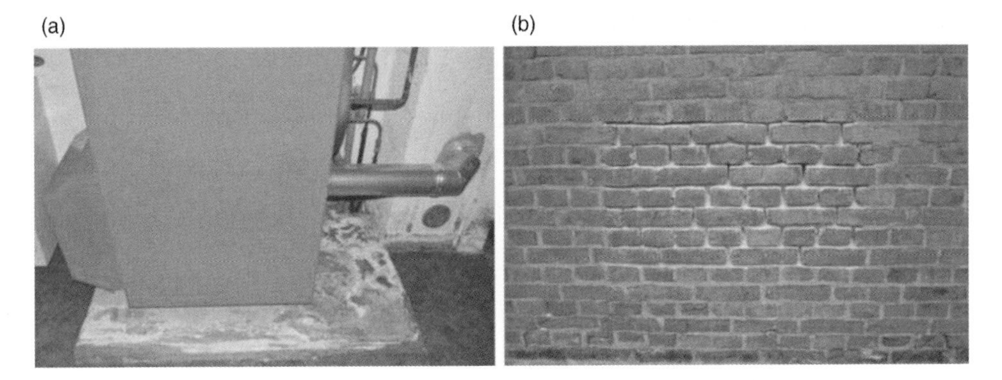

Figure 3.51 (a) $CaCO_3$ efflorescence, (b) damage by crypto-efflorescence.

planned is to wait until all traces have disappeared. Crypto-efflorescence of sodium sulphate (Na_2SO_4) and potassium carbonate (K_2CO_3) stands for type 2. It can cause severe damage, as no effective remedy exists. Called type 3 is the carbonisation of $Ca(OH)_2$ on wall surfaces, which gives a tough lime deposit that degrades the look. It can be removed by brushing, rubbing or scraping, after which the surface is washed using a 10% solution of HCl and rinsed with water. Finally, type 4 includes calcium sulphate attack, which results in swelling and pulverization of the joints. Cutting out the attacked parts and replacing by new masonry is the only remedy possible.

3.4.3.7 Chemical attack

Chemical attack changes the composition of the material. The material-specific and environmental reagents involved use moisture as a vehicle and catalyst. The term environment applies for air loaded with SO_2, chlorides, CO_2 and NO_X but also for acid rain, sewage water and sea water.

Gypsum formation in limestone is one example of chemical attack. Calcite ($CaCO_3$) reacts with sulphuric acid (H_2SO_4) to form gypsum ($CaSO_4$) on the surface and in the surface pores of limestone. The sulphuric acid comes from oxidized sulphur dioxide (SO_2) in the air, after reacting with rainwater:

$$SO_2 + O \rightarrow SO_3$$
$$SO_3 + H_2O \rightarrow H_2SO_4$$
$$H_2SO_4 + CaCO_3 \rightarrow CaSO_4 + CO_2 + H_2O$$

Intact gypsum should protect calcite from further degradation. However, related volumetric expansion and the difference in thermal and hygric movement between gypsum and calcite causes cracking. As a result, the stone draws in more acidulated water and the calcite further dissolves, with gypsum formation continuing, salts being carried with it and frost damage becoming likely. The outcome is a slow but total weathering of the limestone. Eliminating moisture uptake is the way to combat this.

Limestone is therefore often given a water-repellent treatment. This gives the surface layer mechanical, thermal and hygric properties that differ from the stone behind it, but this does not stop wider cracks from showing capillary action, which allows water to accumulate behind the treated surface. Salts then cause crypto-efflorescence, frost damage can develop, and so on. Could limestone last for ever if it weren't for the SO_2 in the air? No. Even CO_2 damages it because it reacts with water to form carbonic acid (H_2CO_3), which transforms calcite into bicarbonates that dissolve in water, thus decalcifying and weathering the stone:

$$CaCO_3 + H_2CO_3 \rightarrow Ca(HCO_3)_2.$$

Also sulphate and alkali granulate reactions figure as chemical attacks. Sulphates diffusing into concrete react with the cement to form Candlot salt which, once present in high enough quantities, causes swelling and cracking. The alkali–granulate reaction in turn indicates that the alkali hydroxides in wet concrete (NaOH, KOH, $Ca(OH)_2$) are reacting with the silicate granulates to form a $CaSiO_3.mH_2O$ gel:

$$Na_2O + H_2O \rightarrow NaOH \qquad 2NaOH + SiO_2 + nH_2O \rightarrow Na_2SiO_3.nH_2O$$

$$Na_2SiO_3.nH_2O + Ca(OH)_2 + H_2O \rightarrow CaSiO_3.mH_2O + 2NaOH$$

That gel is more voluminous than the original granulates, causing the concrete to swell and crack once stresses exceed tensile strength. Unlike carbonation, chlorine and sulphate attack, which are diffusion-induced and act from the surface inwards, the alkali–granulate reaction does it the other way around. The result is a dramatic loss in strength, although a three-dimensional reinforcement can transpose the swelling into a kind of post-stressing. Avoiding the alkali–granulate reaction demands moderate wetness, no use of reactive granulates and no mixing with cement and water that have a high free alkali hydroxide content.

3.4.3.8 Corrosion

Moisture and oxygen trigger corrosion. Both are attracted to the OH ions at the cathode, while wetness acts as ion carrier to the anode. Salts facilitate the process. Some observations have practical importance. Untreated steel starts corroding at 60% relative humidity. Therefore, in temperate climates, even indoors it has to be protected. As ferrous oxide dissolves in water, steel outdoors goes on corroding until complete destruction. Non-ferrous metal oxides instead form a protective layer unless water in contact contains ions (SO_3, Cl and others). Most metals therefore suffer from pitting corrosion under ion-loaded surface condensation. When combining metals, the voltage series defines what happens, see Table 3.40. Water draining from a higher to a lower voltage metal will corrode the lower. The inverse poses no problem.

A specific problem is steel bar corrosion in concrete. The calcium hydroxide ($Ca(OH)_2$) that fresh concrete contains creates an alkaline environment with pH 12.5–13 around the bars, which hinders corrosion. However, as soon as the concrete binds, carbon dioxide CO_2 starts diffusing into the pores, where reaction with the calcium hydroxide in the

Table 3.40 Metals: voltage series compared to a hydrogen electrode.

Metal	Chemical symbol	Voltage
Magnesium	Mg	−1.87
Aluminium	Al	−1.45
Zinc	Zn	−0.76
Iron	Fe	−0.43
Nickel	Ni	−0.25
Tin	Sn	−0.15
Lead	Pb	−0.13
Copper	Cu	0.35
Silver	Ag	0.80
Gold	Au	1.50

concrete, if wet, gives calcite ($CaCO_3$):

$$Ca(OH)_2 + CO_2 \rightarrow CaCO_3$$

Too much moisture turns calcite into the soluble bicarbonate. As the concrete dries, the dissolved bicarbonate slowly moves to the surface where it crystallizes and is washed away by rain. The concrete therefore decalcifies and becomes more porous, while the decreased pH around the bars slowly destabilizes their corroded surface. The increased porosity allows oxygen to diffuse into the pores, and so the pore moisture present then reactivates corrosion. Ferrous oxides now occupy a larger volume than the steel, so the accompanying stresses finally destroy the bar's concrete cover (Figure 3.52).

Figure 3.52 Degraded concrete due to steel bar corrosion.

Chlorines are even more destructive. Some binding accelerators contain NaCl and $CaCl_2$. Near the coast, NaCl is deposited by wind, while concrete in the sea is in direct contact with a saline environment. Chlorines react with the cement in the concrete to form calcium chloride salts. These not only lower the pH but the more voluminous ones, like Friedel salt, also spall the material:

$$Al_2O_3.3CaO.12H_2O + 3CaCl_2.6H_2O \rightarrow \underbrace{Al_2O_3.3CaO.3CaCl_2.30H_2O}_{\text{Friedel salt}}$$

Chlorine ions reaching the steel bars induce pitting corrosion, so slowly destroy the reinforcement.

How can bar corrosion be avoided? The concrete should be well compacted. Depending on the degree of exposure, the cover must be thick enough to retard diffusive flows from reaching the steel bars. Such thicker cover also enlarges the concrete mass, thus requiring more time to decalcify. If necessary, a watertight or sorption-retarding finish should be applied.

3.5 Economics

The cost of better performance is what occupies people the most, while the benefits are often overlooked. Investment in thermal insulation, for example, may result in a better thermal comfort, perhaps an envelope requiring less maintenance, cuts down the heating system needed and will lower the end energy used. A better sound insulation in turn requires investment in tieless cavity party walls, heavy floors with floating screed, sound insulating glass in tight windows, and so on, while the benefits are more privacy and less stress. Except in office buildings where better conditions increase productivity, such comfort gains are difficult to express in terms of money.

3.5.1 Total and net present value

A proven methodology to evaluate the economic benefits of choices consist of evaluating the related total or net present value. The total is calculated as:

$$\begin{aligned} TPV = I_{\text{b+HVAC}} + \sum\nolimits_{j=x,y,z} \left[\frac{I_{\text{b+HVAC}}(1+r_\text{o})^j}{(1+a)^j}\right] + \sum_{i=1}^{n}\left[\frac{K_\text{E}(1+r_\text{E})^i}{(1+a)^i}\right] \\ + \sum_{i=1}^{n}\left[\frac{K_\text{M}(1+r_\text{M})^i}{(1+a)^i}\right] + \sum_{i=1}^{n}\left[\frac{K_\text{P}(1+r_\text{P})^i}{(1+a)^i}\right] + V_\text{end} \end{aligned} \tag{3.96}$$

with $I_{\text{b+HVAC}}$ the initial investment, r_o the mean annual increment in reinvestment costs, K_E the initial annual energy cost, r_E the related mean annual increment, K_M the initial annual maintenance cost, r_M the related mean annual increment, K_P the annual productivity in terms of man-year cost, r_P the related mean annual increment, V_end the salvage value of the building at the end of the present value period, transposed to today, and a the discount factor. The inflation-corrected increments and discount factor vary on a scale of 0 to 1.

The net present value equals:

$$NPV = -\left\{ I_{\text{b+HVAC}} + \sum\nolimits_{j=x,y,z} \left[\frac{I_{\text{b+HVAC}}(1+r_{\text{o}})^{j}}{(1+a)^{j}} \right] \right\} + \sum_{i=1}^{n} \left[\frac{\Delta K_{\text{E}}(1+r_{\text{E}})^{i}}{(1+a)^{i}} \right]$$
$$+ \sum_{i=1}^{n} \left[\frac{\Delta K_{\text{M}}(1+r_{\text{M}})^{i}}{(1+a)^{i}} \right] + \sum_{i=1}^{n} \left[\frac{\Delta K_{\text{P}}(1+r_{\text{P}})^{i}}{(1+a)^{i}} \right] + \Delta V_{\text{end}} \tag{3.97}$$

with Δ the change in annual energy costs, annual maintenance costs, annual productivity in terms man-year costs and salvage value.

Random parameters in both formulas are the annual increments, the discount factor and inflation. All are extrapolated based on past figures. The discount factor accounts, among others, for the perceived risk that investments, for example in energy savings, imply. The time span considered depends on building type, while several factors, ranging from the fame of the designer to building location, impact the salvage value. TPV minimal and NPV maximal allow optimal combinations of variables from a micro-economic point of view to be picked up. To find these, the partial derivatives per variable $x_1 \ldots x_n$, considered as mutually independent, are set to zero:

$$\frac{\partial TPV}{\partial x_1} = 0 \quad \frac{\partial TPV}{\partial x_n} = 0 \quad \frac{\partial NPV}{\partial x_1} = 0 \quad \frac{\partial NPV}{\partial x_n} = 0$$

The solution of that set, with as many variables as equations, gives 'the' optimum. In practice, as analytical solutions mostly fail, finding it requires numerical techniques such as a Monte-Carlo approach, genetic algorithms or stepwise changing each of the variables.

3.5.2 Optimum insulation thickness

The insulation costs (I_{mat}) and its mounting (I_{mont}) define the investment. The first increases more or less linearly with the insulation thickness (d):

$$I_{\text{mat}} = a_{\text{o}} + a_3 d \tag{3.98}$$

with a_{o} debiting the production unit and workers' fees. Mounting costs depend on the number of layers (n) and necessary additional costs (a_2), such as vapour, air and wind barriers needed, higher rafters in a pitched roof or longer ties in cavity walls:

$$I_{\text{mont}} = a_1 n + \max[0, a_2] \tag{3.99}$$

Summing up gives:

$$I_{\text{insul}} = a_{\text{o}} + a_1 n + \max[0, a_2] + a_3 d \tag{3.100}$$

Quantifiable benefits are a cheaper heating system and lower annual heating costs. Suppose, for example, that a central heating system with radiators is installed. Thanks to better insulation less power and thus a smaller boiler, smaller expansion vessel, smaller pump, thinner pipes and less radiator surfaces are needed. As the market only offers a discontinuous series of boilers, pipe diameters and radiators, the investment attributable to 1 m^2 of envelope is a discontinuous function of the thermal resistance. One nonetheless assumes that the investment includes a constant term (b_o) and a term reciprocal to the power needed per m^2 (b_1):

$$I_{\text{heating syst}} = b_o + b_1 \left(\frac{\theta_i - \theta_{eb}}{0.17 + R_o + d/\lambda} \right) \tag{3.101}$$

with R_o the thermal resistance of the non-insulated assembly. Insulating an assembly now impacts the transmission losses. Related energy consumed annually is:

$$E_{\text{use,heat}} = \frac{0.0864 G_{d,\theta_e}^{\theta_i}}{\overline{\eta}_{\text{tot,building}}} B_l (0.17 + R_o + d/\lambda) \qquad (\text{l, m}^3 \text{or kg per annum})$$

with $G_{d,\theta_e}^{\theta_i}$ the transmission degree days and $\overline{\eta}$ the building related mean system efficiency referring to the lower heating value B_l (in MJ/unit). The related annual costs thus become:

$$K_{\text{energy,ann}} = P_E E_{\text{use,heat}} \tag{3.102}$$

with P_E the price of 1 litre of oil, 1 m^3 of gas, 1 kg of coal or 1 kWh_E. Assuming that insulation does not impact maintenance, the total present value over the period considered with insulation thickness d as variable is then equal to:

$$\begin{aligned} TPV = {} & a_o + a_1 n + \max[0, a_2] + a_3 d + b_o + b_1 \left(\frac{\theta_i - \theta_{eb}}{0.17 + R_o + d/\lambda} \right) \\ & + \left(\frac{0.0864\, G_{d,\theta_e}^{\theta_i}}{\overline{\eta}_{\text{tot,building}} B_l (0.17 + R_o + d/\lambda)} \right) P_E \sum_{j=1}^{n} \left[\frac{(1 + r_E)^j}{(1 + a_n)^j} \right] \end{aligned} \tag{3.103}$$

The derivative versus the thickness d zero gives:

$$\underbrace{\left(\frac{d[a_1 n + \max[0, a_2]]}{d(d)} + a_3 \right)}_{A_{\text{insul}}} - \frac{1}{(0.17 + R_o + d/\lambda)^2} \left\{ \underbrace{b_1(\theta_i - \theta_{eb})}_{A_{\text{heating syst}}} + \underbrace{\left(\frac{0.0864\, G_{d,\theta_e}^{\theta_i}}{\overline{\eta}_{\text{tot,building}} B_l} \right) P_E \sum_{j=1}^{n} \left[\frac{(1 + r_E)^j}{(1 + a_n)^j} \right]}_{A_{\text{energy}}} \right\} = 0$$

resulting in following optimum value (d_{opt}) and optimal thermal transmittance (U_{opt}):

$$d_{opt} = \sqrt{\frac{A_{heating\ syst} + A_{energy}}{\lambda A_{insul}}} - \lambda(0.17 + R_o) \qquad U_{opt} = \sqrt{\frac{A_{insul}\lambda}{A_{heating\ syst} + A_{energy}}} \tag{3.104}$$

That result is easily interpreted. The term $A_{heating\ syst}$ represents the investment in the heating system and the term A_{energy} the actual annual energy costs, both for a thermal resistance 1 m^2 K/W. When the two increase, so does the optimal thickness, whereas the thermal transmittance decreases. The same holds for less investment in insulation (term A_{insul}), for example because one layer suffices ($dn/d(d) = 1$). For a higher thermal conductivity (λ) of the insulation, a lower thickness and higher thermal transmittance become optimal. Of course, in reality all quantities considered to be parameters are variable. Happily the square root tempers the impact, which allows quite general conclusions to be formulated. For new construction the optimal thermal transmittance is that, which does not require thicknesses that induce additional costs. The need for a double layer almost always exceeds the optimum. This could motivate manufacturers to produce large thicknesses.

3.5.3 Whole building optimum

3.5.3.1 Methodology

A whole building approach complicates things. Variables are now the thermal insulation, the envelope airtightness, the window area, the type of glazing, the ventilation system, the heating system, PV or not, a solar boiler or not, and so on. Investments in insulation differ between low-slope roofs, pitched roofs, facades and floors, while insulating may generate ample additional costs. Thicker low-slope roof insulation, for example, demands higher parapets and adjusted flashing. Larger wall thicknesses for an equal gross floor area require extra roof surface. Below grade walls, lintels, window edges and sills all demand adjustment. Longer cavity wall ties are needed. Double and triple glazing requires stiffer and deeper window frames, and so on.

Glass systems and insulation thicknesses form a discontinuous set. Energy consumed depends on transmission, ventilation and infiltration losses, usable solar and internal gains and whole heating efficiency. Living habits and the related rebound factor must be accounted for, while the choice of heating and ventilation systems is vast. Genetic algorithms or stepwise changing each variable can take all this into account. The search for the optimum is then a multi-step operation, which includes adapting the building fabric, calculating the power needed, designing and dimensioning the heating system, quantifying the annual net and gross heating demand, related annual end and primary use, annual energy cost, estimating all (re)investments, considering extra maintenance and related costs, guessing increments, inflation and present value factors and calculating either the total or the net present value. Once the set of cases has been evaluated, there follows the search for the lowest total or highest net present value.

3.5.3.2 Example

Given are five non-insulated dwellings, see Figure 3.53 for one of them. All have a preset air leakage at 50 Pa air pressure difference of 12 ach, a purpose-designed natural ventilation system and an 80/60 °C radiator central heating with high efficiency gas boiler, AC-pump on the return in series with a closed expansion vessel and a thermostat in the living room. The boiler also produces the domestic hot water. The construction includes cavity walls plastered inside, single-glazed wooden windows, reinforced concrete decks with plastered ceiling and paved cement screed, heavyweight low-slope roofs with concrete deck, plastered ceiling, lightweight screed and bituminous membrane and, sloped roofs with wooden rafters, underlay, battens, laths and tiled deck. Correct detailing minimizes thermal bridging. Table 3.41 characterizes the five, while Table 3.42 lists all thermal transmittances.

Variables considered are the insulation thickness in the low-sloped and sloped roofs, the cavity walls and the ground floors, the glass type used, airtightness of the dwelling and the ventilation and heating system. Polyurethane foam (PUR) with thermal conductivity 0.023 W/(m.K) is used as an insulation material. Table 3.43 gives the minimum and maximum insulation thicknesses with their stepwise increases and details the glass types considered.

The envelope leakage is first lowered to 3 ach. When balanced ventilation with 80% heat recovery replaces the purpose-designed natural ventilation system, airtightness at 50 Pa is further reduced to 1 ach. The central heating system alternative has 60/40 °C low

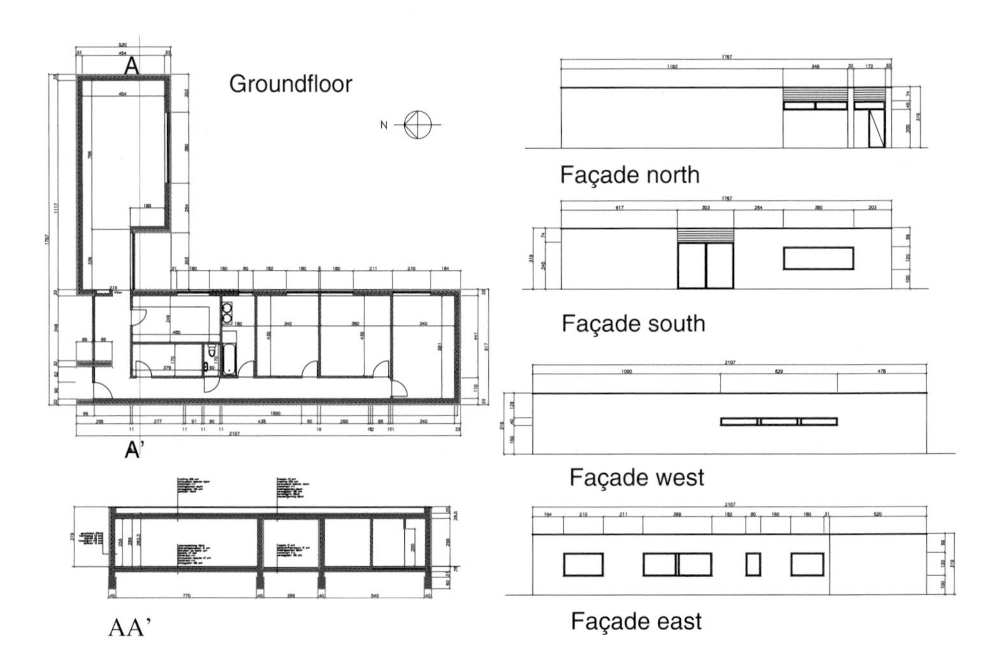

Figure 3.53 Dwelling 1.

Table 3.41 The five dwellings.

Dwelling	House type	Heated volume m³	Envelope area A_T m²	Windows, % of the envelope area	Mean U-value W/(m².K.)	$Q_{prim,ref,T}$ MJ/a
1	Detached, one storey, low-slope roof	472	555	5.2	1.31	126 200
2	Detached, two storeys, sloped roof	625	712	11.4	1.88	159 350
3	End of the row, one storey, inhabited loft, sloped roof	421	317	9.1	1.64	93 480
4	Terraced, two storeys, inhabited loft, sloped roof	435	220	12.7	1.75	83 800
5	Corner apartment, upper floor, low-slope roof	417	256	9.8	1.94	86 000

temperature radiators, a condensing boiler with a PI sensor north outdoors controlling the supply temperature, DC pump and thermostatic valves on each radiator.

Fabric and insulation share the same service life. Double glazing or better has 17% of the total replaced after 20 years and boilers serve for 20 years. Cost data came from catalogues, published price lists and energy suppliers. Investments included all additional costs, also called secondary costs. Heating power was calculated using EN 12831 with −8 °C as design temperature. Figure 3.54 illustrates how the heating system costs change with the mean thermal transmittance of the enclosure for the detached two-storey house 2.

The protected volume of each house is assumed to be constantly heated at 18 °C, a value accounting for night setback but not for all other statistically probable user habits. The reference climate outdoors is that of Uccle. Annual primary energy consumed is

Table 3.42 Non-insulated reference, thermal transmittances.

Part	U-value, W/(m².K)
Floor on grade (EN ISO 13370)	0.54–0.78
Façade	1.4
Low-slope roof	1.2
Sloped roof	1.7

Table 3.43 Minimum and maximum insulation thickness, step applied, glazing.

Insulation	Minimum, mm	Maximum, mm	Step, mm
Envelope part			
Flat roof	30	300	10
Sloped roof	30	300	10
Cavity wall	30	200	10
Floor on grade	30	200	10
Glass	**$U_{central}$, W/(m^2.K)**	**Solar transmittance**	
Double	2.91	0.72–0.75	
Double, low-e, argon filled	1.13	0.61	
Triple, low-e, argon filled	0.80	0.50	

calculated according to EN ISO 13790 with infiltration, dedicated ventilation and internal gains equal to:

$$\dot{V}_{a,infilt} = 0.04 n_{50} V \quad (m^3/h) \tag{3.105}$$

$$\dot{V}_{a,dedic} = \left[0.2 + 0.5 \exp(-V/500)\right] V \quad (m^3/h) \tag{3.106}$$

$$Q_{I,m} = (220 + 0.67V) t_{mo} / 10^6 \quad (MJ/month) \tag{3.107}$$

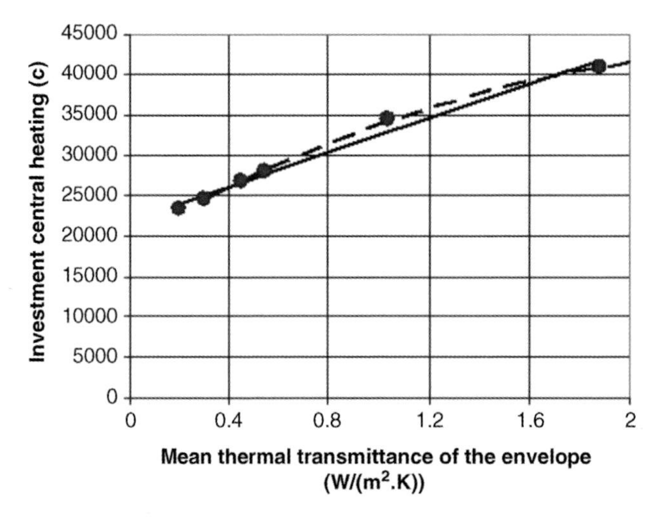

Figure 3.54 Detached two storeys house 2, investment in a hydronic central heating versus mean envelope thermal transmittance (€ of 2004).

Table 3.44 Cases per dwelling.

Case	Glazing	Heating	Ventilation	n_{50}, ach
1	Double (DG)	HE boiler (HEB)	Purpose designed natural (NV)	123
4				
7		C boiler (CB)	Purpose designed natural (NV)	12
13				3
5		C boiler (CB)	Balanced, heat recovery (BVHR)	1
2	Low-e, argon filled double	HE boiler (HEB)	Purpose designed natural (NV)	12
5				3
8		C boiler (CB)	Purpose designed natural (NV)	12
11				3
14		C boiler (CB)	Balanced, heat recovery (BVHR)	1
3	Low-e, argon filled triple	HE boiler (HEB)	Purpose designed natural (NV)	12
6				3
9		C boiler (CB)	Purpose designed natural (NV)	12
12				3
15		C boiler (CB)	Balanced, heat recovery (BVHR)	1

t_{mo} being the seconds per month. The extra investments and total present value are evaluated in relation to the ratio between the calculated primary energy for heating, domestic hot water and auxiliaries ($Q_{prim,T}$) and a reference:

$$E = 100\frac{Q_{prim,T}}{115A_T + 70V + 157.5\dot{V}_{a,dedic}} \quad (\%) \tag{3.108}$$

with E called the level of primary energy consumption.

Fifteen cases per dwelling were considered, see Table 3.44. In each case, graphs illustrate the extra investments and total present value, see Figure 3.55. All show clouds of crosses where each cross represents a result with the lonely cross up at the right fixing the starting combination. Only one cross reaches the absolute minimum, the economic optimum coinciding with the optimal level of primary energy consumption ($E_{optimum}$), see Table 3.45.

As energy for household and lighting is not included, the optima deviate substantially from net zero. In house 3 for example, total primary energy consumed at the optimum

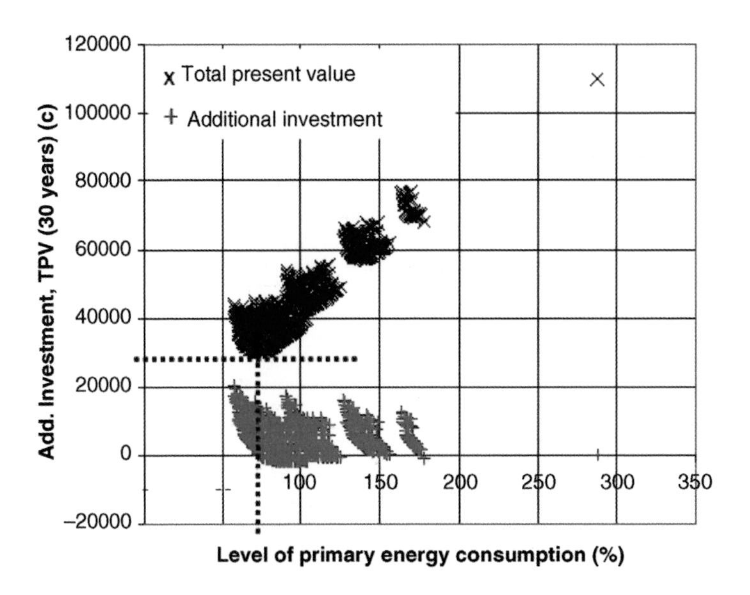

Figure 3.55 House 2, low-e, argon filled double glass, natural ventilation, $n_{50} = 12$ ach, discount factor 5%, increment in energy cost 3.2%, in investment costs 2% (€$_{2004}$).

reaches some 21 000 kWh/a, or, rebound included, 18 400 kWh/a. A run to net zero should demand 113 m^2 of PV, an investment in 2009 exceeding the additional costs that the optimum requires!

3.6 Sustainability

3.6.1 In general

In the past few decades, sustainability has become a prime concern. From the 1970s on, NASA saw the higher ozone layer thinning. Measurements on Mount Kauna in Hawaii gave an increase in atmospheric CO_2 from the 280 ppm before the industrial revolution to 390 ppm today. At the same time, waste, water pollution, air pollution and pressure on the open space coloured the situation locally.

The notion 'sustainable development' was first used in 1987 by the World Council for Economic Development in its Brundtland report 'Our Common Future', defining the concept as 'each generation must use the natural resources in a way the needs and well-being of future generations are not put at risk.' A document published in 1996 by the Civil Engineering Research Foundation (CERF) transposed it to the built environment. In 1998, the International Council for Research and Innovation in Building and Construction (CIB) produced a report on sustainable development and the future of construction, followed by 'Agenda 21 for Sustainable Construction' in 1999.

Table 3.45 Optimal choices.

	Dwelling, optimum	$E_{optimum}$ %	Insulation thicknesses (PUR), cm			
			Low-sloped roof	Pitched roof	Wall	Floor
Present value factor 6.5%. Annual increment in energy cost 0.8%, in investment costs 0%						
1	Low-e, argon DG, $n_{50} = 3$ ach, HEB, NV	**59.4**	7		7	4
2		**60.6**	6	8	8	6
3		**52.8**		8	6	4
4		**55.4**	6	8	6	4
5		**57.1**	10		7	
Present value factor 3.5%. Annual increment in energy cost 3.2%, in investment costs 2%						
1	Low-e, argon DG, $n_{50} = 3$ ach,HEB,NV	**50.8**	10		10	6
2		**57.8**	8	8	10	6
3		**48.6**		10	8	5
4	Low-e, argon 3 DG!	**52.3**	10	8	8	6
5	Same as 1, 2 and 3	**55.0**	10		10	

The links between construction, actual needs, future needs and well-being are manifold. Erecting buildings and infrastructure consumes amazing amounts of non-renewable materials and occupies space for long periods of time. Conditioning the indoors and all other services that buildings are designed for, results in tremendous amounts of primary energy being used, as does related traffic due to the distances between living, working and shopping areas. For some, 'sustainability' still concerns the environment only. Transposed to construction, the issues are then water usage, energy use, material use and waste. A derivative of this vision is life cycle inventory and analysis. However, human needs and well-being require buildings that are functional, accessible for everyone, comfortable, have good indoor air quality, are safe, remain affordable, and so on, in other words have high performance.

3.6.2 Life cycle inventory and analysis

3.6.2.1 Definition

Life cycle inventory and analysis (LCIA) looks to the building's environmental impact from cradle to cradle. Four phases are involved: preparation, construction, service life

Table 3.46 Life cycle inventory at the building level: in- and outflows.

Phase		Inflow	Outflow
1 Preparation	← Recycling	Recycled materials Basic materials Water and other fluids Energy	Building materials and components Noise Solid waste Fluid waste Gaseous waste Environmental damage BUILDING DESIGN
2 Construction		*Building design* Space Building materials and components Water Energy	Hindered traffic Noise Construction waste Liquid waste Gaseous waste BUILDING
3 Service life		*Building* Goods Cleaning products Finishing materials Building materials Energy Water	Noise Solid waste Fluid waste Gaseous waste Construction waste Harming emissions indoors and outdoors
4 Demolition		*Building* Energy Water	Noise Construction waste Solid waste Fluid waste Gaseous waste

and demolition. Preparation concerns the production of materials and the elements needed for building. Construction starts with excavating and ends at delivery. Service life has a 40% probability to take 50 years or more. Sooner or later demolition follows with, if possible, recycling.

LCI counts the inflows and outflows per phase (Table 3.46). Inflows are space, basic materials, fluids and energy, outflows usable materials, components, the building, solid, fluid and gaseous waste and noise. Not all have an environmental impact. That knowledge, creativity, human effort and money don't figure in the list follows from the encasing borders agreed on. Life cycle analysis (LCA) in turn evaluates the outcome of the inventory, looking to criteria such as raw material exhaustion, energy used, global warming potential, ozone depletion potential, eco-toxicity, waste and toxic waste produced, acidification, nitrification, human toxicity, photochemical smog and indoor environment. In most cases only some of these criteria are considered.

3.6.2.2 Some criteria

Global warming is a pressing concern. Burning fossil fuels produces carbon dioxide (CO_2), $\approx$1.75 kg per m^3 of gas, $\approx$2.8 kg per litre of oil or kg of coal. Once in the atmosphere CO_2, like water vapour, methane (CH_4), nitrous oxide (NO_2), SF_6 and all CFCs, absorbs the terrestrial and solar infrared radiation and re-emits both, half of it back to the earth. There, the infrared maintains the moderate temperatures needed for life. However, when these global warming concentrations increase, re-emitting augments, resulting in a slow increase in terrestrial mean temperature, called the global warming effect. Fossil fuel burning ranks as the main cause worldwide. To moderate further worsening, a change to a more CO_2 neutral energy supply is essential. For buildings, less energy means issues such as optimal insulation and excellent airtightness, intelligent use of glass and solar shading, passive measures to abate overheating, high efficiency HVAC systems, PV, heat pumps, CHP and heat and cold storage.

Ozone depletion high up in the stratosphere is also a worry. CFCs are the main culprits here. Until the end of the 20th century their use as refrigerants, blowing agent in insulation materials (PUR and XPS) and firefighting agent (halon) was widespread. The Montreal protocol of 1990 imposed a complete ban on the so-called hard CFCs (Freon 11, 12, 13, 111, 112, 113, 114, 115) and halons between then and 2000 (1997 in the EU), followed by a ban on the soft CFCs in 2012. This forced manufacturers to look for other refrigerants, blowing agents and firefighting agents.

Another more local problem is acidification. Coal and fossil fuels contain small percentages of sulphur. When burnt, the reaction with oxygen produces sulphur dioxide (SO_2), which in a moist atmosphere binds with the water molecules present to form sulphurous acid first and than sulphuric acid. As explained above, acid rain attacks limestone and lime-containing materials, causing accelerated ageing of landmark buildings. Strict legislation on sulphur content in fuels improved the situation drastically.

3.6.2.3 Whole energy use and minimal environmental load

The whole energy consumption includes, besides building usage, the embodied energy, which is the energy needed for manufacturing materials and components, for transporting them to the building site, for construction and demolition, and so on. The sum of usage and embodied energy opens a window for optimization, provided at least one evolves non-linearly. With thermal insulation, the result is a whole-energy optimal thickness. Elements needed are the energy used to produce, transport, mount and demolish 1 m^2 of d metre insulation, that is the embodied part related to LCIA phases 1, 2 and 4, plus, in temperate and cold climates, the energy for heating during phase 3. Embodied is:

$$E_{\text{embodied,insul}} = e_o + e_1 d \tag{3.109}$$

with e_o a thickness-independent embodied part and e_1 the embodied part per m^3 of material. The energy consumed during phase 3 amounts to:

$$E_{\text{use,heat}} = \left[\frac{0.084 G_{\text{d},\theta_e}^{\theta_i}}{\bar{\eta} B_1 (0.17 + R_o + d/\lambda)}\right] T \qquad (3.110)$$

with T the service life in years. The derivative in d of sum zero gives as optimal thickness:

$$d = \sqrt{\frac{T E_{\text{use,heat,an}}}{\lambda e_1}} - \lambda(0.17 + R_o) \qquad (3.111)$$

Higher conductivity of the insulation material, more embodied energy per m^3, shorter service life, lower annual heating use and higher thermal resistance of the part without insulation give a lower optimal value. Again, the methodology is transposable to whole buildings. Such exercises confirm that, compared to passive solar measures such as very high thermal capacity, large sun-orientated glazed surfaces, atriums and Trombe walls, excellent thermal insulation is a smarter decision in temperate climates. Studies also underline that the optima found hardly differ from the economic ones.

Looking to minimal environmental load, again the LCIA phases 1, 2 and 4 shape the embodied part while phase 3 considers the impact that building use generates. Again, as for costs and whole energy consumption, optimal choices exist, which reflect the minimum load cradle to cradle.

3.6.2.4 Recycling

Reprocessing building, insulating and finishing materials from carefully demolished premises in specialized facilities and reusing them closes the loop, so this is at the core of recycling. Reuse can be low or high value: low when the recycled material serves as landfill, for levelling or as an energy source such as synthetic waste used as fuel. High value means that the recycled material is used as an ingredient for new materials and products, as is the case, for example, with crushed old concrete becoming aggregate for new concrete and mineral fibre insulation produced from melted old fibre boards. In the past decade, taxation has accelerated the move to recycling.

3.6.3 High performance buildings

The term 'high performance' makes the link between sustainability and a series of 'fitness for purpose' checks. The UK was the first to adopt a rationale, called BREEAM (Building Research Establishment Environmental Assessment Methodology) that considers the following categories: management (12%), health and well-being (15%), energy (19%), transport (8%), water (6%), materials (12.5%), waste (7.5%), land use and ecology (10%), pollution (10%) and innovation (up to 10%). The percentages reflect the weighting of each category, which differs from the credits per issue in each category. A building gets credits, after which the sum, compared to the maximum possible, is transformed to a percentage that fixes following classes:

Class	%
Outstanding	≥85
Excellent	70–84
Very good	55–69
Good	45–54
Pass	30–44

In the USA, the US Green Building Council (USGBC) developed a rationale called LEED (Leadership in Energy and Environmental Design). The 2014 editions for new construction, existing buildings and specific building typologies consider analogous categories to BREEAM, but ordered differently. Typologies included are data centres, warehouses, distribution centres, hospitality, existing schools, existing retail, mid-rise residences and homes. The categories evaluated are sustainable sites, water efficiency, energy and atmosphere, materials and resources, indoor environmental quality, climate change, human health, water resources, biodiversity, green economy, community and natural resources. Added is innovation in green design and regional priorities. A total of 100 points can be earned, with the following classes:

Class	Points
Platinum	≥80
Gold	60–79
Silver	50–59
Certified	40–49

Germany adopted a rationale based on a 'gesamtnote', with numbers giving a ranking (1 better than 2, 2 better than 3, etc.). France and the Netherlands also have their own systems.

The author helped in developing an instrument for governmental office buildings based on three categories: quality of use, energy, environment. The first relates to the services that an office should offer, the second focuses on the energy needed and the third considers specific issues and durability, a prime element when evaluating sustainability. Each category counts groups of weighted checks: 3 for category 1, 2 for category 2 and 4 for category 3:

Category 1: quality of use	Category 2: energy	Category 3: environment
Accessibility	Building level (net demand, gross demand, end and primary use)	Location and services
Comfort (thermal, acoustics, visual)	Zone and building part level (U-value, damping)	Moisture response
Air quality and ventilation		Water usage
		Material usage

Table 3.47 lists the checks and weights that reward presumed importance of each check.

Each check has five quality levels with level 0 standing for 'what's mandatory just fulfilled' and the levels 1 to 4 given for ever better performance. Checks not in the list, such as functionality, transformability or fire safety, are already subjected to such severe mandatory requirements that gradation makes no sense anymore. Looking to the stepwise stricter mandatory requirements since 1 January 2006, the same will happen with energy. In fact, when in 2021 near zero will be the requirement, still better will become fiction.

Designs that fail at level 0 are excluded from further evaluation. When all checks reach level 4, the result is 80 points per category. Is $S_{j,i}$ the score and $g_{j,i}$ the weight of all checks in a category, the mean scores equal:

Category 1	Category 2	Category 3
$\overline{S}_1 = \frac{\sum_{i=1}^{n} g_{1,i} S_{1,i}}{\sum_{i=1}^{n} g_{1,i}}$	$\overline{S}_2 = \frac{\sum_{i=1}^{n} g_{2,i} S_{2,i}}{\sum_{i=1}^{n} g_{2,i}}$	$\overline{S}_3 = \frac{\sum_{i=1}^{n} g_{3,i} S_{3,i}}{\sum_{i=1}^{n} g_{3,i}}$

The global score is calculated as:

$$S = \left(\overline{S}_1 + \overline{S}_2 + \overline{S}_3\right)/3$$

The mean per category and the global score define the stars that a governmental office building receives:

Scores		Stars	Meaning
Average per category	Global score		
No zeros	$0 < S \leq 1$	1	Acceptable
None below 1	$1 < S \leq 2$	2	Good
None below 2	$2 < S \leq 3$	3	Very good
None below 3	$3 < S \leq 4$	4	Excellent

Defining stars in this way, guarantees that awarded buildings score well in all three categories. High performance in fact should reflect an equilibrated quality. The stars are assigned after testing whether the building complies with the performance levels that the design focused on. Stars also get the year they were awarded added. In fact, each time mandatory performances gain severity, score levels require reviewing. A building with 'old' stars could therefore lose one or more, which was perceived as unjust. Stars and the year awarded are showcased at the building's main entrance.

Table 3.47 Checks with weight per category.

Category	Performance check	Weight
1: Quality of life and well-being	Accessibility for disabled	4
	Thermal comfort, global, cold months	1
	Thermal comfort, global, warm months	1
	Thermal comfort, draft	1
	Thermal comfort, air temperature difference between head and feet	1
	Acoustical comfort: background noise	2
	Acoustical comfort: airborne and contact sound insulation	1
	Vibration control	2
	Visual comfort: illuminance	2
	Visual comfort: daylight factor	1
	Indoor air: CO_2-concentration	1
	Indoor air: perceived quality	1
	Purpose designed ventilation: flows	2/1
	Purpose designed ventilation: peak airing	0/1
2: Energy	Building level	
	Net energy demand: insulation quality	3.5
	Net energy demand: airtightness	1
	Net energy demand: solar screening	2
	Net energy demand: night ventilation?	1
	Gross energy demand: heating, control	1/3
	Gross energy demand: heat emission	1/3| 0
	Gross energy demand: pipe and duct insulation, heating	1/3| 2/3
	Gross energy demand: cooling, control	0.5
	Gross energy demand: pipe and duct insulation, cooling	0.5
	End energy consumption: heat production	0.5
	End energy consumption: cold production	0.5
	Primary energy consumption: renewable energy	2
	Level of primary energy consumption (E) of . . .	5
	Zone and building part	
	Thermal transmittances	2
	Temperature damping	0.5
3: Environment and durability	Location	3
	Facilities	3
	Service life (durability)	5
	Water usage	3
	Material usage	6

Further reading

Adan, O.C.G. (1994) *On the fungal defacement of interior finishes*, Doctoraal Proefschrift, TU/e.

Airaksinen, M., Tuomaala, P. and Holopainen, R. (2007) Modeling human thermal comfort, *Proceedings of Clima 2007 (CD-ROM).*

Ali Mohamed, F. and Hens, H. (2001) *Vermindering van de CO_2 Uitstoot Door Ruimteverwarming in de Residentiële Sector: Rentabiliteit van Bouwkundige Maatregelen*, Eindrapport project 'Kennis van de CO_2-emissies', p. 85 (in Dutch).

Alwaer, H. and Clements-Croome, D.J. (2010) Key performance indicators (KPIs) and priority setting in using the multi-attribute approach for assessing sustainable intelligent buildings. *Buildings & Environment*, **45**, 799–807.

American Hotel and Motel Association (1991) Mold and Mildew in Hotels and Motels, the Survey.

Anon (1979) Temperature without heating, *Proceedings of a CE-symposium, Brussels.*

Anon (1994) *World Energy Outlook*, IEA, Paris, pp. 305.

Anon (1997) *Enhancing the Market Deployment of Energy Technology, a Survey of Eight Technologies*, IEA, Paris, pp. 231.

Anon (2002) Directive 2002/91/EC of the European Parliament and of the council of 16 December 2002 on the energy performance of buildings.

Anon (2004) *Decreet Houdende eisen en Handhavingsmaatregelen op Het Vlak van de Energieprestaties en het Binnenklimaat Voor Gebouwen en tot Invoering van een Energieprestatiecertificaat*, Belgisch staatsblad, 30/07, edn. 3 (in Dutch).

Anon (2007) *Assessment of Office Buildings: Towards A Sustainable Housing of the Flemish Civil Services*, D/2008/3241/080, pp. 126 (in Dutch).

Anon (2005) *Calculation Method for the Characteristic Annual Primary Energy Consumption in Residential Buildings*, Add I to the Flemish EPB-legislation, Belgisch Staatsblad/Moniteur Belge, June, 15 (in Dutch and French).

ANSI/ASHRAE standard 62.2–2013, Ventilation and acceptable indoor air quality in low-rise residential buildings, pp. 50

ASHRAE (1998) Field Studies of Thermal Comfort and Adaptation, *A Collection of Papers from the ASHRAE Winter Meeting in San Francisco.*

ASHRAE (2001) *Handbook of Fundamentals.*

ASHRAE (2007) *Handbook of HVAC Applications*, Chapter 36, Atlanta.

ASHRAE (2008) *Advanced Energy Design Guide for K-12 School Buildings*, ISBN 1-931862-41-9, pp. 171.

Barron, E. (1990) Ongewisse warmte. *Natuur en Techniek*, **2**, 96–105 (in Dutch).

Benzinger, T.H. (1978) *The Physiological Basis for Thermal Comfort*, First International Symposium on Indoor Climate, Copenhagen.

Bogaerts, F. (1987) *Syntheserapport Energie Audits*, Rapport RD-Energie (in Dutch).

BRE (2010) BREEAM Assessment, Zwin Bespoke.

Caluwaerts, P. (1976 –1985) *De Warmtehuishouding in Gebouwen 's Winters en 's Zomers*, Cursus Thermische isolatie en vochtproblemen in gebouwen, TI-KVIV (in Dutch).

CEN (2002) EN ISO 13790 *Thermal Performance of Buildings*, Calculation of Energy Use for Heating.

CEN (2003) EN 12831, Heating Systems in Buildings, Method for Calculation of the Design Load.

CEN/TC 89 (1992) Working draft 'Thermal performance of buildings-Calculation of energy use for heating – Non-residential buildings'.

CEN/TC 89 (1993) Working draft 'Thermal performance of buildings-Calculation of energy use for heating – Residential buildings'.

CERF (1996) Construction Industry, Research Prospectuses for the 21st Century, Report #96-5016.T, Civil Engineering Research Foundation, Washington DC, pp. 130

CIB (1998) *Sustainable Development and the Future of Construction*, Publication 225, Rotterdam.

CIB (1999) *Agenda 21 on Sustainable Construction*, Publication 21, Rotterdam, pp. 120.

Commission Energy 2030 (2006) *Belgium's Energy Challenges Towards 2030*, report commissioned by the federal minister of energy.

Cziesielski, E. (1991) Energiegerechte Sanierung von Korrosionsschäden bei Stahlbetongebäuden. *Bauphysik*, **5**, 138–143 (in German).

D'haeseleer, W. (redactie) (2005) *Energie Vandaag en Morgen, Beschouwingen Over Energievoorziening en –Gebruik*, ACCO, Leuven, pp. 292 (in Dutch).

Davis, M. and Vernon, M. (2013) 4 ways LEED w4 will change business, http://www.greenbiz.com/blog/2013/10/18/4.

De Bie, C. and Deschrijver, F. (1988) Aantasting van Natuursteen: gevallenstudie Jeruzalem, eindwerk KU Leuven (in Dutch).

De Grave, A. (1957) *Bouwfysica 1*, Uitgeverij SIC, Brussel, (in Dutch).

Desmyter, J., Potoms, G., Demars, P. and Jacobs, J. (2001) *De alkali-silicaat reactie. basisbegrippen en belang voor belgië*, WTCB Tijdschrift, 2^{e} trimester, pp. 3–16 (in Dutch).

DOE (1997) Office of Energy Efficiency and Renewable Energy, Scenarios of US carbon reductions.

Du Plessis, C. (2002) Agenda 21 for Sustainable Construction in Developing Countries, published for CIB and UNEP by CSIR Building and Construction Technology, Pretoria, pp. 82

Environment International (1991) *Special Issue Healthy Buildings*, Pergamon Press, New York.

Erhorn, H. (1998) Fördert oder schadet die europäische Normung der Niedrigenergiebauweise in Deutschland. *Gesundheitsingenieur*, **119**, Heft 5, pp., 236–239 (in German).

Fachvereinigung Mineralfaserindustrie e.V., Umgang mit Mineralwolle-Dämmstoffen (in German).

Fagerlund, G. (1974) *Critical Moisture Contents at Freezing of Porous Materials*, Symposium CIB-RILEM on Moisture Problems in Buildings, Rotterdam.

Fanger, P.O. (1972) *Thermal Comfort*, Mc Graw-Hill Book Company, New York.

Fanger, O. (1988) The olf and the decipol. *ASHRAE Journal*, 35 382.27.

Fanger, O. (1988) Hidden olfs in sick buildings. *ASHRAE Journal*, 40–43.

Fanger, O. (1988) Introduction of the olf and decipol units to quantify air pollution perceived by humans indoors and outdoors. *Energy and Buildings*, **12**, 1–6.

Fanger, P.O. (2002) Prediction of thermal sensation in non-air-conditioned buildings in warm climates, *Proceedings of the Indoor Air 2002 Conference, Montery, California*, vol. 3, pp. 23–28.

Fanger, P.O., Lauridsen, J., Bluyssen, P. and Clausen, G. (1988) Air pollution in offices and assembly halls. *Quantified by the Olf Unit, Energy and Buildings*, **12**, 7–19.

Federaal Planbureau (2006) *Het Klimaatbeleid na 2012, Analyse van de Scenario's Voor Emissiereductie Tegen 2020 en 2050*, Studie in opdracht van de federale minister van leefmilieu, pp. 246 (in Dutch).

Feist, W. (1997) *Lebenszyklus-Bilanzen im Vergleich: Niedrigenergiehaus, Passivhaus, Energieautarkes Haus*, Wksb, Heft 39, 42 Jahrgang, Juli, pp. 53–57 (in German).

Fisk, W., Satish, U., Mendell, M. *et al.* (2013) Is CO_2 indoor pollutant. *ASHRAE Journal*, March 2013, 84–85.

Fisk, W., Satish, U., Mendell, M. *et al.* (2013) Is CO_2 an indoor pollutant. *REHVA Journal*, October 2013, 63.

Flannery, T. (2005) *The Weather Makers: The History and Future Impact of Climate Change*, Text Publishing, Melbourne.

Flannery, T. (2006) *De Weermakers*, Uitgeverij Atlas, Amsterdam, pp. 324 (in Dutch).

Gagge, A.P. (1980) *The New Effective Temperature ET∗*, An index of human adaption to warm environments, Environmental Physiology, Elsevier, Amsterdam.

Gertis, K. (1991) Verstärkter baulicher Wärmeschutz–ein Weg zur Vermeidung der bevorstehenden Klimaveränderungen? *Bauphysik*, **5**, 133–137 (In German).

Gertis, K. (1991) Verstärkter baulicher Wärmeschutz–ein Weg zur Vermeidung der bevorstehenden Klimaveränderung. *Bauphysik*, **13** (Heft 5), 132–137 (in German).

Gertis, K., Hauser, G., Sedlbauer, K. and Sobek, W. (2008) Was bedeuted "Platin"? Zur Entwicklung von Nachhaltigkeitsbewertungsverfahren. *Bauphysik*, **30** (4), 244–256 (in German).

Geurts, H. and Van Dorland, R. (2005) *Klimaatverandering*, Teleac KNMI, Kosmos-Z&K uitgevers, Utrecht/Antwerpen, pp. 128 (in Dutch).

Gesellschaft für rationelle Energieverwendung (GRE) (1996) *Heizenergieeinsparung im Gebäudebestand*, BAUCOM, Böhl-Iggelheim, pp. 99 (in German).

Grumman, D. (ed.) (2003) *ASHRAE GreenGuide*, Atlanta, pp. 163.

Guerra-Santin, O. and Itard, L. (2010) Occupants' behaviour: determinants and effects on residential heating consumption. *Building Research & Information*, **38** (3), 318–338.

Hendriks, L. and Hens, H. (2000) *Building Envelopes in a Holistic Perspective*, Final report IEA-ECBCS Annex 32, ACCO, Leuven, pp. 102 + ad.

Hens, H. (1982) *Bouwfysica 2: Warmte en Vocht, Praktische Problemen en Toepassingen*, vol. **1** ACCO, Leuven (in Dutch).

Hens, H. (1985) *Studie van een Woning Met Laag Energiegebruik*, Rapport RD-Energie (in Dutch).

Hens, H. (1991) Bouwen en gezondheid: mythe of werkelijkheid. *Onze Alma Mater*, 1991/4, 285–304 (in Dutch).

Hens, H. (1991) *Analysis of Causes of Dampness, Influence of Salt Attack*, L'umidita ascendente nelle murale: fenomenologia e sperimentazione, Bari.

Hens, H. (1993) *Toegepaste*, Bouwfysica 1, Randvoorwaarden en Prestatie-eisen, ACCO, Leuven, (in Dutch).

Hens, H. (1993) Thermal Retrofit of a Middle Class House: a Monitored Case, *Proceedings of the International Symposium on Energy Efficient Buildings, Stuttgart, 9–11 March.*

Hens, H. (1996) *Toegepaste Bouwfysica 3, Gebouwen en Installaties*, ACCO, Leuven, pp. 329 (in Dutch).

Hens, H. (1999) Fungal defacement in buildings: a performance related approach. *International Journal of HVAC&R Research*, **5** (3), 265–280.

Hens, H. (2003) *Bouwen, Gebouwgebruik en Milieu, Geen Problemen of Toch?* Wetenschap op nieuwe wegen, Lessen van de 21^{e} eeuw, Raymaekers en Van Riel, ed., Universitaire Pers, Leuven, pp. 219–243 (in Dutch).

Hens, H. (2004) *Cost Efficiency of PUR/PIR Insulation*, Report 2004/14, written on demand of the European PU-industry, Laboratory of Building Physics, pp. 76.

Hens, H. (2006) *Duurzaam Bouwen*, Francqui leerstoel VUB, pp. 80 (in Dutch).

Hens, H. and Ali Mohamed, F. (1993) Thermal Quality and Energy Use in the Existing Housing Stock, *Proceedings of the International Symposium on Energy Efficient Buildings, Stuttgart, 9–11 March.*

Hens, H. and Rose, W.B. (2008) The Erlanger House at the University of Illinois-A Performance-based Evaluation, *Proceedings of the Building Physics Symposium, Leuven, 29–31 October*, pp. 227–235.

Hens, H. and Standaert, P. (1980) *Energiebesparing in de Huishoudelijke Sector, Gezien Door Een Bouwfysische Bril*, Studiedag 'Feiten en mythen in zake energiebesparing in huishoudelijke en gelijkgestelde sectoren', Studiedag TI-KVIV (in Dutch).

Hens, H. and Verdonck, B. (1997) *Wonen, Verwarmen: Energie en Emissies*, CO_2-project Electrabel-SPE, pp. 52 (in Dutch).

Hens, H., Verbeeck, G., Van der Veken, J. *et al.* (2007) *Ontwikkeling via Levenscyclusanalyse van Extreem Lage Energie- en Pollutiewoningen (EL^2EP)*, Eindrapport GBOU-project (in Dutch).

Heron (2002) Special Issue on ASR, Vol 47, no 2.

Huber, J.W., Baillie, A.P. and Griffiths, I.D. (1987) *Thermal Comfort as a Predictive Tool in Home Environments*, CIB-W77 meeting, Holzkirchen.

Hui, Z. (2003) *Human thermal sensation and comfort under transient and non-uniform thermal environments*, PhD thesis, University of California, Berkeley.

IEA (1995) *Solar Heating and Cooling Programme*, Solar low energy houses of IEA Task 13, James & James Publishers, London.

IEA-Annex 14 (1991) *Condensation and Energy*, Final Report, volume 1: Source Book, ACCO, Leuven.

IEA-Annex 14 (1991) *Condensation and Energy*, Final Report, volume 3: Case Studies, ACCO, Leuven.

IEA-EXCO ECBCS Annex 8 (1987) *Inhabitants' Behaviour with Regard to Ventilation*, Final report.

Itard, L. (2012) Werkelijk energiegebruik en gebruikersgedrag, lezing Nationale experten workshop, VITO, 1-6-2012.

Iyer-Raniga, U. and Pow Chew Wong, J. (2012) Evaluation of whole life cycle assessment for heritage buildings in Australia. *Buildings & Environment*, **47**, 138–149.

Johannesson, G. (2006) Building energy – a design tool meeting the requirements for energy performance standards at early design – validation, in *Research in Building Physics and Building Engineering* (eds. P. Fazio, H. Ge, J. Rao and G. Desmarais), Taylor & Francis, London, pp. 627–634.

Johansson, T., Kelly, H., Reddy, A. and Williams, R. (1993) *Renewable Energy*, Sources for Fuels and Electricity, Island Press, pp. 1160.

KU Leuven, Laboratorium Bouwfysica (1982–1988) Rapporten over de Haram-al-Shariff te Jeruzalem, Het Oude Gerechtshof te Brugge, Arenberg kasteel te Heverlee e.a. (in Dutch and English).

KU Leuven, Laboratorium Bouwfysica (1994–2003) Rapporten over het thermisch comfort in grote verkoopsruimten, trading zalen en woningen (in Dutch).

Laboratorium Bouwfysica (1997) Databases STUD9497.xls, ANE9497.xls, ANAF9497.xls.

Labs, K., Carmody, J., Sterling, R. *et al.* (1988) *Building Foundation Design Handbook*, Chapter 11.1 on Termite Control, ORNL.

Lee, Y.S. and Guerin, D.A. (2010) Indoor environmental quality differences between office types in LEED-certified buildings in the US. *Buildings & Environment*, **45**, 1104–1112.

Lévêque, F. (1996) *Modélisation de la Prévision de la Consommation D'électricité du Secteur Résidentiel Belge*, Memoire, UCL, pp. 82 + annexes.

Liddament, M. (1996) *A Guide to Energy efficient Ventilation*, AIVC, pp. 254.

Long, M. (2013) *LEED v4, the Newest Version of LEED Green Building Program launched at USGBC's Annual Greenbuild Conference, posted in Media, November 20.*

Mayer, E. (1990) Untersuchung der physikalischen Ursachen von Zugluft. *Gesundheitsingenieur*, **1**, 17–30 (in German).

Mayer, E. (1990) Thermische Behaglichkeit, Einfluss der Luftbewegung auf das Arbeiten in Reinraum. *Reinraumtechnik*, **4**, 30–34 (in German).

Meadows, D., Randers, J. and Behrens, W. (1973) *Rapport van de Club van Rome*, Uitgeverij het Spectrum (in Dutch).

Ministerie van Economische Zaken Administratie voor Energie, Energiebalansen 1984–1985–1986 (in Dutch).

Mohamed, F.A. and Hens, H. (2001) *Vermindering van de CO_2 Uitstoot Door Ruimteverwarming in de Residentiële Sector: Rendabiliteit van Bouwkundige Maatregelen*, Rapport Kennis CO_2-emissies, Electrabel/SPE (in Dutch).

Mumovic, D. and Santamouris, M. (2009) *A Handbook of Sustainable Building Design & Engineering*, Earthscan Publishing Company, pp. 423.

Muzzin, G. (1982) *Uitbloeïingen op Baksteen*, WTCB-Tijdschrift, **4**, pp. 2–17 (in Dutch).

Nationaal Programma RD-Energie (1984) *Ontwerp en Thermische Uitrusting van Gebouwen, Deel 1*, DPWB (in Dutch).

Nationaal Programma RD-Energie (1984) *Ontwerp en Thermische Uitrusting van Gebouwen, Deel 2*, DPWB (in Dutch).

NBN B 62–100 (1974) *Hygrothermische Eigenschappen der Gebouwen: Thermische Isolatie, Winteromstandigheden*, BIN, (in Dutch).

NBN B62-003 (1986) Berekening van de warmteverliezen van gebouwen (in Dutch and in French).

NBN B62-002 (1987) Berekening van de warmtedoorgangscoëfficiënten van wanden (in Dutch and in French).

NBN EN 12831 (2003) *Verwarmingssystemen in Gebouwen*, Methode voor de berekening van de ontwerpwarmtebelasting, pp. 76 (in English).

NIBS (2008) High Performance Buildings: assessment to the US Congress and US Department of Energy in Response to Section 914 of the Energy Policy Act of 2005 (Public Law 109-058), pp. 27.

Norgard, J. (1989) Low Electricity Appliances-Options for the Future, in *'Electricity, Efficient End-Use and New Generation Technologies and Their Planning Applications'*, Lund University Press.

Norm ISO 7730 (1984) *Moderate Thermal Environments-Determination of the PMV and PPD Indexes and Specification of the Conditions of Thermal Comfort.*

Olesen, B.W. (1975) *Thermal Comfort Requirements for Floors*, Technical University of Denmark.

Opfergeld, D. (1985) *Guide de la Gestion Énergétique des Établissements Scolaires*, RD-Energie, Rapport E3/III.1.8 (In French).

PATO (1986) Sectie Bouwkunde, Syllabus van de leergang: Leidt energiebesparing tot vochtproblemen (in Dutch).

Poffijn, A. (1989) *Binnenluchtverontreiniging Door Radon*, TI-KVIV studiedag Gezond Bouwen, fabel of realitei (in Dutch).

prEN 12831 (2000) *Heating Systems in Buildings, Method for Calculation of the Design Heat Load*, Final Draft, August, pp. 77.

Raatschen, W. (1990) *Demand Controlled Ventilating System, State of the Art Review*, Report IEA, Annex 18.

Radon, J. and Werner, H. (1992) Quantifizierung des Solar-Ausnutzungsgrades zur Berechnung des Heizenergiebedarfs von Gebäuden. *Bauphysik*, **14**, Heft 1, 7–11 (in German).

RD-Energie (1980–1985) *Rationeel Energiegebruik in de Gebouwen*, Integratieoefeningen (in Dutch).

REHVA (2005) *Clima 2005 8th World Congress, Proceedings, Lausanne (CD-ROM).*

Rousseau, E. (1992) *Bouwproducten: Asbest en Andere Probleemstoffen*, WTCB-Beroepsvervolmaking: Luchtkwaliteit in Gebouwen (in Dutch).

Rudbeck, C. (1999) *Methods for Designing Building Envelope Components Prepared For Repair and Maintenance*, Report R-035, Technical University of Denmark.

Sachverständigengremium 'Gesundes Bauen und Wohnen' (1986) *Gesundes Bauen und Wohnen: Antworten auf Aktuelle Fragen*, Bonn, (in German).

Sedlbauer, K. (2001) *Vorhersage von Schimmelpilzbildung auf und in Bauteilen*, Doktor-Ingenieur Abhandlung, Universität Stuttgart (in German).

Seppänen, O. and Fisk, W. (2005) Indoor climate and productivity, *Proceedings of the International Conference on the Energy performance of Buildings, AIVC Brussels, 21–23 September.*

Shum, M. (2002) An Overview of the health effects due to mold exposure, *Proceedings of the Indoor Air 2002 Conference, Montery, California*, Vol 3, pp. 17–22.

Spengler, J., Burge, H. and Su, J. (1991) *Biological Agents and the Home Environment*, Bugs, Mold & Rot 1, Proceedings pp. 11–18.

Stevens, W.J. (1992) *Het Sick Building Syndrome, een Air Conditioning Ziekte*, WTCB-Beroepsvervolmaking: Luchtkwaliteit in Gebouwen (in Dutch).

Stichting Bouwresearch (1974) *Voorstudie Energiebesparing*, Rapport BII-46 (in Dutch).

Sunikka-Blank, M. and Galvin, R. (2012) Introducing the prebound effect: the gap between performance and actual energy consumption. *Building Research & Information*, **40** (3), 260–273.

TI-KVIV (1985) Kursus Thermische Isolatie en Vochtproblemen in Gebouwen: Teksten (in Dutch).

TU-Delft (1981) *Afdeling der Bouwkunde*, Research in Energy Savings, Eindrapport.

Tuomaala, P., Airaksinen, M. and Holopainen, R. (2004) A concept for utilizing detailed human thermal model for evaluation of thermal comfort, *Proceedings of Clima 2007 (CD-ROM)*.

USGBC (2009) LEED 2009 for New Construction and Major Renovations, 88 pp.

Uyttenbroeck, J. (1989) *Thermisch Comfort*, TI-KVIV studiedag Gezond Bouwen, fabel of realiteit (in Dutch).

Vaes, F. (1984) *Hygrothermisch Gedrag van Lichte Geventileerde Daken*, Eindrapport onderzoek TCHN-KUL-IWONL (in Dutch).

Van Bronswijk, J.E.M.H. (1992) Biologische agentia en bouwfysica. *Bouwfysica*, **1**, 7–12 (in Dutch).

Van Cauteren, Prof. Ir. R. Leerstoel (1987) Restauratie in de civiele techniek: tekstboek (in Dutch).

Van den Ham, E.R. and Ackers, J.G. (1992) Radon in het binnenmilieu. *Bouwfysica*, **1**, 21–25 (in Dutch).

Van der Veken, J., Hens, H., Peeters, L. *et al.* (2006) Economy, energy and ecology based comparison of heating systems in dwellings, in *Research in Building Physics and Building Engineering* (eds. P. Fazio, H. Ge, J. Rao, G. Desmarais), Taylor & Francis, London, pp. 661–668.

Van der Wal, J.F. (1992) Belasting van de binnenlucht door emissie uit (bouw)materialen en producten. *Bouwfysica*, **1**, 12–21 (in Dutch).

Vandermarcke, B. (1982) *Thermisch Comfort: Basis Voor Het Ontwerpen van een Zwembad, Eindwerk*, KU Leuven (in Dutch).

Verbeeck, G. and Hens, H. (2002) *Energiezuinige Renovaties: Economisch Optimum, Rendabiliteit*, Rapport Kennis CO_2-emissies, Electrabel/SPE, (in Dutch).

Verbeeck, G. (2007) *Optimisation of extremely low energy residential buildings*, PhD thesis, K.U.Leuven.

Verbeeck, G. and Hens, H. (2006) Development of extremely low energy dwellings through life cycle optimization, in *Research in Building Physics and Building Engineering* (eds. P. Fazio, H. Ge, J. Rao and G. Desmarais), Taylor & Francis, London, pp. 579–586.

Verbeeck, G. and Hens, H. (2007) Life cycle inventory of extremely low energy dwellings. *Proceedings Plea 2007, Singapore, 22–24 November.*

Verbruggen, A. (1988) Investeringsanalyse, Rationeel Energiegebruik in Kantoorgebouwen, Wilrijk (in Dutch).

Verbruggen, A. (ed.) (1994) *Leren om te Keren*, Milieu en Natuurrapport Vlaanderen, Galant, Leuven, pp. 823 (in Dutch).

Verbruggen, A. (ed.) (1996) *Leren om te Keren, Milieu en Natuurrapport Vlaanderen*, Galant, Leuven, pp. 585.

Vlaamse regering (2005) Besluit tot vaststelling van de eisen op het vlak van de energieprestaties en het binnenklimaat in gebouwen met bijlagen (in Dutch).

Vliet-SENSIVV (1995 1999), Onderzoek naar de bouwfysische kwaliteit van woningen, gebouwd na 1993, WTCB (in Dutch).

Vos, B.H. and Tammes, E. (1973) *Thermal Insulation, Seen from an Economic Viewpoint*, Bouwcentrum, Rotterdam.

WCED (1987) The Brundtland Report.

Weiss, J.S. and O'Neill, M.K. (2002) Health effects from stachybotris exposure in indoor air, a critical review, *Proceedings of the Indoor Air 2002 Conference, Montery, California*, vol. 3, pp. 23–28.

Werner, H. (1986) *Berechnung des Heizenergiebedarfs von Gebäuden nach der Methode ISO-DP 9164*, HLH 37, Heft 11, pp. 541–545 (in German).

Winnepenninckx, E. (1992) *Het Vlaamse Isolatie- en Ventilatiedekreet, Hoofdstuk Over Isoleren en Milieu*, Eindwerk IHAM (in Dutch).

Wouters, P. (1989) *De Mogelijkheden en Beperkingen van Ventileren Met Betrekking tot Gezond Wonen'*, TI-KVIV, studiedag Gezond Bouwen, fabel of realiteit (in Dutch).

Wouters, P. (2000) *Quality in Relation to Indoor Climate and Energy Efficiency. An analysis of trends, achievements and remaining challenges*, Doctoraal proefschrift, UCL.

Wouters, P. and Hens, H. (1981) *Isolatie en Haar Impact Op Het Bouwproces*, Ronde Tafel 'Isolatie' (in Dutch).

Wouters, P. and Hens, H. (1982) *De Verbetering van Bestaande Gebouwen*, Studiedag ATIC over energiebesparing in flatgebouwen (in Dutch).

WTCB (1972) *Onderzoek Naar de Vorstvastheid van Bouwmaterialen en Bepaling van de Eisen Voor Keuring der Materialen*, Studie- en Researchrapport 15 (in Dutch).

4 Envelope and fabric: heat, air and moisture metrics

4.1 Introduction

Chapter 3 discussed a performance cluster at the building level. Here we will step one level down, analysing the heat, air and moisture metrics at the envelope and fabric level. For opaque parts the checks include airtightness, thermal transmittance, transient thermal response, moisture tolerance, thermal bridging and hygrothermal loading. Airtightness comes first as it impacts the next three. If it is lacking, then thermal transmittance and insulation quality uncouple, transient response degrades and moisture tolerance becomes more of a problem. For transparent parts mastering solar gains replaces the transient response, while light transmittance and view to the outside add to it. With floors, the contact coefficient of the finish requires consideration. For structural, noise and other important metric clusters at the envelope level, reference is made to both volumes on 'Performance-based Building Design'.

4.2 Airtightness

4.2.1 Air flow patterns

Lack of airtightness not only impairs the thermal transmittance, transient thermal response and moisture tolerance, but it also decreases sound insulation and degrades qualities at the building level, such as draught, end energy use for heating and cooling and fire safety. When evaluating airtightness as an overall performance, seven patterns of interaction are involved:

Pattern	Cause	Consequences
Outflow In → Out	Assembly not airtight Temperature differences between indoors and outdoors Situated at the leeward side Overpressure indoors	Thermal transmittance no longer reflecting insulation quality Higher interstitial condensation risk and larger deposits, faster drying to outdoors Uncontrolled infiltration
Inflow In ← Out	Assembly not airtight Temperature differences between indoors and outdoors Situated at the windward side Under pressure indoors	Thermal transmittance no longer reflecting insulation quality Worse transient thermal response Increased mould and surface condensation risk, faster drying mainly to indoors Drop in sound insulation against airborne noise from outdoors Uncontrolled infiltration

Applied Building Physics: Ambient Conditions, Building Performance and Material Properties, Second Edition. Hugo Hens.

Cavity venting 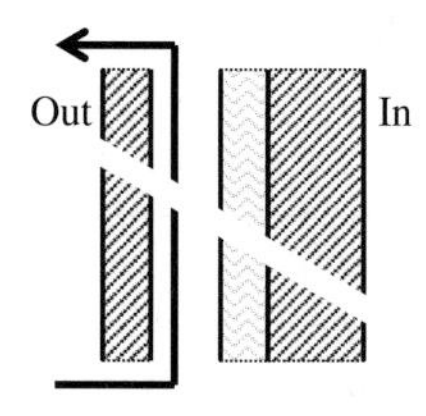 	Cavity at the outer side of the thermal insulation, air permeable or airtight outside cladding with weep holes up and down Wind pressure differences along the outside surface Temperature difference between cavity and outdoors	Hardly any increase in thermal transmittance Considered beneficial for moisture tolerance though condensation by under-cooling against any cladding more likely Worse sound insulation against airborne noise from outdoors
Wind washing 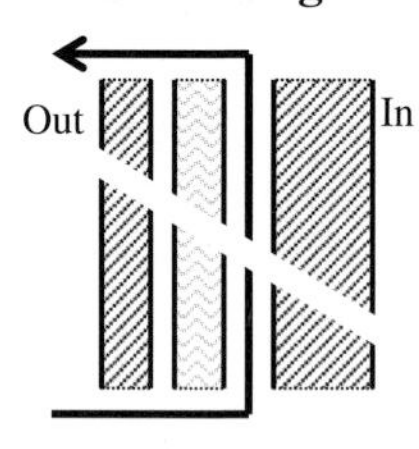 	Air layer at the inside of the thermal insulation at different heights disclosed to outdoors or, cavity filled with air-permeable insulation Wind pressure differences along the outside surface, temperature difference between the cavity and outdoors	Large increase in thermal transmittance Worse transient thermal response Increased risk of mould and surface condensation on the inside surface of the inner leaf Worse sound insulation against airborne noise from outdoors
Indoor air venting 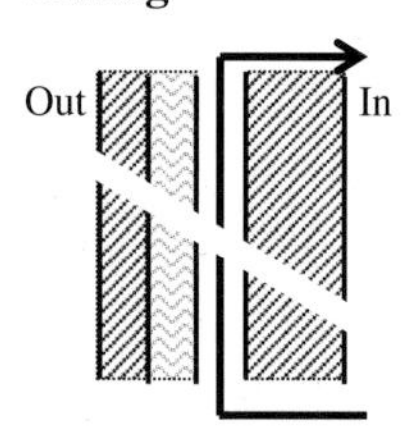 	Air layer at the inside of the thermal insulation at different heights disclosed to the indoors Temperature differences along the inside surface Air pressure differences along the inside surface	Small increase in thermal transmittance Worse sound insulation against airborne noise from outdoors
Indoor air washing 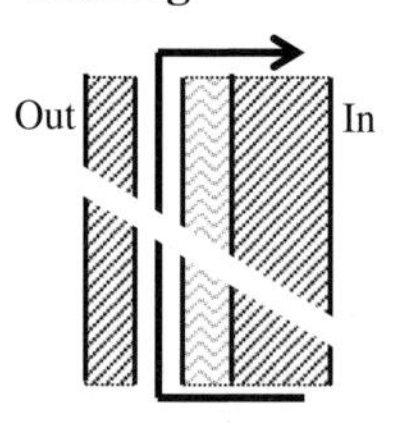 	Cavity at the outside of the thermal insulation at different heights disclosed to the indoors or air permeable insulation Temperature differences along the inside surface Air pressure differences along the inside surface	Large increase in thermal transmittance High interstitial condensation risk, larger deposits Worse sound insulation against airborne noise from outdoors
Air looping 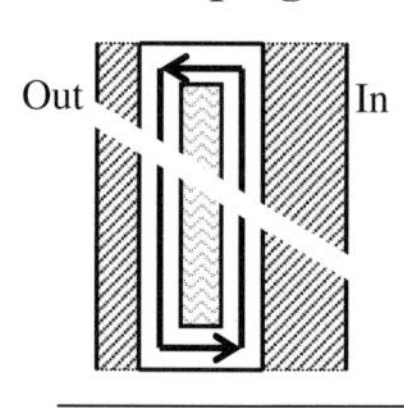 	Cavity at both sides of the thermal insulation, leaks at different heights in the insulation layer or air permeable insulation Temperature difference over the insulation layer	Large increase in thermal transmittance Somewhat higher interstitial condensation risk

Limiting air inflow and outflow demands inclusion of an air barrier somewhere in the envelope. If at the inner side, inside air washing is prevented. If at the outer side, it also acts as a wind barrier controlling wind washing, albeit indoor air venting and washing might not be stopped. Air barriers must be continuous, properly fastened and supported at both sides to resist wind loads.

Eliminating inside air and wind washing is best done by mounting an air barrier inside and a wind barrier in front of the thermal insulation. An impermeable insulation layer of course may ensure both. No air space anywhere in front of such insulation precludes outside air venting, while no air space at its back side prevents wind washing, though open joints between boards may still allow indoor air venting. In leaky envelopes, the seven patterns combine.

4.2.2 Performance requirements

4.2.2.1 Air infiltration and exfiltration

In principle, the answer to 'how airtight should an envelope be?' is 'perfectly', but in practice, complete airtightness remains a fiction. Even when the design aims at perfectness, difficulties in execution will induce misfits that turn 'perfectly tight' into a fairy tale. As a consequence, even a little inside air ingress will short-circuit the diffusion resistance between indoors and the interfaces where interstitial condensation may happen. Whether linked moisture deposit, if any, will be acceptable, depends on the assembly's layout and the ambient conditions with the air and vapour pressure differentials with outdoors as the main factors.

A realistic approach consists of linking the airtightness needed to the indoor climate class and the air pressure differentials expected. Avoiding concentrated leaks such as cracks, perforations, open joints and keeping the area-averaged air permeance coefficient below 10^{-5} kg/(m^2.s.Pa^b) ensures tolerance in the indoor climate classes 1, 2 and 3:

$$\bar{a} = 10^{-5} = \frac{G_a}{A\Delta P_a^b}$$

4.2.2.2 Inside air washing, wind washing and air looping

These three must stay moderate enough to ensure that the area-averaged heat flow entering any assembly divided by the difference between the inside and outside reference temperature, giving a so-called equivalent thermal transmittance, hardly increases compared to transmission only (U). This is also written as the transmission thermal transmittance multiplied with the Nusselt number (Nu):

$$U_{eq} = U^* Nu = U(1 + x/100) \quad (\mathrm{W/(m^2.K)}) \tag{4.1}$$

For an increase (x) not greater than 10%, the Nusselt number cannot exceed a value of 1.1. Take a cavity wall with 9 cm thick brick veneer, 13 cm wide cavity filled with 10 cm thick PUR boards and a plastered, 14 cm thick lightweight inside leaf, thermal resistance

0.88 m^2.K/W. The wall is 2.7 m high. Two weep holes per metre run perforate the veneer at bottom and top. Owing to lazy workmanship the fill sits somewhere in the cavity with its underside above the bottom and its top below the top weep holes. Wind washing induces air flows in the cavities left in front (1) and behind the fill (2), proportional to the third power of their widths:

$$G_a = G_{a,1} + G_{a,2} \quad G_{a1} = G_a \frac{d_1^3}{d_1^3 + d_2^3} \quad G_{a2} = G_a \frac{d_2^3}{d_1^3 + d_2^3}$$

As the veneer has a negligible thermal resistance compared to the insulation, the temperature in the cavity in front will hardly differ from the one outdoors, which allows the temperature in the cavity behind to be written as:

$$\theta_2 = \theta_{2,\infty} + \left(\theta_e - \theta_{2,\infty}\right) \exp\left(- \frac{R_1 + R_2}{c_a G_{a,2} R_1 R_2} z \right)$$

with R_1 and R_2 the thermal resistances, respectively from outdoors and indoors to the cavity behind and $\theta_{2,\propto}$ the temperature without wind washing in that cavity. The Nusselt number, the equivalent thermal transmittance and the insulation efficiency then become:

$$Nu = \left\{ 1 - c_a G_{a,2} \frac{U R_1^2}{H} \left[\exp\left(- \frac{1}{U R_1 R_2 c_a G_{a,2}} H \right) - 1 \right] \right\}$$

$$U_{\text{eff}} = U \times Nu$$

$$\eta_{is} = 100 \frac{d_{\text{ins,eq}}}{d_{\text{ins}}} = 100 \frac{U}{U_{\text{eq}}} \left(\frac{1 - U_{\text{eq}} R_o}{1 - U R_o} \right)$$

with U the clear wall thermal transmittance, H the distance between bottom and top weep holes and R_o the thermal resistance of the non-insulated wall. Figure 4.1 illustrates how the efficiency and Nusselt number depend on the cavity width behind the insulation. Clearly, $Nu = 1.1$ demands careful workmanship as the efficiency drops quickly for a width beyond 8 mm.

4.3 Thermal transmittance

4.3.1 Definitions

4.3.1.1 Opaque envelope assemblies above grade

When airtight, the whole wall thermal transmittance of such assembly is calculated as:

$$U = U_o + \left(\sum_j (\psi_j L_j) + \sum_k \chi_k \right) \Big/ A \quad (\text{W}/(\text{m}^2.\text{K})) \tag{4.2}$$

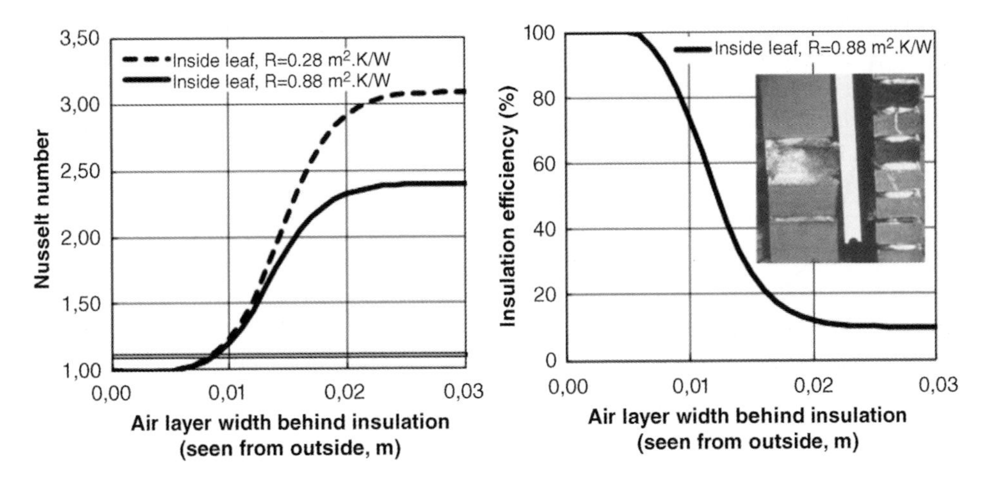

Figure 4.1 Cavity wall, partial fill (10 cm PUR) with cavities (or air layers) at both sides. Wind washing at 4 m/s wind speed (open terrain, 10 m high). Nusselt number and insulation efficiency.

with U_o its clear wall thermal transmittance, A its surface, ψ_j the linear thermal transmittances (W/(m.K)) and L_j the lengths of all linear and χ the local thermal transmittances (W/K) of all local thermal bridges within the surface A. Floors on grade, floors above basements, floors above crawl spaces and transparent parts have whole wall thermal transmittances that are quantified differently, see both volumes on 'Performance-based Building Design'.

4.3.1.2 Whole envelope

The envelope, the sum of all the opaque facade parts, the lowest floor parts, the opaque roof parts and the transparent parts enclosing the protected volume in parallel, included all linear and local thermal bridges, has as mean thermal transmittance:

$$U_{\mathrm{m}} = \frac{a_{\mathrm{fl}} U_{\mathrm{fl}} A_{\mathrm{fl}} + \sum_{\mathrm{opaque}} aUA + \sum_j \left(a\psi_j L_j\right) + \sum_k a\chi_k + \sum_w U_w A_w}{A_{\mathrm{T}}} \quad (\mathrm{W/(m^2.K)}) \tag{4.3}$$

with U_{fl} the thermal transmittance, A_{fl} the surface area of the lowest floor, U the thermal transmittance and A the surface area of all other opaque building parts (facade and roofs), U_{w} the thermal transmittance and A_{w} the surface area of all transparent parts and the a's, reduction factors for on and below grade ($a_{\mathrm{fl}} < 1$) parts and party walls with neighbour buildings ($a = 0$) or partitions with unheated spaces outside the protected volume ($a < 1$).

4.3.2 Basis for requirements

4.3.2.1 Envelope parts

Values advanced must keep mould likelihood in edges and corners below 5%. In temperate climates this requires thermal transmittances (U_o) below 0.46 W/(m^2.K).

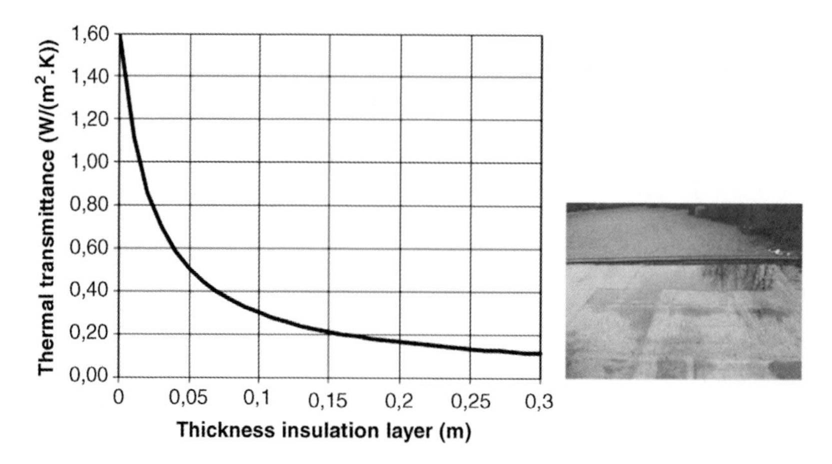

Figure 4.2 Low-sloped roof, insulation thickness versus thermal transmittance.

Also, the optimum in terms of total present value may be aimed for, resulting in a range of 0.2–0.46 W/(m^2.K). Of course one could also target minimum whole energy use, included the embodied energy, or minimal whole environmental load, a track leading to clear wall values of 0.15 W/(m^2.K) or lower. But do not forget that the property is inversely proportional to the insulation thickness, so ever larger thicknesses will generate ever lower additional benefits, see Figure 4.2.

4.3.2.2 Whole envelope

The preferred approach is to look for the optimum from a total present value perspective.

4.3.3 Examples of requirements

Years before the EU's Energy Performance Directive of 2002 and the recast of 2010 came into force, countries had already mandated maximal thermal transmittances (U_{max}). Some even limited the envelope mean, for example in relation to building compactness. Today, mandatory prevails, due to the long service life that characterizes insulation measures.

4.3.3.1 Envelope parts

Table 4.1 gives the maximum thermal transmittances as required in a few countries and regions.

4.3.3.2 Whole envelopes

Mandating maximum thermal transmittances per envelope part has its disadvantages. Very large glass surfaces remain applicable. Even with low-e, gas-filled double and triple glass, such abundance does not offer real benefit in temperate climates because winter solar gains remain low, while overheating in summer becomes more of a problem. Also poor

Table 4.1 Thermal transmittance requirements (U_{max}).

Element	U_{max} (W/(m^2.K))	
	New construction	Retrofit
Belgium (Flanders, from 1/1/2016)		
Walls	0.24	0.24
Roofs	0.24	0.24
Floors above grade	0.24	0.24
Floors above basements and crawlspaces	0.24	0.24
Floors above outdoor spaces	0.24	0.24
Walls contacting the ground	0.24	0.24
Party walls and floors	1.0	1.0
Glass	1.1	1.1
Windows	1.5	1.5
Doors and gates	2.0	2.0
Post filling existing cavity walls		0.55
Existing massive walls, insulation inside		—
Germany (normally heated buildings)		
Walls	0.24	
Roofs	0.24	
Low sloped roofs	0.20	
Floors above grade	0.30	
Floors above basements and crawlspaces	0.30	
Floors above outdoor spaces	0.30	
Walls contacting the ground	0.30	
Glass	1.10	
Windows	1.30	
UK		
Walls	0.30/0.35	
Roofs	0.16	
Floors	0.25	
Windows	2.00	
Sweden	**Oil or gas heating**	**Electrical heating**
Walls	0.18	0.10
Roofs	0.13	0.08
Floors	0.15	0.10
Windows	1.30	1.10
Outer doors	1.30	1.10

compactness is not penalized. One possible approach consists of limiting the transmission losses, which is easily controllable during design as it only requires calculating the product of the envelope's surface (A_T) by the mean thermal transmittance (U_m):

$$Q_{\mathrm{T,ann}} \div U_m A_T \quad (\mathrm{MJ/a}) \tag{4.4}$$

Imposing, for example, proportionality to the protected volume allows to write:

$$U_m = \alpha V / A_T = \alpha C$$

with C the compactness in m. The result is a straight line with slope α passing through the origin in a $[C,U_m]$ ordinate system. The lower that slope, the less the transmission losses, see Figure 4.3(a).

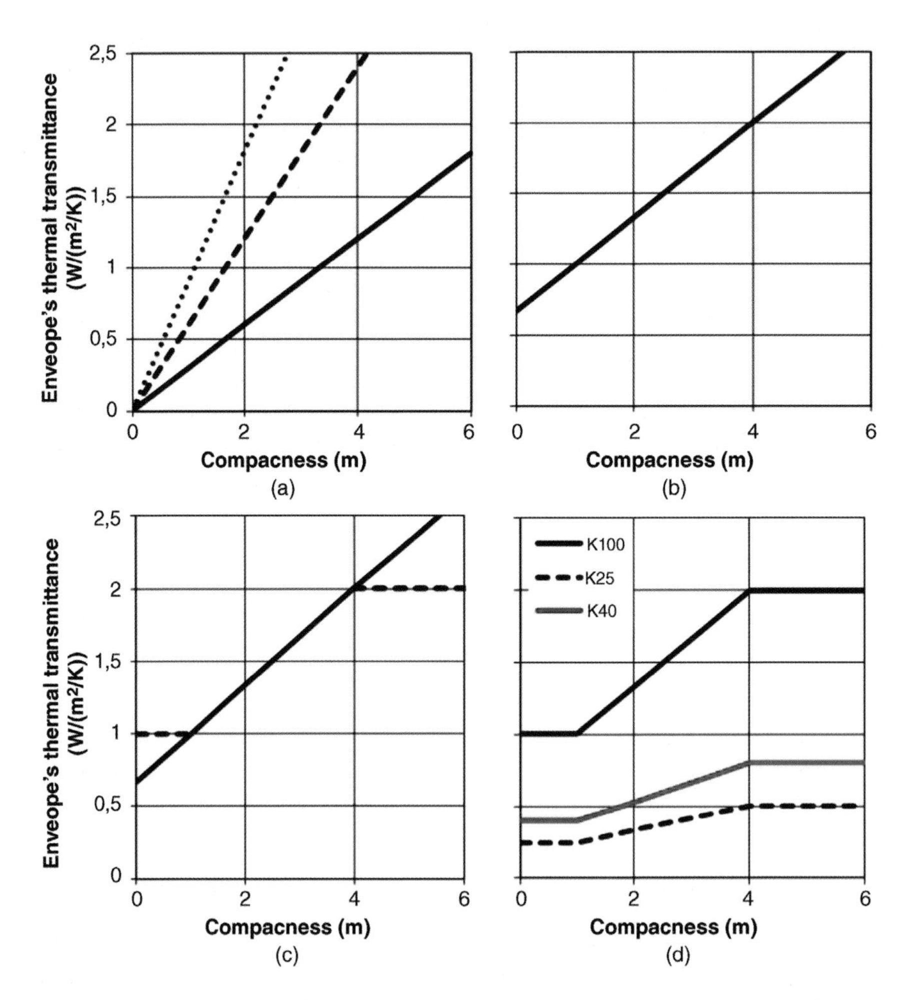

Figure 4.3 Compactness (C) versus the envelope mean thermal transmittance (U_m).

However, keeping that straight line under all circumstances is pointless. At high compactness, the mean thermal transmittance required looks so weak that mould growth, surface condensation and comfort complaints may prevail. At very low compactness, requirements in turn become so severe that investments explode and buildability becomes a real question mark. Glazed surfaces make a through-the-origin line anyway fictitious. In fact, one has:

$$\begin{aligned} U_\mathrm{m} &= U_\mathrm{m,op} + \left(A_\mathrm{T,w}/A_\mathrm{T}\right)\left(U_\mathrm{w} - U_\mathrm{m,op}\right) \\ &\approx U_\mathrm{m,op} + \left[A_\mathrm{T,w}/V\left(U_\mathrm{w} - U_\mathrm{m,op}\right)\right]C \\ &= a + bC \text{ with } a = U_\mathrm{m,op} \text{ and } b = A_\mathrm{T,w}/V\left(U_\mathrm{w} - U_\mathrm{m,op}\right) \end{aligned}$$

That is a straight line with $U_\mathrm{m,op}$ the mean thermal transmittance of the opaque parts, U_w the mean thermal transmittance of the glazed ones and $A_\mathrm{T,w}$ the total area that these take up. As the protected volume can also be written as $A_\mathrm{fl}h$ with A_fl the total floor area in m^2 and h the floor-to-floor storey height in m, the slope b in the formula expresses proportionality with the ratio between window and floor area but inverse proportionality with storey height, see Figure 4.3(b). Fixing a and b, for example a equal to 2/3 and b to 1/3, delivers a basis for formulating envelope requirements. The straight line then intersects compactness in the point (−2, 0). Weakening the still-too-strict requirements at low and strengthening the still-to-poor at high compactness is done by keeping a constant, for example below compactness 1 m and above compactness 4 m, see Figure 4.3(c). With $a = 1$ W/(m^2.K) at $C = 1$ m and a = 2 W/(m^2.K) at $C = 4$ m, both for b = 0, the straight line becomes trapped and could for example be called ‘level of thermal insulation K100’. Any building with a (C, U_m) couple on that line gets that level. All other levels then correspond with trapped lines proportional to K100, see Figure 4.3(d):

$$\begin{array}{ll} C \le 1\,\mathrm{m} & K = 100U_\mathrm{m} \\ 1 < C < 4\,\mathrm{m} & K = \dfrac{100U_\mathrm{m}}{2/3 + C/3} \quad (\mathrm{W/(m^2.K)}) \\ C > 4\,\mathrm{m} & K = 50U_\mathrm{m} \end{array} \tag{4.5}$$

Imposing requirements is simple now. Low energy, for example, requires K25. New construction in Belgium must have a K40 envelope. Needed, of course, are rules on how to calculate the protected volume, the enclosing envelope surface, all building part surfaces, and so on.

Compactness can also be defined as A_T/V in m^{-1}. Lower numbers then mean more and higher mean less compact, while the straight line turns into a hyperbola. Figure 4.4 shows the former German envelope requirements. Also here, the same corrections apply as already explained: constant below compactness 0.2 m^{-1} and above compactness 1.05 m^{-1}, hyperbolic in between:

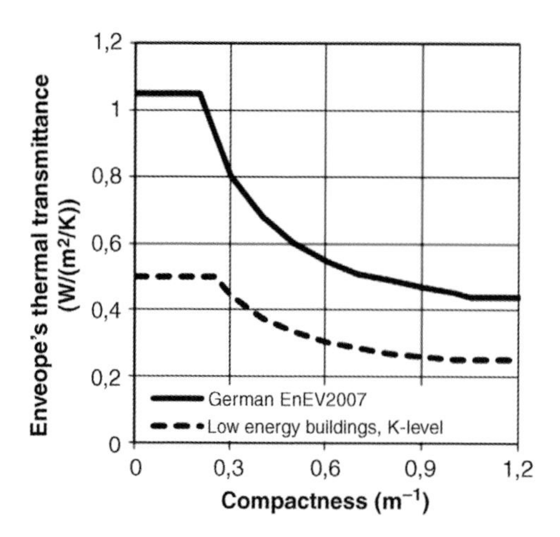

Figure 4.4 German envelope mean thermal transmittance requirements.

$$\begin{array}{ll} C' \leq 0.2\,\text{m}^{-1} & U_\text{m} = 1.05 \\ 0.2 < C' \leq 1.05\,\text{m}^{-1} & U_\text{m} = 0.3 + \dfrac{0.15}{C'} \quad (\text{W}/(\text{m}^2.\text{K})) \\ C' > 1.05\,\text{m}^{-1} & U_\text{m} = 0.44 \end{array} \tag{4.6}$$

4.4 Transient thermal response

4.4.1 Properties of importance

In regions with a temperate climate a highly dampening transient thermal response of the building fabric and the envelope is one of the means to enhance summer comfort. In regions with warm but dry climates, energy use for active cooling benefits from such dampening response. Characteristics used to quantify that response in case of opaque flat envelope parts are:

Harmonic (period: 1 day)	Others
Temperature damping D_θ	Time constant τ
Dynamic thermal resistance D_q	
Admittance Ad	

The advantage of the harmonic properties is that they are analytically calculable, see 'Building Physics, Heat, Air and Moisture'. Simple modelling in turn allows estimating the time constant, but a correct calculation demands a numerical approach. One such simple model pictures the opaque part as a resistance–capacitance–resistance circuit.

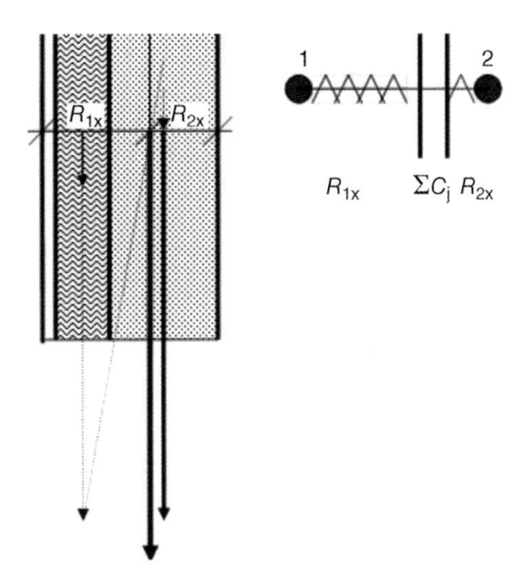

Figure 4.5 The <thermal resistance R_1/capacitance ΣC/thermal resistance R_2> circuit.

Consider the thermal capacity per layer ($C=\rho c d$ in J/K) as a vector at each layer's centre. The vector field so formed has a resultant (ΣC_j) passing through the point of action of the assembly. With as ordinate x, the distance between that point and environment 1, related thermal resistance equals R_{1x} while the one between x and environment 2 is R_{2x}. The heat balance for the circuit formed is (Figure 4.5):

$$\frac{\theta_1 - \theta_x}{R_{1x}} + \frac{\theta_2 - \theta_x}{R_{2x}} = \left(\sum C_j\right)\frac{\mathrm{d}\theta_x}{\mathrm{d}t}$$

with θ_x the temperature at the point of action and θ_1 and θ_2 the temperatures respectively in environments 1 and 2.

A step change at time zero by one of these two ambient temperatures gives in x:

$$\theta_x = \theta_{x,\infty} + \left(\theta_{x,o} - \theta_{x,\infty}\right)\exp\left(-\frac{t}{\overline{R}\sum C_j}\right)$$

with $\overline{R}\sum C_j$ the time constant and $\overline{R} = R_{1x}R_{2x}/(R_{1x}+R_{2x})$ the harmonic mean of the thermal resistances R_{x1} and R_{x2}. Temperature $\theta_{x,\infty}$ in turn equals:

Step change $\Delta\theta_1$	*Step change* $\Delta\theta_2$
$\theta_{x,\infty} = \overline{R}\left(\frac{\theta_{1,o}+\Delta\theta_1}{R_{1x}} + \frac{\theta_{2,o}}{R_{2x}}\right)$	$\theta_{x,\infty} = \overline{R}\left(\frac{\theta_{1,o}}{R_{1x}} + \frac{\theta_{2,o}+\Delta\theta_2}{R_{2x}}\right)$

The time constant clearly increases, together with the capacity of the assembly and the harmonic mean of both resistances.

4.4.2 Performance requirements

Formulating requirements for the harmonic properties is not evident. The reciprocal of the mandated thermal transmittance (U_{max}),for example, fixes the value of the dynamic thermal resistance amplitude for an assembly without thermal capacity. For all others, following relation holds:

$$\left[D_{\text{q}}\right] > 1/U_{\text{max}} \quad (\text{m}^2.\text{K/W})$$

The mandated thermal transmittance also fixes the admittance amplitude for an assembly without thermal capacity, whereas for an assembly with infinite thermal capacity not shielded by insulation that amplitude becomes equal to the thermal surface film coefficient indoors (h_{i}). For real assemblies the value obeys:

$$U_{\text{max}} < [Ad] < h_{\text{i}}.$$

Or, a low thermal transmittance and a high surface film coefficient indoors opens a large interval for the admittance amplitude. A requirement could be:

$$[Ad] \geq h_{\text{i}}/2 \quad (\text{W}/(\text{m}^2.\text{K}))$$

In theory, temperature damping may vary from 1 to infinity. The relevance of very high values, however, is relative. If for example on a daily basis the sol-air temperature fluctuates between 10 and 80 °C, the amplitude of the temperature indoors, if the whole envelope forms one assembly and the temperature outdoors is the same at all faces, will equal 35/[D_{q}]:

$[D_\theta]$	$\hat{\alpha}_i$, °C
4	8.8
8	4.4
16	2.3
32	1.1

A 2.5 °C difference indoors between day and night is hardly problematic for thermal comfort. So, a lowest value metric suffices, for example $[D_\theta] \geq 15$.

4.4.3 Consequences for the building fabric

How to achieve high damping? Following simple model gives an answer. Take a two-layer assembly, one insulating without capacitance, thermal resistance R, the other heavy, capacitance C and thermal resistance zero. The thermal surface film coefficient at one side is h_1, at the other side h_2. The heat balance is:

$$\frac{\theta_1 - \theta_x}{R + 1/h_1} + \frac{\theta_2 - \theta_x}{1/h_2} = C\frac{\theta_x}{\text{d}t} \tag{4.7}$$

with θ_x the central temperature in the capacitance, θ_1 the ambient temperature at the insulation and θ_2 the ambient temperature at the heavy layer side. Assume 1 is outdoors ($h_1 = h_e$, θ_1). The temperature there fluctuates harmonically with period T. The temperature indoors (θ_2) will show the same periodicity, but damped and time shifted, with as complex value:

$$\theta_2 = \alpha_2 \exp\,(\mathrm{i}2\pi t/T)$$

Temperature damping presumes a heat flux zero at the inside face 2. It is then also zero between the inside and the capacitance, meaning that the complex amplitude at the point of action (α_x) equals α_2. The heat balance becomes ($\exp(\mathrm{i}2\pi t/T)$ eliminated):

$$\frac{\alpha_1 - \alpha_2}{R + 1/h_1} = \frac{\mathrm{i}2\pi C}{T}\alpha_2 \tag{4.8}$$

giving as damping amplitude ($R_1 = R + 1/h_1$):

$$D_\theta^{1,2} = \alpha_1/\alpha_2 = 1 + \mathrm{i}2\pi R_1 C/T \rightarrow \hat{D}_\theta^{1,2} = \sqrt{1 + \left(\frac{2\pi R_1 C}{T}\right)^2} \tag{4.9}$$

a value close to a linear function of the time constant R_1C. Let's reverse the situation with 2 outdoors and 1 indoors, θ_2 becoming the driver now . The complex heat flux at the inside face 1 yet becomes zero, changing the balance into ($h_2 = h_e$):

$$h_e(\alpha_2 - \alpha_1) = \frac{\mathrm{i}2\pi C}{T}\alpha_2$$

giving as damping amplitude:

$$D_\theta^{2,1} = \alpha_2/\alpha_1 = 1 + \mathrm{i}2\pi C/(Th_2) \rightarrow \hat{D}_\theta^{2,1} = \sqrt{1 + \left(\frac{2\pi C}{Th_2}\right)^2} \tag{4.10}$$

As the surface resistance $1/h_2$ has a value far below the sum of the thermal resistances $1/h_1 + R$, the damping amplitude $D_\theta^{1,2}$ must be much larger than $D_\theta^{2,1}$. Neglecting the 1 under the square root gives for their ratio:

$$\hat{D}_\theta^{1,2}/\hat{D}_\theta^{2,1} = 1 + Rh_e$$

that is a value 38.5 ($h_e = 25$ W/(m^2.K)) for an insulation with thermal resistance 1.5 m^2. K/W. Achieving high temperature damping thus requires assemblies with a massive layer inside and the insulation outside. A damping amplitude 15 relates to a time constant of 205 800 s or 2.38 days. By the same argument, a massive layer inside also gives the highest admittances, at least if not screened by a thin insulation layer or a high inside surface film resistance ($1/h_i$).

4.5 Moisture tolerance

4.5.1 In general

Several wetness sources put moisture tolerance at stake. Under the heading liquid water stay construction moisture, rain, rising damp, water heads and accidental leaks. Hygroscopic moisture, surface and interstitial condensation instead are vapour related. Chapter 3 already showed that moisture figures as prime damaging actor looking to durability and well-being.

4.5.2 Construction moisture

4.5.2.1 Definition

The term construction moisture applies to all excess wetness in the fabric before service life starts. This includes rain and snow during construction (Figure 4.6), more water in mortar, concrete and gypsum plaster than chemically needed, bricks wetted to facilitate laying, timber arriving on site too humid, aerated concrete containing up to 250 kg production moisture per m^3, carbonisation generating water, and too much hygroscopic moisture in apparently dry materials. The amounts in a massive building fabric are quite impressive. Tests on low-sloped roof screeds gave some 120 kg/m^3. Drawbacks due to drying of that moisture are extra heating energy used and higher vapour pressures indoors during early service life, included increased mould and surface condensation risk and durability issues.

4.5.2.2 Performance requirements

Construction moisture must dry without causing unacceptable interstitial condensation in the thermal insulation or other moisture sensitive layers composing the enclosure, while the envelope and fabric as constructed might not unnecessarily retard drying. This does not mean that a correctly composed assembly will dry fast. The time needed

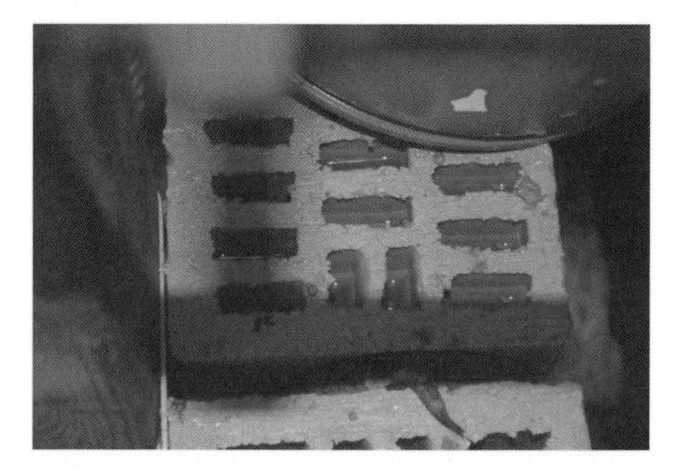

Figure 4.6 Construction moisture due to rain during construction.

depends on the amounts of moisture involved, the thickness and hygrothermal properties of all the layers and the ambient conditions: temperature outdoors, solar irradiation, under-cooling, precipitation, relative humidity indoors and outdoors, operative temperature indoors and air pressure differences.

4.5.2.3 Consequences for the building fabric

Drying and no interstitial condensation elsewhere generates specific rules. Never sandwich moist layers and dry ones between vapour retarding foils. With the wet layers at the warm side, always insert a vapour retarder directly behind any vapour permeable thermal insulation or moisture sensitive layer at the cold side. Wait before painting, wallpapering and flooring until the fabric is air-dry, and so on.

In temperate climates, ventilation and heating accelerate drying. Consider a humid envelope layer facing indoors. The vapour flux during the first drying stage is approximately:

$$g_{v,d} = \beta(p_{sat,s} - p_i) = \frac{h_{c,i}}{N\lambda_a}(p_{sat,s} - p_i) \quad (kg/(m^2.s)) \tag{4.11}$$

with β the surface film coefficient for diffusion, $h_{c,i}$ the convective surface film coefficient (both increasing with air movement), N the diffusion constant, l_a the thermal conductivity of air, $p_{sat,s}$ the vapour saturation pressure on the inside surface and p_i the vapour pressure indoors. Fast drying presumes a large difference between these two vapour pressures and a high convective surface film coefficient. Otherwise said, it requires the surface to be heated, the air speed to be raised and a low relative humidity indoors to be maintained. With G_a the ventilation flow and A the drying surface, the mean vapour pressure indoors will be:

$$p_i = p_e + \frac{g_{v,d}A}{6.21\ 10^{-6}G_a} \quad (Pa) \tag{4.12}$$

Combining that equation with the previous one gives as vapour flow:

$$g_{v,d} = \frac{h_{c,i}(p_{sat,s} - p_e)}{N\lambda_a + h_{c,i}A/(6.21\ 10^{-6}G_a)} \tag{4.13}$$

If the inside surface temperature θ_s and related vapour saturation pressure ($p_{sat,s}$) is known, then the flow could be quantifiable. For partition walls with equal operative temperature at either side, the surface temperature (θ_s) during the first drying stage obeys:

$$\theta_s = \theta_i - \frac{l_b g_{v,d}}{h_{c,i} + 5.4\, e_L} \tag{4.14}$$

with e_L the longwave emissivity of both faces and l_b the heat of evaporation. Combination with the vapour flux equation gives a first relation between surface temperature and vapour saturation pressure. The equation of state offers a second one. Together, they make the drying flux calculable as Figure 4.7 illustrates for a plastered room with volume $4 \times 4 \times 2.7 = 43.2\ m^3$ and $56.8\ m^2$ wet surface (four walls minus the window and the

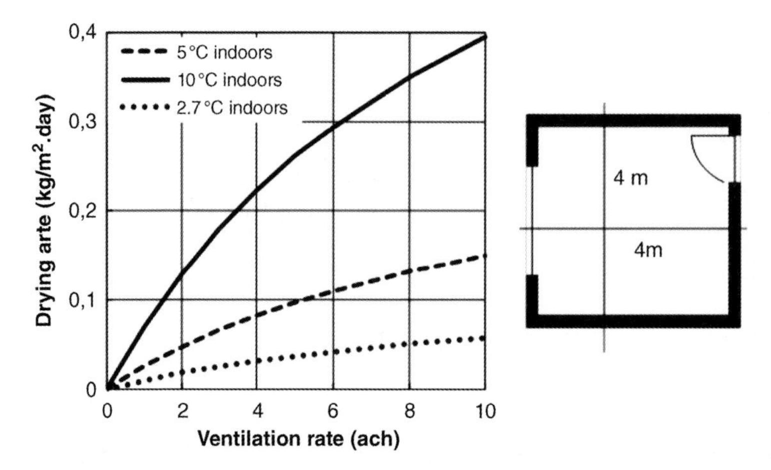

Figure 4.7 First stage drying rate in a newly plastered 4 × 4 × 2.4 m^3 room.

ceiling). Outdoors is the mean January temperature 2.7 °C. Parameters are the ventilation rate and the temperature indoors. Both look effective, though the added benefit shrinks with increasing ventilation rate.

However, faster drying during the first stage causes the second to start at higher transition moisture content:

$$w_{\mathrm{tr}} = w_{\mathrm{cr}} + g_{\mathrm{vd}} d/(3D_{\mathrm{w}}) \tag{4.15}$$

During that second stage, the flux drops to:

$$g_{\mathrm{v,d}} = \frac{p_{\mathrm{sat},x} - p_{\mathrm{i}}}{N\lambda_{\mathrm{a}}/h_{\mathrm{c,i}} + \mu N x} = \frac{p_{\mathrm{sat},x} - p_{\mathrm{i}}}{Z_x} \tag{4.16}$$

with $p_{\mathrm{sat,x}}$ the vapour saturation pressure at the moisture front at a distance x from inside in the partition wall and μ the vapour resistance factor along x. Air velocity at the surface so quickly loses impact because drying increasingly depends on the diffusion resistance to the indoors, a value that painting inside faces too early will enlarge, thus retarding drying even more and allowing moisture to re-humidify the already dry part (Figure 4.8). Happily, temperature and relative humidity indoors keep their impact.

4.5.3 Rain

4.5.3.1 The problem

It's not the rain itself that is the problem, but rain leakage. This occurs when precipitation and wind-driven rain wet layers in the enclosure, whose function or location requires dryness (Figure 4.9).

Take the thermal insulation and the layers on its inside. Capillary insulation in contact with outside cladding may suck in penetrating rain. Solar driven flow from wet outside

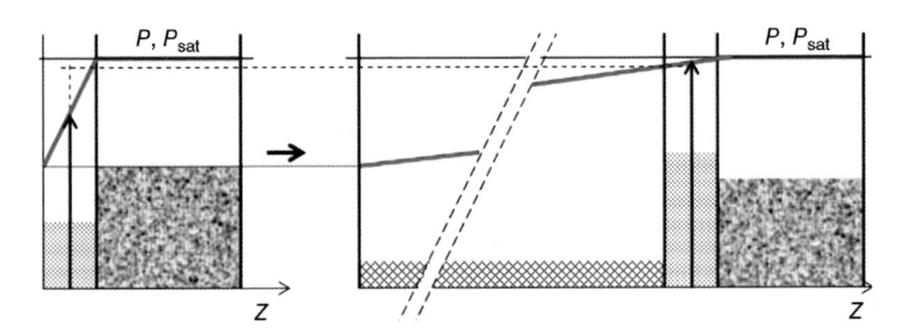

Figure 4.8 Re-humidification of the dry thickness, vapour tight finish applied too early.

finishes to indoors can humidify layers on the inside of the insulation. If this is the inside finish, its drying will release vapour indoors while absorbing latent heat of evaporation, which tends to cool the interior. How much extra heating will be required to keep the set point temperature indoors and how important can that evaporative cooling effect be? To know that, consider a ventilated room with volume V and A m^2 of identical exterior walls. Where all are air-dry, the heating power (Φ_H) to keep the room comfortable when

Figure 4.9 Rain leakage.

neither solar nor internal gains play and all adjacent rooms are at that temperature becomes:

$$\Phi_H = (UA + 0.34\, nV)(\theta_i - \theta_e) \quad (W)$$

If all indoor wall faces are wet, the surface balance changes to (per m^2):

$$h_i(\theta_i - \theta_{si}) + \frac{\theta_e - \theta_{si}}{R_T + 1/h_e} = l_b g_{vd}$$

with g_{vd} the drying flux, l_b the heat of evaporation and R_T the wall's thermal resistance. At room level, the heat balance then turns into:

$$h_i A(\theta_i - \theta_{si}) + \Phi'_H + 0.34\, nV\theta_e = 0.34\, nV\theta_i$$

giving as heating power:

$$\Phi'_H = \left(\frac{h_i/(R_T + 1/h_e)}{h_i + P} A + \frac{l_b g_{vd}}{\theta_i - \theta_e} + 0.34\, nV\right)(\theta_i - \theta_e) \quad (4.17)$$

If the thermal resistance R_T overwhelms the surface film coefficient h_i, that equation simplifies to:

$$\Phi'_H = (UA + 0.34\, nV)(\theta_i - \theta_e) + l_b g_{vd} A$$

Or, keeping the set point requires delivering the heat of evaporation absorbed by the kilograms of moisture drying per unit of time, which means extra heating. If that extra cannot be transmitted, the indoor temperature will drop to:

$$\theta_i = \theta_e + \frac{\Phi_H - l_b g_{vd} A}{UA + 0.34\, nV}$$

Evaporative cooling then is a fact. Rooms with outside walls partly wet by rain thus either require more energy to heat or become cooler. Relative humidity indoors increases anyway, which may degrade perceived indoor air quality and could induce mould and sometimes surface condensation.

Rain penetration as defined does not mean that all layers in an assembly must remain dry. Nobody, except manufacturers of frost-sensitive or salt-containing bricks, worries about a frost-resisting, low-salt brick veneer sucking rain, on condition that solar-driven vapour flow in warm weather does not cause harm. In temperate climates tiles become wet in autumn, stay wet in winter and dry during the spring. This again troubles no-one, except those still manufacturing frost-sensitive tiles.

4.5.3.2 Performance requirements

These follow from the problems mentioned. Each exposed envelope part must be protected or assembled in such a way that extreme precipitation and wind-driven rain humidifies neither the thermal insulation nor the layers at its inside, while claddings and veneers that suck rain should suffer neither from detrimental functional nor unacceptable aesthetic degradation.

4.5.3.3 Modelling

During rainy weather horizontal and inclined surfaces always collect water. Vertical surfaces instead only do when rain and wind act together, giving as wind-driven impingement rate:

$$g_{r,v} = (0.2C_r v_w \cos\theta) g_{r,h}$$

θ being the angle between wind direction and normal to the surface. C_r is called the wind driven rain factor, a function of building location, the surroundings, the spot on the facade, local detailing, and so on. The product $0.2C_r v_w \cos\theta$ represents the catch ratio. Its value follows from measurement or calculations combining computerized fluid dynamics (CFD) with droplet tracing. Catch ratios are highest at corners and the top of facades facing the wind, see Figure 4.10 and Table 4.2.

Looking to how to control rain impingement and seepage, the easiest measure is to shield, which is quite efficient and easily done above and along facades by using overhangs. As designers often dislike it for aesthetic reasons, handling drainage, storage and transmission is the other way out (Figure 4.11). Drainage develops along outside surfaces and, where gravity and wind transmit water, between layers in free contact. Buffering becomes a reality when capillary layers suck and retain rain, while transmission happens each time capillary layers contact each other or touch thin air layers that act as drainage planes.

Looking to buffering in more detail, capillary finishes first suck the impinging droplets until their surface becomes capillary saturated. At that stage, a water film forms that adsorption retains for a while until run-off fingers down and wets the areas below, while further buffering until capillary saturation of the finish occurs. How long suction takes

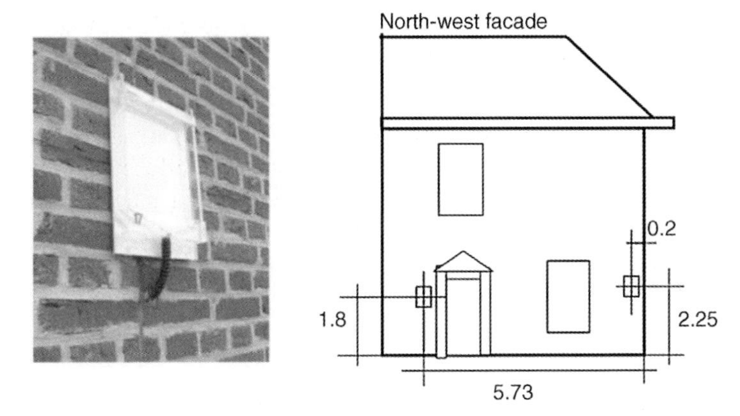

Figure 4.10 End of the row house, north-west facade, wind-driven rain gauges.

Table 4.2 Measured catch ratios, see the house of Figure 4.10.

Location	NW 5.73/1.8 m	NW 0.2/2.25 m	SW 0.36/2.25 m	SW 6.53/1.8 m	NE 1.72/2.25 m	NE 4.38/2.7 m
Catch ratio	0.016	0.26	0.23	0.047	0.008	0.014

(t_f) before run-off begins, depends on the wind driven rain intensity ($g_{r,v}$) and the finish's capillary water absorption coefficient (A):

$$\frac{A}{2g_{r,v}^2} \le t_f \le \frac{\pi^2 A}{16 g_{r,v}^2} \tag{4.18}$$

When that coefficient is large and rain intensity low, sucking postpones film formation for quite some time, allowing the layer to buffer lots of rain. When that coefficient is low, film formation proceeds rapidly, quickly turning the outside surface into a drainage plane. Bricks are moderately to highly capillary active ($A = 0.2$–1 kg/(m^2.s$^{0.5}$)). A brick veneer therefore acts as buffering volume, which minimizes the rain load on joints and facade protrusions. Conversely, concrete, concrete blocks, sand-lime stone, water-repellent stuccoes, paints and timber claddings are hardly capillary active ($A \le 0.02$ kg/(m^2.s$^{0.5}$)). They mainly act as drainage planes. Non-porous materials, such as glass, plastics and metals, give quasi instant run-off.

At first sight, run-off along non-sucking surfaces should give an increased water flow from the zone hit by rain and a constant flow on the non-hit surfaces below. If, as for high-rises, the wetted zone is large, wind driven rain should generate significant run-off, but observation contradicts this. Only the top storeys of high-rises catch rain, while friction, obstruction by facade reliefs and evaporation greatly reduces the run-off on its way down.

As stated, capillary outside finishes act as buffering volumes. However, when in suction contact with a capillary layer behind, rainwater will be transmitted. Gravity intervenes when run-off creates water puddles above facade protrusions or fills leaky joints, which then empty at the rear. Wind does the same by squeezing run-off through cracks wider than 0.5 mm, while the kinetic energy of the impinging droplets supports transmission

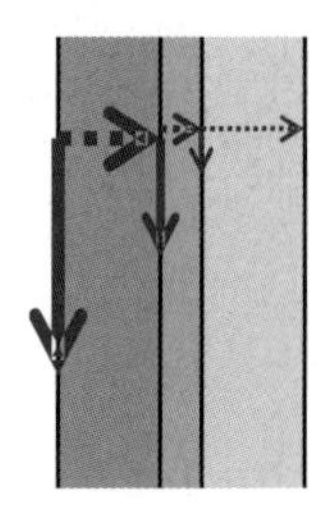

Bold vertical arrows = drainage
Dark to light grey = buffering
Dotted horizontal arrows = transmission

Figure 4.11 Rain control: drainage, storage and transmission.

by pushing run-off through yawning fissures, cracks and joints. Run-off entering a crack or leak seeps through at a rate:

$$G_w = -83.3 b^3 \text{grad}\, P_w/\eta \quad (\text{kg/s}) \tag{4.19}$$

with η the dynamic viscosity of water (0.000 15 kg.s/m^2), b the 'equivalent' width and grad P_w the pressure gradient across that crack or leak, gravity-related equal to $\rho_w g z$, with g acceleration by gravity (9.81 m/s^2) and z the depth of the water puddle facing or built up in the crack or leak.

4.5.3.4 Consequences for the building envelope

When designing envelopes, shielding, buffering, exterior surface drainage (one-step control), or combined drainage and buffering with exclusion of unwanted transmission (two-step control) are the measures that, correctly executed, ensure rain-tightness.

The principle behind shielding is simple. What's higher protects what's lower, Take roof overhangs, sills and coping stones (Figure 4.12). Sills require a slope to the outside, a kerf at the underside, end dams and a back dam, principles that help to shape window sashes and frames.

Buffering in turn combines outside drainage with rain storage. To function properly, a buffering wall must be thick enough to keep the moisture away from inside, even after

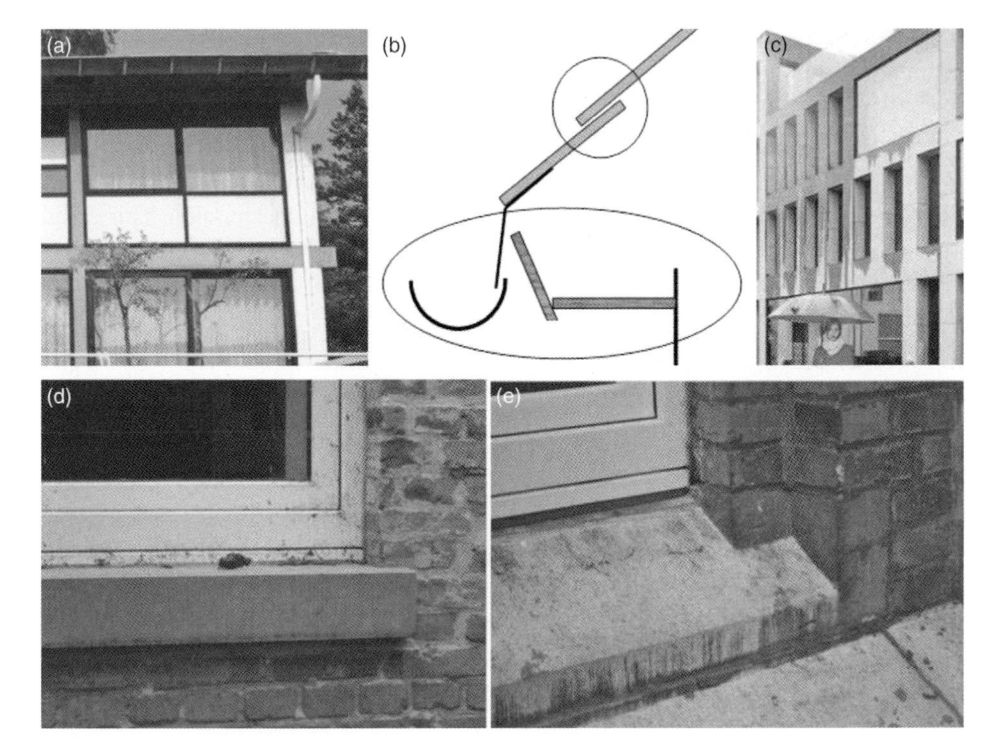

Figure 4.12 (a) and (b) Roof overhang, (c) no overhang and sills, (d) sill without end and back dam, (e) sill with end and back dam.

long-lasting rain events. In temperate but humid climates, 30 hours of uninterrupted rain can happen, whereas ‘keep away’ means not passing the wall’s midline. Take an aerated concrete wall, capillary water absorption coefficient 0.08 kg/(m^2.s$^{0.5}$), capillary moisture content 350 kg/m^3. The thickness needed is:

$$d = 2\left(\frac{A}{w_c}\right)\sqrt{T} = 2\left(\frac{0.08}{350}\right)\sqrt{2.5 \times 24 \times 3600} = 0.21\ \text{m}$$

With masonry, head and bed mortar joints often act as short-circuits, which is why buffering requires one and half brick walls, where-in a continuous mortar joint with air voids that splits the wall in two parts acts as a rain stop, see Figure 4.13. Of course, the bed joints still short-circuit that stop to some extend

In a one step control, the only drainage plane is the outside surface (Figure 4.13). To function properly the outer finish must be watertight, water-repellent or fine-porous. Watertight ones neither buffer nor transmit water. Water repellent ones have a contact angle below 90°, while fine-porous choices must be thick enough to prevent rainwater from reaching the substrate:

$$d_{pl} = B\sqrt{T} \quad \text{(m)} \quad \text{with} \quad B \approx A/w_c$$

where B is the water penetration coefficient and T rain duration. Thin ones thus need a really low capillary water absorption coefficient but a high capillary moisture content, which means very fine pores but high enough porosity and tortuosity. If so, they largely prohibit substrates with wider pores from sucking rain. Of course, fine pores and high tortuosity makes these thin finishes vapour-retarding. A drawback of one-step solutions is damage sensitivity. Once the finish is perforated or cracked, rain can penetrate.

When going for a two-step control, the outside and reverse side of the outer finish act as drainage planes, while a capillary break behind prohibits rain transmission. This could be a cavity or a water repellent thermal insulation. Examples include brick veneers having a tray at the bottom of the cavity behind with weep holes to the outside just above (Figure 4.14), facade elements with open joints and a cavity behind, metal claddings with a cavity behind, and the rebates between hinged window sashes and the frames. The

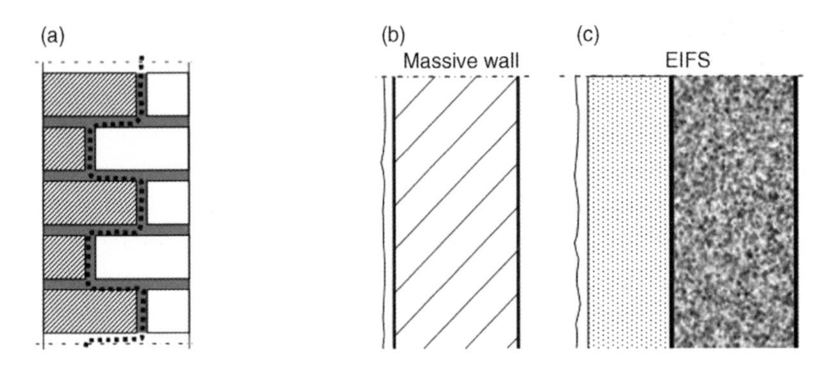

Figure 4.13 (a) rain control by buffering, (b) and (c) one-step control, two examples.

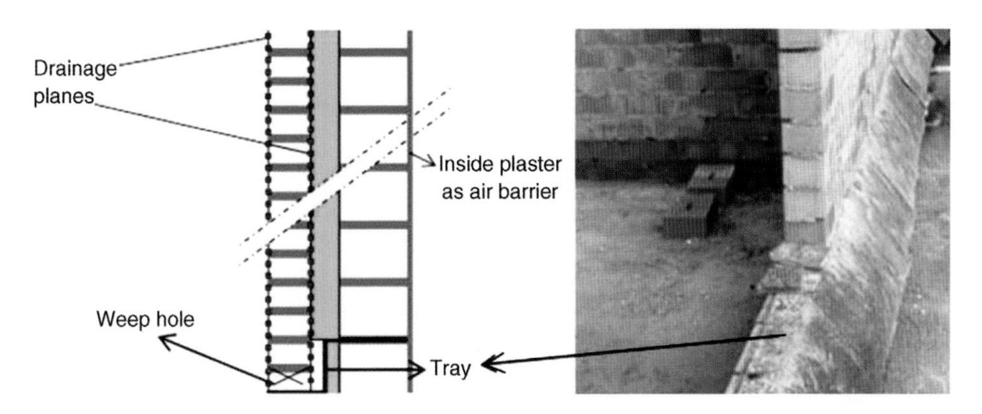

Figure 4.14 Filled cavity wall, two-step rain control.

layers at the inside of the capillary break in turn must be airtight, meaning that rain-tightness and air/wind-tightness are split.

Cavities extending three-dimensionally behind a veneer wall or outside cladding don't level out the wind pressure differences, a necessary condition to avoid heavy wind from blowing reverse side run-off across the cavity to the inside leaf. Compartmentalization of these cavities is often the only way out.

One major advantage of these two step solutions is damage insensitivity. A cracked or perforated outside finish doesn't kill its function. However, professionals often overlook the air and wind barring function of the inside leaf. When closing joints, for example, the best seal, wind- and airtight, should sit on the inside. Many put it outside.

4.5.4 Rising damp

4.5.4.1 Definition

'Rising damp' refers to walls wetted from below, which can happen when a wall or some of its layers are capillary and contact the water table, capillary moist soil, sink water or collected rainwater. This cause of dampness is often seen as a problem that only occurs in massive walls, although also the inside leaf of a cavity wall may suck the veneer's reverse side run-off when the cavity misses a correctly installed tray at the bottom.

4.5.4.2 Performance requirements

Rising damp must be avoided because large areas can become and stay wet, whereas drying to the inside either increases energy use for heating or invokes evaporative cooling, while raising relative humidity indoors and increasing mould risk (see Figure 4.15).

Moisture content in affected walls can reach high values ($w_{cr} \leq w \leq w_c$), resulting in plaster, wallpaper and paint damage. In many cases rising damp carries dissolved salts that crystallize where drying happens. At higher relative humidity they rehydrate. Such

Figure 4.15 Rising damp, causing extensive mould growth.

salts also increase hygroscopicity and retard, even exclude, drying when it is humid and warm outdoors.

4.5.4.3 Modelling

The height at which damp will stop in a homogeneous wall depends on the balance between capillary suction and evaporation at the wall's surfaces. In contact with the water table or with ponding rainwater, rising starts from capillary saturated. When contacting capillary moist soil, the difference in suction between wall material and soil fixes the moisture content to start off. If it is the soil that sucks the most, damp will hardly rise. If it's the wall, then damp will rise but at a lower rate and a lower amount than if in direct contact with water. The ambient conditions at both wall faces and the diffusion resistance of the finishes determine the evaporation. The larger that resistance, the less vapour will evaporate and the higher the damp will rise up the wall for a given suction. A simple model helps clarifying that balance.

In a homogeneous wall, rising per unit of time equals:

$$v_{\mathrm{m}} = r\left(\frac{\sigma \cos \Theta}{4\eta}\right)\left(\frac{1}{h} - \frac{1}{h_{\max}}\right) \tag{4.20}$$

with h the height that is damp and h_{max} the maximum height dampness would reach without evaporation:

$$h_{max} = 2\sigma \cos \Theta/(r\rho_w g)$$

The viscosity (η) and the surface tension (σ) of the rising moisture, the contact angle Θ and the mean radius r of the pores in the material fix the water penetration coefficient as:

$$B = \sqrt{r \frac{\sigma \cos \Theta}{2\eta}} \tag{4.21}$$

or:

$$\frac{B^2}{2} = \frac{r\sigma \cos \Theta}{4\eta}$$

With $B = A/w_c$ and the sucked water flow G_m equal to $v_m w_c d$ with d wall thickness and w_c the material's capillary moisture content, without evaporation the flow and damp height write as:

$$G_m = \frac{ABd}{2}\left(\frac{1}{h} - \frac{1}{h_{max}}\right) \quad h_{max} = \frac{5.5 \times 10^{-4}}{B^2} \tag{4.22}$$

Once at equilibrium, what's sucked must equal what's evaporated (G_e):

$$G_m = G_e = h\left[\frac{p_{sat,1}(1-\phi_1)}{1/\beta_1 + Z_1} + \frac{p_{sat,2}(1-\phi_2)}{1/\beta_2 + Z_2}\right] \tag{4.23}$$

where β_1 and β_2 are the surface film coefficients for diffusion and Z_1 and Z_2 the diffusion resistances of the finishes at both wall faces. Combining both flow equations gives:

$$h^2 - \left[\frac{ABd}{2h_{max}\left(g_{d,1} + g_{d,2}\right)}\right]h + \frac{ABd}{2\left(g_{d,1} + g_{d,2}\right)} = 0$$

a quadratic expression with as positive root:

$$h = \frac{ABd}{4h_{max}\left(g_{d,1} + g_{d,2}\right)}\left[\sqrt{1 + \frac{4h_{max}^2\left(g_{d,1} + g_{d,2}\right)}{ABd}} - 1\right] \tag{4.24}$$

The moisture profile in the wall follows from:

$$\frac{d}{dz}\left(D_w \frac{dw}{dz}\right) = \frac{g_{d,1} + g_{d,2}}{d}$$

If moisture diffusivity (D_w) and the drying rate stay constant along the damp height (h), the solution is ($z=0$: $w=w_c$; $z=h$: $w=w_{cr}$):

$$w = \left(\frac{g_{d,1}+g_{d,2}}{2dD_w}\right)z^2 - \left[\frac{w_c - w_{cr}}{h} + \frac{(g_{d,1}+g_{d,2})h}{2dD_w}\right]z + w_c \tag{4.25}$$

a parabola with highest moisture content in the contact plane with the water and lowest at final damp height. For a diffusivity that changes with moisture content ($D_w = f(w)$), the profile becomes more rectangular with a smaller gradient along the damp height.

Decreasing moisture content with height is a rising damp characteristic, although dissolved salts can obscure this picture. Figure 4.16 gives the final height in a 30 cm thick, joint free partition wall, whose basement contacts the water table, for 20 °C and 50% relative humidity indoors. The abscissa is the product of capillary water penetration and absorption coefficient (AB). When this product becomes larger – which is the case for ever more course-porous materials with an ever lower pore volume – dampness first climbs to a maximum height to drop beyond. Very limited heights typify materials with ultrafine or really coarse pores.

Figure 4.17 shows the final damp height after painting such a homogeneous wall, which added 1 m of diffusion thickness to both wall faces. Decreased drying speed now increases that height by a factor of 6! In other words, hiding damp walls behind vapour retarding finishes is no solution. After a few years, wetness will appear again, this time above the vapour retarding finish.

Walls contacting moist soil, brick walls with mortar joints and walls containing dissolved salts show different behaviour. Moist soil still fits with the model on condition that a lower capillary water absorption coefficient and capillary moisture content are assumed. To what extent needs to be tested. Mortar joints in brick walls in turn act as capillary breaks. If the mortar used is coarsely porous, the joints will suck very little

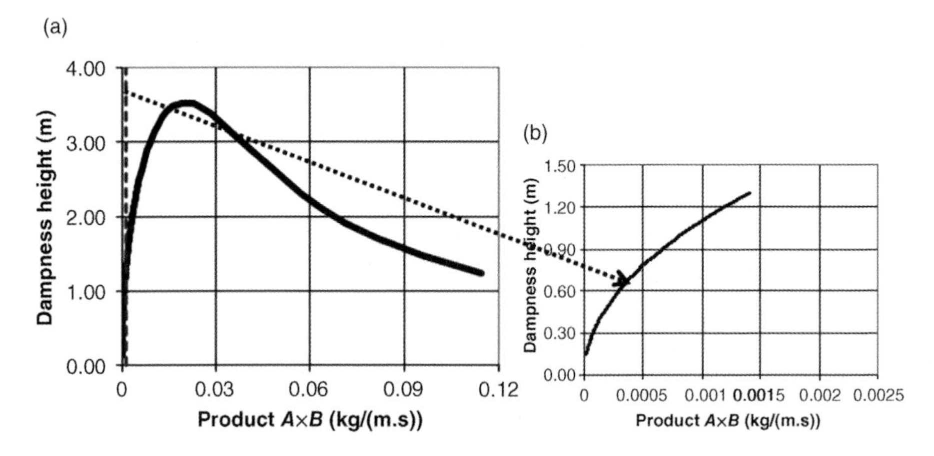

Figure 4.16 Unpainted 30 cm thick partition wall, no joints: damp height at 20 °C, 50% RH and $\beta_i = 2.6\times10^{-8}$ s/m. (a) The curve, which details the part between the damp height axis and the dashed vertical in the curve left, represents bricks. (B) the wall.

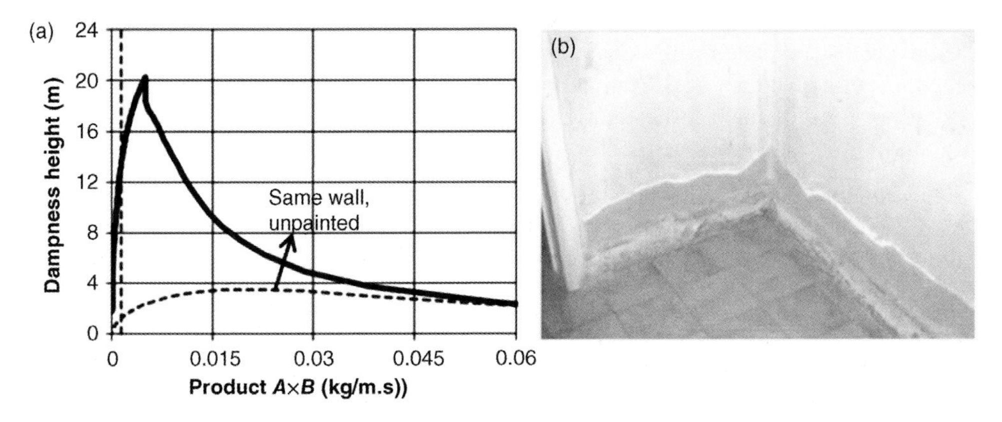

Figure 4.17 (a) The homogeneous wall painted (diffusion thickness 1 m), damp height at 20 °C, 50% RH (b) gypsum plaster short-circuiting the mortar joints.

moisture out of the bricks. In the opposite case, the bricks will pick up very little wetness from the joints. Of course, head and bed joints may turn wet by suction without humidifying the bricks. But even then they retard moisture uptake and limit dampness to a few brick layers, though the plaster often forms a short-circuit as Figure 4.17 shows. Even so, salts mitigate that joint effect, while high concentrations at both wall surfaces and at the moisture front change hygroscopicity to the extent that high moisture content persists even after curing.

4.5.4.4 Avoiding or curing rising damp

In a new construction, insertion just above grade of a section wide watertight membrane in all walls stops rising damp. The same applies above locations where run-off collects, see Figure 4.18.

For retrofit. a first possible cure consists of eliminating suction by inserting watertight membranes or steel plates just above grade in all damp walls. An alternative is filling or

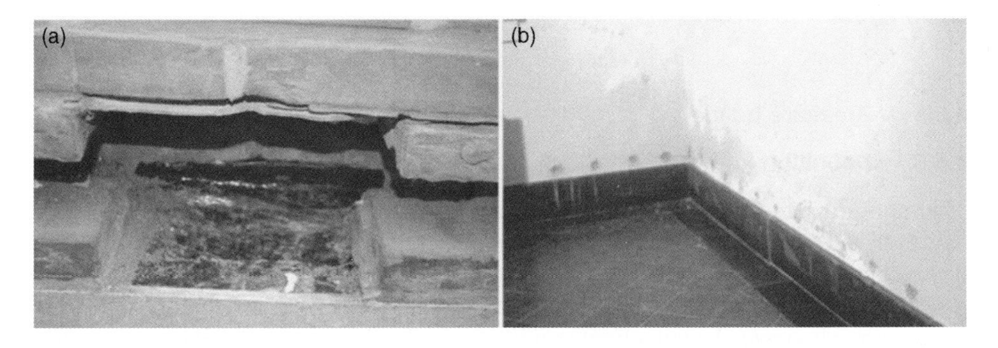

Figure 4.18 (a) Tray avoiding rising damp in the inside leaf of a cavity wall, (b) curing by injection.

making the material pores water repellent by injection or infusion, see Figure 4.18. Unless walls are loaded with salts, in which case hardly anything helps, these are by far the most effective measures. Activating drying is a second possibility. For that, all the retarding finishes have to be removed. This is much less effective than eliminating suction but it sometimes helps.

Repelling dampness could be a third. Electro-kinesis does this, at least in theory. Capillary suction in fact induces a voltage difference between the water in contact and the pore walls. Reversing it should turn suction into repulsion, forcing the water out. This, however, demands energy. In passive electro-kinesis, the electrodes used link a conductor that is embedded above grade in all damp walls to the earth. Corrosion in the contacts between the two delivers the energy needed, though only until breakage. Then, suction restarts. With active electro-kinesis energy is supplied to the conductor by a voltage of less than 2 volts, thus excluding conductor corrosion by avoiding the dissolution of the dampness present in oxygen and hydrogen. But, being active, it consumes electrical energy! And, neither form of electro-kinesis really guarantees drying, merely because dissolved salts make rising damp electrically conducting. So electro-kinesis is a fairly useless measure.

Fourth possibility but also useless is inserting drying pipes above grade in the damp walls. They lack any effect, mainly because the air they contain becomes 100% humid, limiting the drying area to the pipe section. The only conductance left is the surface film coefficient for diffusion at the pipe's aperture, though at high ambient relative humidity, temperature and vapour pressure outdoors give a really weak driving force:

$$G_{\mathrm{v,d,pipe}} = \beta \left[p_{\mathrm{sat}}(\theta_{\mathrm{e}}) - p_E\right] \left(\frac{\pi d_{\mathrm{pipe}}^2}{4}\right)$$

Equally without any benefit is hiding damp walls behind vapour retarding finishes.

A final thing which should not be done when retrofitting is applying any of the effective measures but without first investigating salt presence, especially when it concerns former stable walls, walls that have been in contact with cesspools or walls that have sucked salt-loaded meltwater. No cure for rising damp is of any help then. Hiding behind a non-capillary insulation or a brick veneer is often the only option left.

4.5.5 Pressure heads

4.5.5.1 Definition

The term pressure heads applies to moisture driven by pressure gradients. The differentials that stack, wind and fans generate are generally too small. Instead, the pressure heads constructions below the water table, swimming pool walls (Figure 4.19) and water reservoirs face are often large and constant. The same holds for those that water wells and rain ponds create. To put a figure on it, 10 cm of water gives 1000 Pa, 1 m of water 10 000 Pa.

Figure 4.19 Natatorium, leaking swimming pool wall.

4.5.5.2 Performance requirements

Wetting and leakage should not disturb the function of the spaces whose enclosure faces pressure heads. Thermal insulation, if present, must remain dry, while moistening from outside must stop at the water barrier in the assembly.

4.5.5.3 Modelling

Pressure flow is calculated using Darcy's law for saturated water displacement. The resulting moisture content in open-porous materials mostly exceeds capillary with frost damage for assemblies subjected to temperature swings below 0 °C as one of the risks.

4.5.5.4 Protecting the building fabric

A first possible measure consists of providing drainage around below grade constructions that face rain ponds. Drains collect water using the pressure heads that the pond builds up in the soil, the zero pressure head in the porous pipes and the high water conductivity of the surrounding fill. Between rain events, the ponding water is curving comparable to the temperature field in soils contacting a colder wall (Figure 4.20). The result is a moisture flow per metre run, equal to:

$$G_{\mathrm{w}} = \sqrt{\frac{w_{\mathrm{b}} k_{\mathrm{w,b}}}{\pi t}} \Delta P_{\mathrm{w}}$$

with w_{b} the saturation moisture content in, and $k_{\mathrm{w,b}}$ water conductivity of, the soil, t the time and ΔP_{w} pressure head. Drains function properly as long as the head they face remain below some 0.5 m.

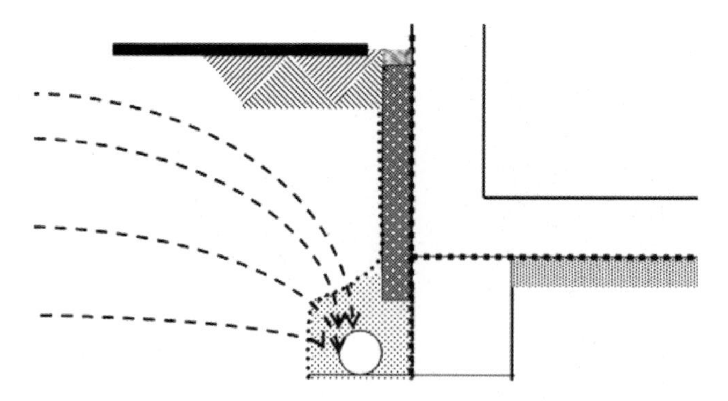

Figure 4.20 Drainage.

A logical second measure is ensuring water tightness of the below grade enclosures. If the related spaces require dryness, the damming construction should form a water barrier at the water head side of the thermal insulation. If usage allows, an alternative is retarded permeation so that the water front in the single-layer damming construction stabilises away from the inside. For that, diffusion from that front to indoors should equal water inflow, giving a steady state relation between the distance (x) from inside and the product of water conductivity (k_a) and vapour resistance factor (μ), assuming the surface film resistance for diffusion is negligible:

$$x = \frac{d}{\frac{\Delta P_w N(k_w \mu)}{p_{sat,x} - p_i} + 1}$$

with ΔP_w water head, $p_{sat,x}$ the saturation pressure at the water front in the wall and p_i the vapour pressure indoors. To function properly, the wall material should combine a low water conductivity with a low vapour resistance factor, which is contradictory as the last requires a high open porosity, large pore diameters and low tortuosity whereas the first needs the inverse.

4.5.6 Accidental leaks

The word accidental embraces uncommon and low probability, but the consequences of such leaks could be disastrous. Figure 4.21 shows a timber-framed dwelling, where a hot water pipe, built into an outer wall, leaked for a couple of years before the amazing amount of damage inflicted led to its detection.

4.5.7 Hygroscopic moisture

4.5.7.1 Definition

The terms ‘hygroscopic’ and ‘sorption’ are used to denote the moisture content in a material in equilibrium with the relative humidity in the pore air (ϕ). Sorption curves are S-shaped, with quite an increase at low relative humidity, a tilted platform between

Figure 4.21 Damage caused by a leaky hot water pipe.

30% and 80% and again a steep rise beyond 80%. Hysteresis compels desorption to stay above the sorption curve – see Figure 4.22 which shows a (de)sorption curve measured between 33% and 98% relative humidity, thus missing the sharp increase below 33%.

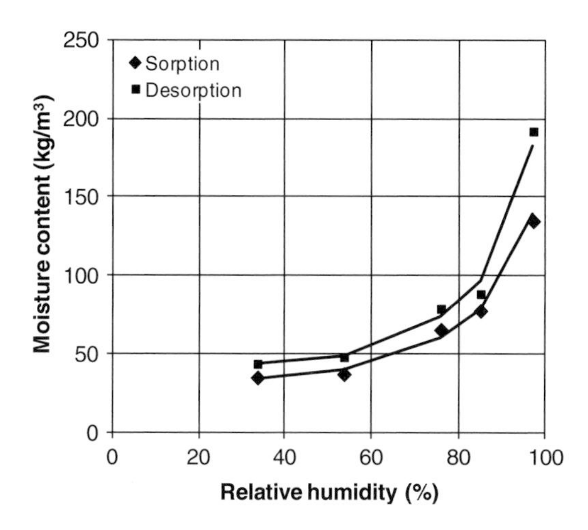

Figure 4.22 Sorption/desorption isotherm.

4.5.7.2 Performance requirements

Hygroscopic moisture reflects an equilibrium state. However, a too high or too low value, which means a too high or too low relative humidity, causes problems. Timber, textiles and paper for example shrink at a relative humidity below 30%. Massive wood panels then crack, paintings are damaged, and so on. A too high relative humidity in turn fosters mould.

4.5.7.3 Modelling

Evaluating cracking risk requires a complex heat–moisture/stress–strain model. Instead, fixing the temperature factor needed to prevent mould in temperate and cold climates is based on a simple rationale. First, knowing the four-week mean vapour pressure indoors (p_i) during the coldest month of the year and assuming 80% as the maximum allowable surface relative humidity, transpose the related vapour saturation pressure into a dew point temperature (θ_d), not to be exceeded indoors during these four weeks:

$$p_{sat,si} = p_i/0.8 \quad p_{sat,si} \rightarrow \theta_d\left(p_{sat,si}\right)$$

Then derive the related mean temperature factor:

$$f_{h_i} \geq \frac{\theta_d\left(p_{sat,si}\right) - \theta_e}{\theta_i - \theta_e}$$

4.5.7.4 Consequences for the building fabric

The model showed that in temperate climates envelope design should guarantee a temperature factor 0.7 or higher everywhere indoors on all opaque parts. Very complex structural thermal bridges require control using appropriate software, included a surface film coefficient inside as mandated by the standard (e.g. 4 W/(m^2.K)) and linked to the operative temperature at the room's centre, 1.7 m above the floor. If 0.7 is respected everywhere, the design complies. Otherwise, adjustment and new controls are needed.

4.5.8 Surface condensation

4.5.8.1 Definition

The term 'surface condensation' commonly covers water vapour condensing on the inside surfaces of any envelope or building fabric part (Figure 4.23). Of course, condensate is also deposited on outside surfaces but this rarely causes trouble.

4.5.8.2 Performance requirements

That on a daily basis drying must win, when outdoors the design temperature for heating reigns, is often the metric advanced, be it that lasting surface condensation on single glass will apparently have no consequences. Anyhow, run-off will moisten the window sashes. So, if made from softwood, they may gradually rot. Reveals that suck condensate deposited on aluminium window frames could turn mouldy. Surface condensation along

Figure 4.23 Surface condensation.

the perimeter of double or triple glass is a sign of a too high vapour release or failing ventilation indoors. And opaque envelope parts that are a little warmer than these glasses may turn mouldy if the four-week mean relative humidity at their inside surface passes 80% or when occasional surface condensation followed by drying pushes relative humidity temporarily well above 80%.

4.5.8.3 Modelling

Modelling starts with gathering information on the most likely daily mean vapour pressure indoors. If this value exceeds the daily mean inside surface saturation pressure somewhere, condensate will be deposited there:

$$g_c = \beta\left(p - p_{sat,s}\right) \approx 7.4 \times 10^{-9} h_c\left(p - p_{sat,s}\right) \quad (\mathrm{kg/(m^2.s)}) \tag{4.26}$$

with β the surface film coefficient for diffusion and h_c the convective surface film coefficient. Heat transfer involved is:

$$q = g_c l_b \approx 2.5 \times 10^6 g_c$$

For 1 m^2 of flat assembly, the steady-state thermal balance at the condensing side then becomes:

$$-2.5 \times 10^6 g_c = h_1(\theta_1 - \theta_s) + \frac{\theta_2 - \theta_s}{1/U - 1/h_1}$$

resulting in the following surface temperature:

$$\theta_s = \frac{h_1\theta_1 + U'\theta_2 + 2.5 \times 10^6\, g_c}{U' + h_1} \text{ with } U' = \frac{1}{1/U - 1/h_1} \tag{4.27}$$

In these formulas 1 denotes the condensation and 2 the other side of the assembly with θ_1 and θ_2 the related temperatures, operative indoors and sol-air outdoors. To quantify surface condensation, the system formed by the condensing flux and the temperature equation above is solved iteratively. When after a period of condensation the surface

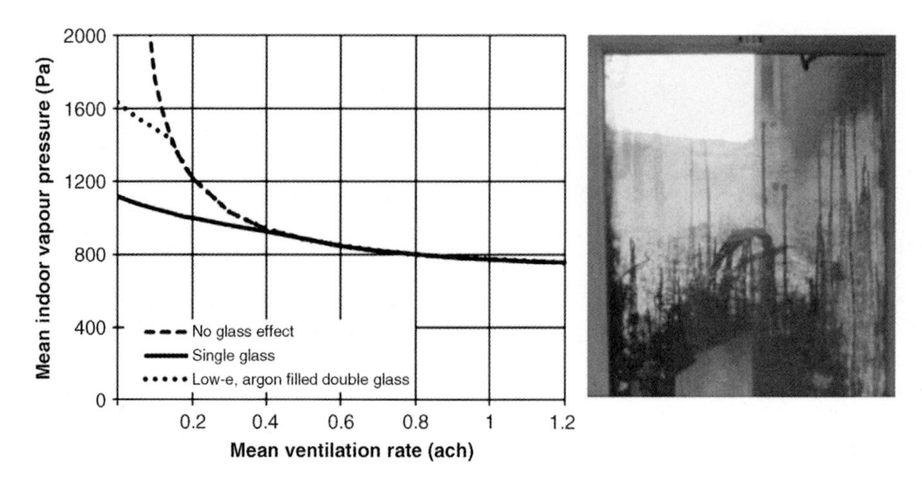

Figure 4.24 Two persons bedroom $4 \times 4 \times 2.5\,m^3$, $2.52\,m^2$ of glass. Monthly mean vapour pressure indoors depending on glass type and ventilation rate where the room is occupied 8 hours a day (monthly mean outdoors 2.7 °C, 663 Pa, indoors 13.9 °C).

warms up and sees its saturation pressure passing vapour pressure in the air close by, drying starts and goes on until all condensate has evaporated:

$$\int_{o}^{t} g_{\mathrm{d}} \mathrm{d}t \leq m_{\mathrm{c}} \tag{4.28}$$

Too high vapour pressure or too low inside surface temperature are the culprits indoors, with the first, as explained, pointing to high vapour releases or/and bad ventilation and the second to poor insulation and/or details acting as thermal bridges. Of course, condensation alternating with drying will stabilize vapour pressure indoors at levels that depend on the thermal resistance of the surfaces involved, usually the glazing, and the ventilation rate, see Figure 4.24.

Under-cooling or sudden weather changes, from cold to warm and humid, invoke surface condensation outdoors. The results are icy roads, frost on cars, rime on roofs and on argon-filled low-e double and triple glazing outside. In temperate climates, the winter amounts on well-insulated facades compare with what wind-driven rain deposits.

4.5.8.4 Consequences for the envelope

Insulate well and avoid thermal bridges with too low a temperature factor.

4.5.9 Interstitial condensation

4.5.9.1 Definition

Interstitial condensation refers to vapour deposited as liquid in building assemblies. For a long time, vapour released indoors was assumed to be the only wrongdoer. More

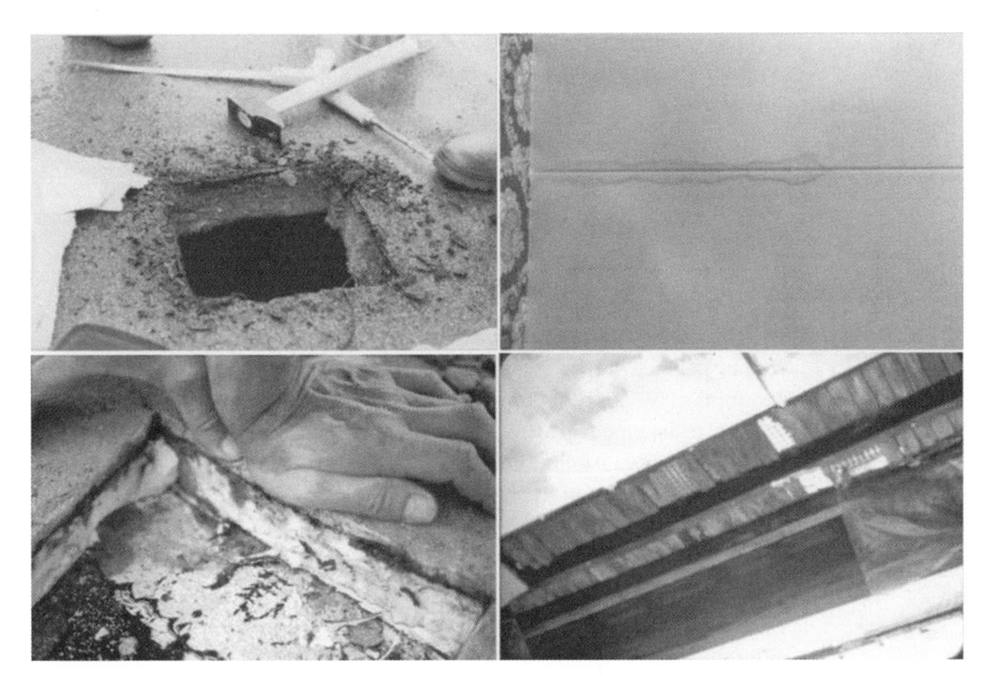

Figure 4.25 Damage caused by interstitial condensation.

recently it has been shown that temperature gradients can force construction moisture, hygroscopic moisture, rising damp and absorbed or drained rain to evaporate and deposit as condensate elsewhere in an assembly. Driving forces are diffusion and, more powerful, air ingress. Design flaws and workmanship errors, albeit often hidden for a long time, are usually the cause of the problems (Figure 4.25).

4.5.9.2 Modelling

In the late 1950s, H. Glaser advanced a simple rationale for checking interstitial condensation in cold storage walls. The method did well because most materials in such enclosure were neither capillary nor hygroscopic, while the assemblies were airtight, the differences in temperature and relative humidity only fluctuated modestly and vapour diffusion was the driver. In the 1960s, some started to use his rationale to judge building assemblies on moisture tolerance, which was a step too far. Building enclosures in fact face transient conditions, airtightness is not guaranteed, the materials used are often capillary and hygroscopic, gravity and pressure heads intervene, and so on. In later years more realistic boundary conditions, capillary redistribution and air egress as a driving force were included, giving birth to the many transient heat, air, moisture models that exist today.

Until the 1990s the Glaser method remained the reference for checking interstitial condensation tolerance of building assemblies, regardless of being far too simple because only airtight parts with non-hygroscopic inside finish, neither a capillary nor a hygroscopic outside cladding and closed cell insulation in between convened. A sad result was a vapour barrier mania that survived for decades. That diffusion only method includes many assumptions.

Geometry	1. Flat assembly composed of plane-parallel layers
Moisture flow	2. Vapour only. False for capillary-porous materials beyond critical moisture content
	3. Diffusion sole driving force. Valid as long as the assemblies and their layers are airtight. Testing and practice show that this is mostly untrue
	4. Hygroscopicity not considered. Eliminates hygric inertia and makes the vapour balance steady state under all circumstances. Softened as an assumption by applying monthly mean boundary conditions
	5. Vapour resistance factors constant. Does not fit with reality but makes calculation easy. An alternative is using different values depending on whether a layer sits at the inside or at the outside of the insulation
Heat flow	6. Equivalent conduction only. Excludes any form of enthalpy flow
	7. Latent heat release far too low to have impact
	8. Thermal inertia not involved. Heat flow is steady state, even under varying boundary conditions. Softened as an assumption by applying monthly mean boundary conditions
	9. Thermal conductivity of all materials constant. Does not fit with reality but makes calculation easy.

Regarding the boundary conditions, slope, orientation and shortwave absorptivity of the outside surface have to be known. The European standard presumes an outside surface protected from sun and under-cooling, so the monthly mean outdoor air temperature applies. A better approach is to use the equivalent temperature for condensation and drying (see Chapter 1). Also needed are the monthly mean vapour pressures outdoors for the location considered and the temperature and vapour pressure indoors. If no measured data is available, following annual mean and amplitude for the temperature indoors can be used:

Building type	Annual mean, °C	Annual amplitude, °C
Dwellings, schools, office buildings	20	3
Hospitals	23	2
Natatoriums	30	2

The monthly mean vapour pressure (p_i) indoors can be calculated if the monthly mean outdoors (p_e), the mean ventilation rate (n in ach), the hourly mean vapour release indoors ($G_{v,p}$), and the built indoor volume (V) are known:

$$p_i = p_e + \frac{R(273.15 + \theta_i)G_{v,p}}{nV}$$

Otherwise, the pivot vapour pressure for the indoor climate class that the building belongs to is used. For the temperature and that climate class related vapour pressure indoors the monthly means follow from:

$$= annual\ mean + annual\ amplitude \times C(t)$$

with $C(t)$ the time function, in Chapter 1 only given for Uccle (Brussels).

Consider first an assembly, which is dry at the start. Fix the thermal and diffusion resistances of all layers. Redraft the assembly in a [diffusion resistance/vapour pressure] axis system and calculate the temperature and saturation pressure in all interfaces (p_{sat}) for the coldest month of the year, typically January (southern hemisphere: July). Coupling the successive interface saturation pressures by straight lines mostly suffices as curve. Draft the vapour pressure line. If intersecting saturation, interstitial condensate is a fact. In this case, trace the tangents from the vapour pressures indoors (face $Z=Z_T$) and outdoors (face $Z=0$) to the saturation curve. A coinciding point of contact means condensate deposits in that one interface. Otherwise a tangent scan from highest to lowest point of contact fixes all interfaces or zones where deposit happens. The difference in slope per interface or zone between incoming and outgoing tangent then gives the monthly amount condensed:

$$m_c = 86400 d_{mo} g_c \quad (kg/m^2) \tag{4.29}$$

with g_c the condensing flux in kg/(m^2.s) and d_{mo} days per month.

With one condensing interface, add the successive amounts that during the colder months condense there and subtract the amounts that during the warmer months dry. Once the deposit left drops below zero, keep that zero during the months that follow until intersection with saturation restarts. Unless annual drying excludes accumulation, condensing will continue until a limit state is reached.

Two or more condensing interfaces complicate the counting. The annual total follows from the algebraic sum of the twelve incoming and outgoing tangent slopes, assuming vapour saturation at their contact points. Assessing the allowance requires the annual total to be redistributed over all condensing interfaces, according to the difference in slope between the arriving and departing tangent in each of them. If condensation zones were found, the derivative of the saturation curve defines the distribution of the deposit in them.

Move now to assemblies where-in one, sometimes more layers contain construction moisture, rain or rising damp. The calculation becomes more complex then. Take one wet layer. Again, redraft the assembly in a [diffusion resistance/vapour pressure] axis system and assume vapour saturation pressure in the wet layers. As evaporation happens in the interfaces with neighbouring layers, connect the wet interface at the inside straight on with the vapour pressure indoors and the wet interface at the outside straight on with the vapour pressure outdoors. If either one intersects saturation, wetness from the moist layer will condense elsewhere in the assembly. The next steps then reflect the dry wall case. Replace the intersecting straight line by the tangents coming from the wet interfaces and from the vapour pressure indoors or outdoors. The points of contact fix the condensation interface or zone, and so on.

More advanced but still not truly realistic is the combined diffusion and convection method, which handles the following assumptions:

Geometry	1. Assembly composed of plane-parallel layers
Moisture flow	2. Vapour only
	3. Exfiltration/infiltration and diffusion as driving forces
	4. No sorption/desorption
	5. Diffusion resistance factor constant
Heat flow	6. The result of equivalent conduction and enthalpy flow
	7. Latent heat release too low to be considered
	8. No thermal inertia
	9. Thermal conductivity constant

The calculation is often restricted to a reference cold week, for Uccle:

Temperature, °C	Relative humidity, %	Radiation, hor. surface, W/m^2	Surface film coefficient, W/(m^2.K)	Wind velocity (free field), m/s
−2.5	95	−30	17	3.8

Slope, orientation, composition, radiant properties of the outside surface, thickness, air permeance, equivalent thermal conductivity and equivalent vapour resistance factor of all layers must be known. First quantify the air flow rate through the assembly. For that, wind, stack and fan induced air pressure differences are needed. Then calculate the thermal and diffusion resistances of all composing layers. Temperature and vapour pressure indoors have to be known. If measured data or best guesses fail, the indoor climate class that the building or room belongs to fixes the values to be used.

If the assembly starts air-dry, calculate the temperature, saturation and vapour pressure in it, the last curving exponentially in the [diffusion resistance/vapour pressure] axis system. Interstitial condensation happens when the vapour pressure curve crosses saturation. If so, calculate the incoming and outgoing tangent exponentials from the vapour pressure indoors and outdoors to the interface where the distance from the vapour pressure exponential to the saturation curve is largest. If none of these two exponentials crosses saturation, condensate deposits in that one interface. Otherwise, a tangent exponentials scan must fix all intermediate condensing interfaces or zones. Quantify the amounts deposited (G_c in kg/(m^2.week)). For assemblies with wet layers, the evaluation runs according to the methodology for diffusion.

4.5.9.3 Performance requirements

Allowance is typically expressed in terms of the annual deposit per m^2 that has no harmful consequences:

Belgium	**Allowance, kg/m^2**
Accumulating deposit	*Limit state*
– Problem-free drainage possible	—
– Frost-resisting, d metre thick stony materials without vapour-tight finish outside	$m_c \leq w_{cr}d$
Winter deposit, no accumulation	*Maximum*
– Stony materials that are frost resisting and have a vapour-tight finish at the outside, stony materials that are not frost resisting (both d metre thick)	$\leq 0.05 w_{cr}d$
– Wood and moisture proof wood-based materials (plywood, particle board, fibre board, OSB)	$\leq 0.03 \rho d$
– Non-moisture proof wood based materials	≤ 0.05
– Foils with slope(s) below 15°	$\leq 0.4 - 0.3s/15$
– Foils with slope(s) beyond 15°	≤ 0.1
– Slightly moisture sensitive insulation materials (mineral fibre, EPS, XPS, PUR)	$\leq \max\left[\frac{12.5\lambda}{U_o(0.6-\lambda)}, 0.5\right]$

Switzerland	**Allowance, kg/m^2**
Accumulating deposit	Not allowed
Winter deposit, no accumulation	*Maximum*
– Stony materials	≤ 0.8
– Wood and moisture proof wood-based materials	$\leq 0.03 \rho d$
– Non-moisture proof wood based materials	0
– Non-capillary foils with any slope	0.02
– Slightly moisture sensitive insulation materials	$\leq 10d$

In both tables, w_{cr} is critical moisture content in kg/m^3, ρ density in kg/m^3, λ thermal conductivity and U_o the clear wall thermal transmittance of the assembly controlled.

4.5.9.4 Consequences for the building envelope

Most important is a good airtightness of all envelope parts. If this is the case, then a vapour retarder with the right diffusion resistance can still be needed at the warm side of the insulation, unless the diffusion resistance of the composing layers drops from the warmer, more humid to the colder, drier side. As a reminder, the warmer side is indoors in temperate and cold climates and outdoors in hot, humid climates. Of course, conflicts with other moisture sources are not unlikely. Rainproof, vapour-tight outside finishes that are combined with a vapour retarder inside may box up the insulation with layers containing construction moisture. Solar driven vapour flow can force rain buffered in outside finishes to condense against the vapour retarder's insulation side, and so on.

4.5.9.5 Remark

What's called interstitial condensation often stands for a local increase in sorption moisture or for wetness that suction has redistributed. In fact, condensate only gives droplets when deposited against non-porous layers such as metal claddings, glass, synthetic foils and finishes. Sucking layers instead must first become capillary wet before droplets form, but even then secondary uptake may still absorb some of the deposit.

4.5.10 All moisture sources combined

4.5.10.1 Modelling

Construction moisture, rain, rising damp, sorption/desorption, surface and interstitial condensation, all interact, so only combined models can simulate how assemblies really respond.

The geometry can be one-, two- or three-dimensional. Most models that consider airflow use the hydraulic circuit analogy. Properties needed are air permeability of all materials and the air permeances of cavities, leaks, cracks, composite layers, and so on. A few use computational fluid dynamics (CFD). Heat flow combines conduction, convection, radiation, enthalpy flow and latent heat. Thermal inertia is accounted for. Material properties needed are density, specific heat capacity as a function of moisture content, equivalent thermal conductivity as a function of moisture content and temperature and the radiant surface properties.

Moisture migrates as vapour and liquid. Below 0 °C transformation to ice occurs. Models that only deal with vapour flow simulate reality quite well as long as moisture content remains hygroscopic. Driving forces then are equivalent diffusion and convection. Many models, however, only consider diffusion, which excludes air leaky assemblies from simulation. Driving forces for liquid flow are temperature, suction, gravity and pressure heads, although most models only include suction. Hygric inertia is involved through sorption/desorption, with air transfers, if considered, as short-circuits. Material properties needed are the moisture retention curve, water vapour permeability and moisture conductivity, all as functions of the driving forces involved.

As boundary conditions, hourly, or even ten-minute values of all ambient parameters, including wind and rain, are needed. The contact conditions could include suction, diffusion, mixed or drained contact, albeit many models only consider suction between capillary materials and diffusion between a capillary and non-capillary or between two non-capillary materials

4.5.10.2 Performance requirements

The combined heat, air, moisture reality impacts end energy use, sometimes indoor environmental quality and durability. For insulated buildings the increase in thermal transmittances by humidification and air patterns developing should not be more than 0.1 U. For low energy buildings the limit is 0.05 U. Hygrothermal cracking harming rain-tightness, airtightness or aesthetics and favouring dirt accumulation must be avoided. Moisture content in layers subjected to freeze/thaw cycles should stay below the critical value for frost ($w < w_{cr,f}$), while relative humidity in moisture-proof wood-based materials must not reach 95%. In salt-laden stony materials relative humidity has to remain low enough to exclude hydration of the salts present. Changes in moisture content should also not activate problematic crystallization. Relative humidity in the interfaces between materials must remain under the mould germination value:

$$\phi_{\text{crit}} \geq min\left\{100;\ \left(0.033\theta_s^2 - 1.5\theta_s + 96\right)[1.25 - 0.0734\ \ln(T)]\right\}$$

with T the period considered in days. Relative humidity on wood should obey:

Temperature, °C	Mould probability on a scale of 0 to 1		
	RH ≤ 75%	RH 75–90%	RH > 90%
8–20	0	0–0.5	0.5–1
0–8	0	0–0.1	0.1–0.8
<0	0	0–0.02	0.02–0.5

while moisture content must not exceed fibre saturation. On floor coverings time-averaged relative humidity should remain below the values listed in Table 4.3.

Premature perforation of metallic finishes must be prevented. To give an example, after four years outdoors, weight loss by corrosion of an aluminium sheet will total (Δm in g/m^2):

$$\Delta m = 0.85 + 0.029 \times TOW \times [SO_2] \times [O_3] + 80 \times g_R \times \left[H^+\right]$$

with TOW the time fraction on a scale of 0 to 1 the sheet is staying wet and g_R the average precipitation in m per year. The terms between brackets represent gas concentrations in the outdoor air (μg/m^3). For an exponential increase with weight

Table 4.3 Acceptable time-averaged relative humidity in floor coverings.

Covering	RH, %
Synthetic carpet with mould sensitive lower layer	80
Non-alkali-resisting floor covers glued on a cement screed	
Six months periods	
Composite products	90
Homogeneous synthetic carpets	85
Shorter periods	
Composite products	95
Homogeneous synthetic carpets	90
Cork floor cover without synthetic protective layer	80
Cork floor cover with synthetic protective layer	85
Linoleum	85–90
Self-levelling screed	90–95
Synthetic carpet without mould sensitive layer, needle felt	99

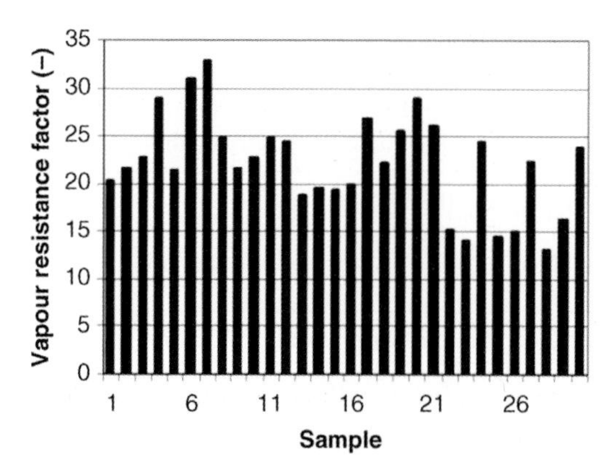

Figure 4.26 Dry cup vapour resistance factors measured for 30 bricks from the same batch.

loss zero at time zero, the loss after four years calculated with the formula given and assuming that half that loss is reality after one year, loss after x years can be quantified.

4.5.10.3 Why models still have limitations

In theory combined models could evaluate the moisture tolerance of building assemblies in its entire complexity, but there are many reasons why this still is not the case.

Looking to the material properties that models use, the embedded continuity approach assumes that each building and insulation material behaves as a collection of identical infinitesimally small representative elementary volumes (REVs). In reality, most of them are too inhomogeneous and too anisotropic for that. As a consequence, the single-valued properties and property relationships used show uncertainty, which Figure 4.26 illustrates by showing the dry cup vapour resistance factor measured for 30 bricks from the same batch.

The deviations are large. So, what value should be used for calculation? The arithmetic average? The harmonic average? For a half-brick wall, the latter wins. Water vapour diffusion through a square metre of bricks without joints in between in fact can be written as:

$$g = \sum_{i=1}^{67} \left(\frac{A_{\text{brick}} \Delta p}{\mu_i dN} \right) = \frac{\Delta p}{\mu_{\text{m}} dN}$$

giving:

$$\frac{1}{\mu_{\text{m}}} = \sum_{i=1}^{67} \frac{A_{\text{brick}}}{\mu_i} \tag{4.30}$$

For a one or a one and a half brick wall things are less clear.

The brick properties are often considered as representative for masonry. Table 4.4 compares the harmonic mean of the 30 measured with the vapour resistance thickness of a veneer wall brick-laid with these bricks. The difference is striking. Table 4.5 compares their air

Table 4.4 Bricks and masonry: vapour diffusion thickness.

Material	Thickness, m	Density, kg/m^3 %→	Vapour diffusion thickness, μd in m				
			$\phi=28$	$\phi=54$	$\phi=73$	$\phi=82$	$\phi=91$
Brick		1900	2.89		2.31		
Half-brick veneer	0.09			1.20		0.51	

permeance. Whereas the bricks are airtight, the half-brick veneer is not. Reasons for these deviating values include badly filled mortar joints and micro-cracks around the bricks.

Characteristics defining the moisture response of open-porous materials are specific moisture capacity and moisture permeability. The first is the derivative of a spot-wise measured water retention curve. Differentiating starts from a fitted approximate function, which makes the derivative even more approximate. Hysteresis also affects that capacity. The second is calculated from the diffusivity, which follows from a [position/time] scan using γ-ray, x-ray or NMR of successive moisture profiles in a material sample during suction or drying. Most profiles include a flat part at high and low moisture content with a steep front in between. The weak slope and scatter in measured values in both flat parts make correct calculation of the diffusivity difficult. The steep part indicates that diffusivity increases suddenly there. How suddenly is difficult to assess, which is why a Boltzmann transform is used that unifies all moisture profiles into one, although this is again loaded with uncertainty.

There are real differences between the assemblies as built and as modelled. Their geometry as built, including joints, cracks, voids, leaks and air spaces, is never known, which is why models consider idealized sections, see the cavity wall of Figure 4.27. In this way, air inflow and outflow, air looping, wind washing, water ingress by gravity and pressure heads, water run-off and drainage are too often overlooked.

The contact conditions between layers are mostly unknown. Most models therefore assume perfect hydraulic contact, which in reality is the exception. More likely is that a thin air space is left. Only air and vapour are transmitted then, except if condensation or water ingress from outside fills that space causing gravity-activated water run-off. The air space might also belong to a network of cracks, joints and voids that cross the assembly

Table 4.5 Bricks and veneer wall: air permeance ($K_a = C\Delta P_a^{n-1}$).

Material	Thickness m	Density kg/m^3	Air permeance K_a $kg/(m^2.s.Pa)$	
			C	n
Brick		1900	0	1
One-brick veneer	0.09	135	0.4×10^{-4}	0.81

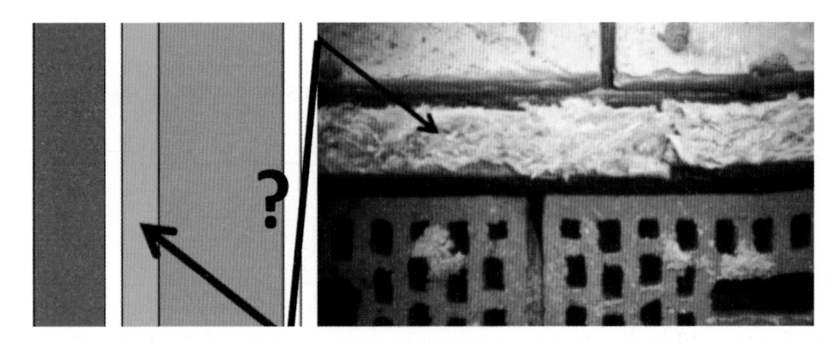

Figure 4.27 Filled cavity walls: real assembly versus dummy, used in modelling.

and activate air flow, often with detrimental heat, air and moisture effects. Likely is natural contact, characterized by a random distribution of air voids that includes all the inconveniences that thin air spaces have and perfect hydraulic contact spots. Also possible are real contacts with inter-penetration of materials moulded together, such as bricklaying and rendering. These interpenetrations behave as an additional layer with unknown properties

Regarding the boundary conditions, predicting transient wind driven rain patterns on a building enclosure remains hard. And even if doable with the actual combined CFD/droplet tracing models, major questions still remain. What happens when rain droplets hit a surface? How much water splashes away? How important is evaporation? Does a surface suck until capillary saturated before run-off, wetting parts underneath, starts, or does each single drop activates suction and run-off? Why does run-off go into fingering? How does rain seep across veneer walls (Figure 4.28). These are all important questions because wind-driven rain is often the main moisture load.

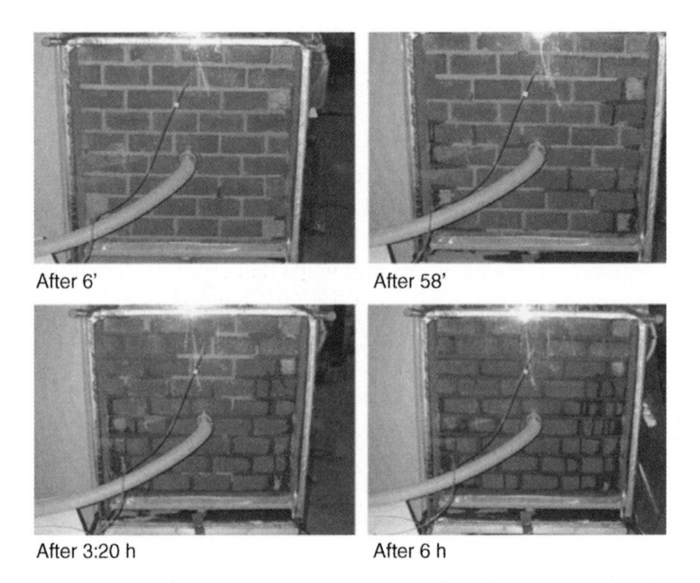

Figure 4.28 Concrete block veneer wall: seeping of the head joints.

Most models handle the ambient conditions as time-dependent only. This is untrue. Temperatures change with height. Surface film coefficients differ spot wise. Parts of an enclosure stay shadowed by parapets and overhangs while others see the sun for hours. Wind-driven rain mainly hits the upper and corner parts, and so on. This overall randomness affects the enclosure's response.

The indoor conditions used in turn often combine a constant temperature with given vapour concentration and constant air pressure differences. However, night setback and excessive solar gains subject the indoor temperature to daily periodicity. Vapour pressure differences with outdoors, though dampened by all sorption-active surfaces, furniture and furnishings indoors, continue oscillating due to changes in vapour release, ventilation rate and air exchanged between spaces. Air pressure differences not only depend on wind and temperature differences but also on building layout, the ventilation system and air leak distribution over enclosure and fabric. Air flow patterns couple the hygrothermal load of each assembly to the whole building response and the user habits. This randomizes the response of all enclosure parts.

Whereas heat transmission takes hours and moisture movement days, or even weeks, air flow takes seconds. This induces numerical problems, which is why many models overlook air as a carrier of heat and vapour.

Gravity and pressure flow drive water through cracks, voids, open joints, and so on. Activation demands a water film or water head. Rain run-off along the exterior surface may induce random seepage across cracks in the rain screen. Head joints in brick veneers so become water filled, giving backside run-off. This all-sided brick wetting leads to faster and more rain stored than most models calculate (Figure 4.29). Where run-off sticks to mouldings, water heads develop that further facilitate weeping, a fact that no model predicts. Consequences in terms of leakage are often problematic.

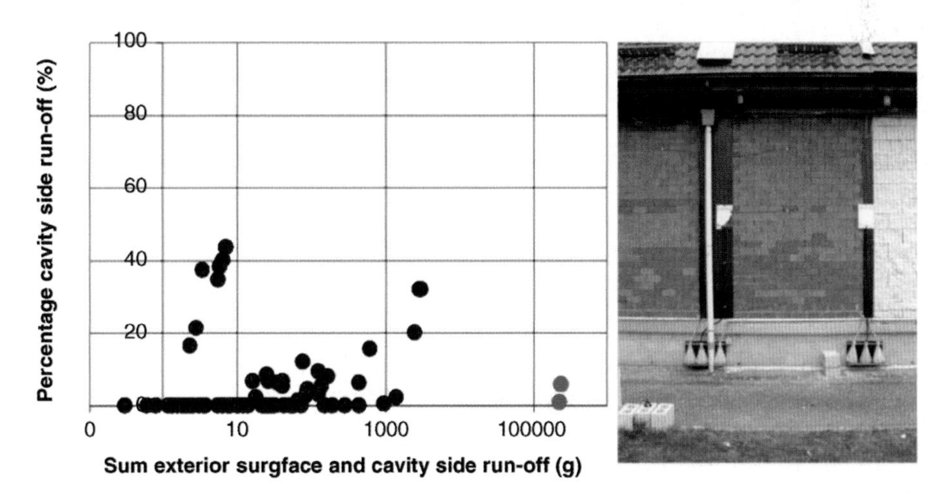

Figure 4.29 Cavity run-off as measured on site for a cavity wall with concrete block veneer and a mineral wool full cavity fill.

Finally, practice everytime again shows that moisture impacts durability. Yet, many tools neither predict which deficiencies the heat, air, moisture response may cause nor do they evaluate tolerability. In general, this depends on the construction's limit state properties and load severity. Loads are variable, while the actual properties vary considerably and degrade with time, meaning that for a given load, safety slowly decreases. As a consequence, decay and service life are random facts.

4.5.10.4 Three examples where full models were hardly of any help

Interesting is to look to what extend the actual models are of help in solving moisture issues. Therefore, three buildings where problems surfaced are discussed in some depth: a university building, an office and a villa

The university building housed a wide range of facilities: underground parking, lecture theatres, library, seminar rooms and individual offices. The design team proposed a volume that narrowed from basement to top, with the lecture theatres just above the parking, the next floor containing the library and seminar rooms and all offices filling the upper storeys. The result was a building with oblique outer walls, composed of a brick veneer, a cavity partly filled with PUR-boards and a concrete inside leaf with brick finish (Figure 4.30).

The complaints included large moisture spots formed on the inside brick finish and rain penetrating along the window sills (Figure 4.30). What were the causes? A couple of years earlier, the veneer was given a water repellent treatment. Because of that intervention the oblique facades functioned as drainage planes with concentrated run-off at their sides (Figure 4.30). That water film penetrated the cracks between bricks and mortar, leaked into the cavity, where it dripped on the insulation, ran off, seeped across the joints between boards and wetted the concrete inside leaf. There shrinkage cracks directed the water to the inside brick finish. At the same time, lack of upward sill-folds caused rain penetration under the windows (Figure 4.30). None of this was predicted by any full heat, air, moisture tool. In an attempt to cure this, a storey-wise regressing veneer replaced one of the oblique ones (Figure 4.31). The appearance was awful, while the repair induced thermal bridging, lifting the whole wall thermal transmittance from 0.49 to 0.64 W/(m^2.K). Lacking cavity trays at the bottom of each storey-high part also gave opportunity for further leakage. The final solution proposed therefore was to replace all brick veneers by zinc sidings fixed on timber joists, the bays in between filled with insulation, and exchanging the zinc sills below the windows by new ones with a back dam. The whole repair, however, was so expensive that the university decided to vacate the building.

The five storeys high office building had a south-west-facing rear facade, which consisted of aluminium window strips alternating with cavity wall strips, consisting of a brick veneer, a cavity filled with 7.5 cm glass fibre boards and a precast concrete inside leaf. The windows were mounted flush with the veneer (Figure 4.32(a)). The main complaint was that, when the sun shone after a few rainy days, the rear side's highest floor had water dripping down the windows onto the inside sills. The lime that this water contained soiled the glass and sills to such extend that the renting company left the floor (Figure 4.32(b)).

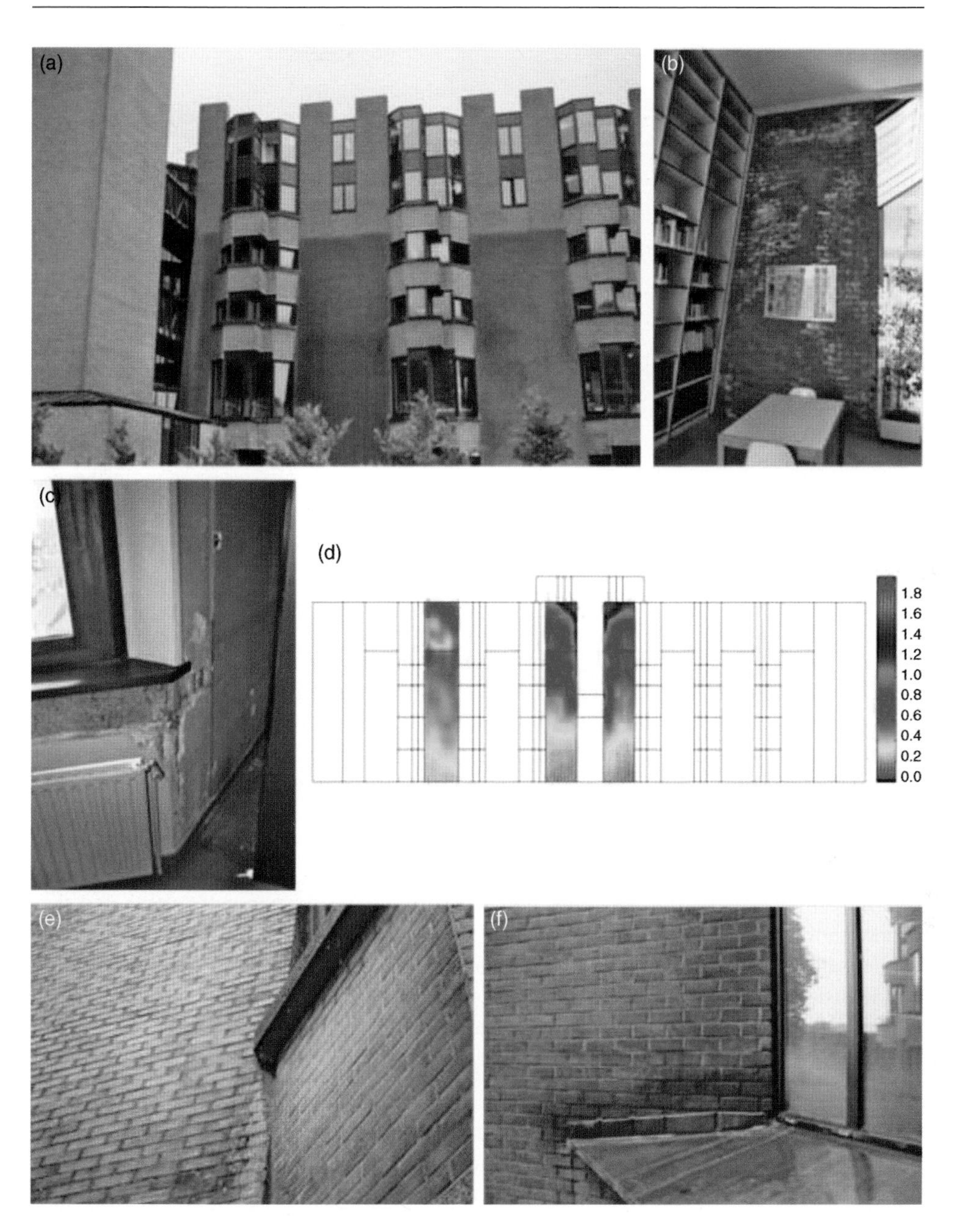

Figure 4.30 (a) The building, (b) moisture spots, (c) leakage along window sills, (d) catch ratio, (e) run-off, (f) no upward sill folds.

Figure 4.31 Solution, first trial.

Why did this happen? In north-west Europe wind-driven rain mainly comes from south-west. A CFD droplet tracing simulation showed that the rear side's highest floor was hardest hit (Figure 4.32). The capillary veneer now acted as a rain buffer. But the south-west is also the sunny side, which is why the veneer suffered from solar-driven vapour flow to the inside each time the sun appeared. The expected result was first hygroscopic loading of the concrete leaf, followed by sucking any further deposit. But dripping inside? A cavity tray contacting the underside of the concrete should have drained the water film formed, if any, to the outside. However, a detailed check revealed that the tray was not glued against the concrete, while that concrete was so well compacted that condensation produced droplets. As the windows sat flush with the veneer, the condensate could not do anything other than run off behind the tray and wet the glass on the inside and the sills below. Although combined models predicted the solar driven vapour flow, with the properties listed in catalogues the concrete should have buffered the related condensation deposit. Moreover, the tray details that caused dripping could not be modelled.

The best cure was to tear down the upper part of the veneer, remove the insulation, glue the cavity tray against the inside leaf, remount the insulation, rebuild the veneer and repair the windows. A less effective alternative consisted of rendering the building facade with water repellent stucco and repairing the windows. The least solid but cheapest solution was brushing the veneer with a water-repellent product and repairing the windows. This last repair method was the one chosen.

The villa had filled cavity walls, double-glazed timber windows, an insulated tiled roof, underfloor heating and a garden and terrace sloping towards the rear facade (Figure 4.33).

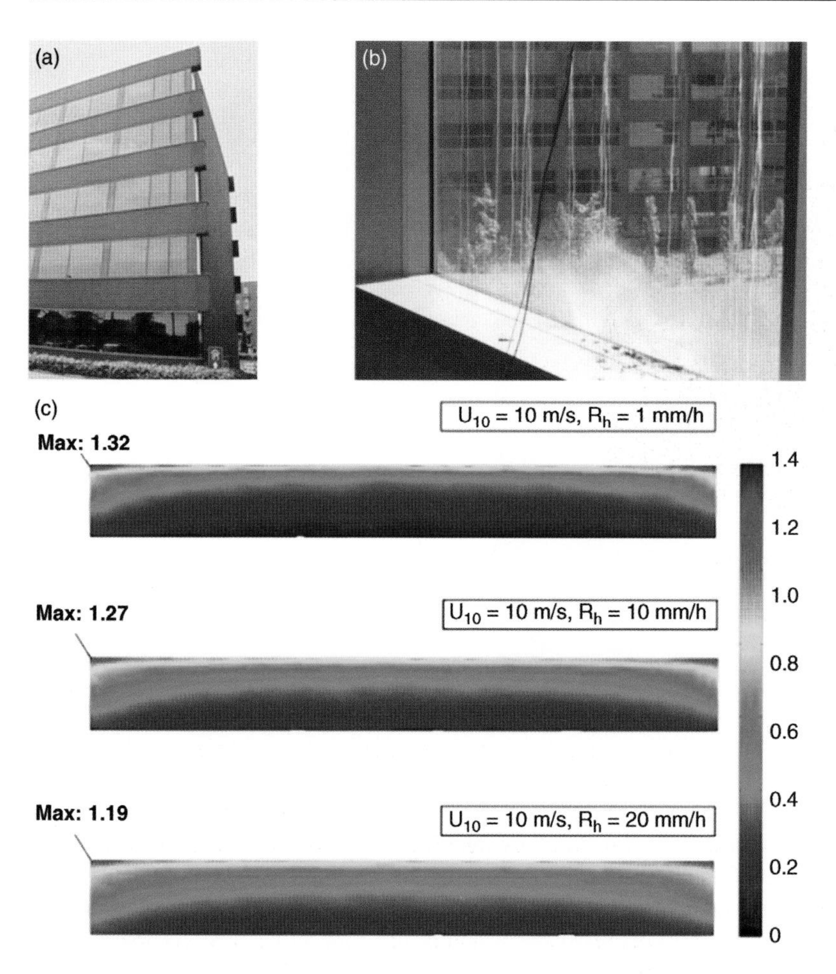

Figure 4.32 (a) The building at the street side, (b) lime deposit on the glazing, (c) results of a wind-driven rain calculation.

Remarkably, the inside leafs and partition walls on the groundfloor suffered from rising damp (Figure 4.33(c)). An on-site analysis showed that cavity and partitions walls had a waterproof insert at floor level. Digging along one of the sidewalls proved that there was no ground water. Previous damage cases, however, showed that gardens and terraces sloping towards a building drain rainwater straight to the outer walls, where it seeps through when both the waterproofing and cavity tray sit above the floor deck, and that was the case here. The thermal insulation below the underfloor heating facilitated the water spread over the concrete deck and suction by inside leafs and partition walls from below the waterproof membrane. Also the outlet of the washing machine that was concreted in the screed leaked, as an analysis of the salt efflorescence on nearby partitions proved. They mainly consisted of sodium and sulphates, with sodium sulphate

Figure 4.33 (a) Villa, (b) sloping garden and terrace, (c) rising damp in a partition wall.

being a corrosion inhibitor in washing powder. Again, none of the combined heat, air, moisture tools was of any help. Curing included water-tightening the below grade walls up to the cavity tray and installation of a foundation deep drain all around the villa, coupled to the street sewerage. Also the washing machine outlet was repaired.

4.6 Thermal bridges

4.6.1 Definition

The term 'thermal bridge' refers to envelope spots where heat transfer develops two- or three-dimensionally with as a result in temperate and cold climates a larger heat flow and lower inside surface temperatures than across and on adjacent opaque parts. The consequences of this are more heat lost and often very low surface temperatures. The first happens thanks to the increase in the envelope's mean thermal transmittance, while the second is evaluated using as property the temperature factor (f_{hi}):

$$f_{h_i} = \frac{\theta_{si,min} - \theta_e}{\theta_i - \theta_e} \quad (-) \tag{4.31}$$

The lower its value, the more sensitive a thermal bridge is to dust deposit, mould, surface condensation and local crack formation.

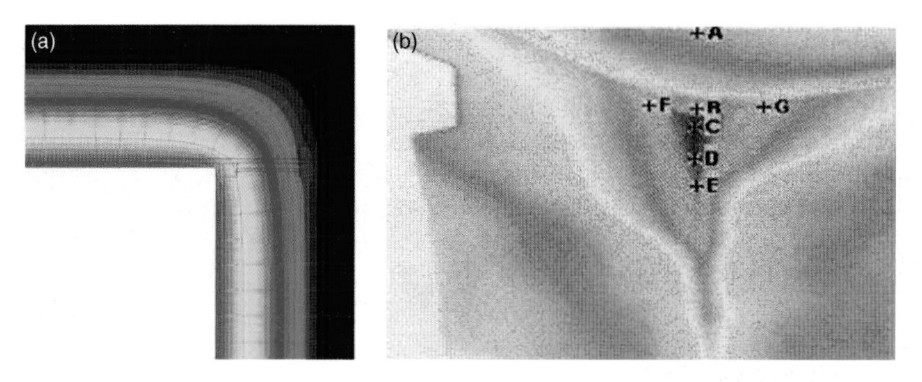

Figure 4.34 Geometric thermal bridges, (a) an edge between two facade walls (calculated temperature field), (b) a corner between two facade walls and a floor (IR-picture).

Structural thermal bridges differ from geometric ones. The first reflect the need for structural integrity, the second are a consequence of the building's geometry (Figure 4.34).

4.6.2 Performance requirements

In temperate climates the lowest acceptable temperature factor, 0.7, limits mould risk to some 5%, provided ventilation is normal and the daily mean temperature indoors is more than 12 °C. If both conditions are maintained, double glazing will show surface condensation rather than thermal bridges will suffer from mould.

4.6.3 Consequences for the envelope

There are two principles involved here. First, eliminate structural thermal bridges as much as possible by applying thermal breaks. This requires a continuous insulation. For that, it must be possible to trace a line on the drawings along the envelope in section and floor plan without leaving the insulation. Of course, structural safety or buildability often makes complete avoidance impossible. Tricks then are long conduction paths (less heat loss), and having the inside surface larger than the outside one (higher surface temperature). Secondly, neutralize geometric thermal bridges thanks to a correct insulation of the envelope.

4.7 Contact coefficients

Contact coefficients are important for foot comfort. Preferred for floor finishes is a low value. This is less important for walls, although high values turn each contact into a cold experience. Aluminium inside finishes, for example, never gained market share because of their high contact coefficient ($b = 23\,380$ J/(m^2.K.s$^{0.5}$)). The same holds for cast concrete. Inhabitants call it cold, not by lack of insulation and thus a low inside surface temperature but because of its high contact coefficient ($b = 2440$ J/(m^2,K.s$^{0.5}$).

4.8 Hygrothermal stress and strain

That materials deform and become stressed and strained under humidity and temperature changes cannot be avoided. What to exclude are cracks that degrade performance and aesthetics. Therefore, stresses should at no point exceed tensile strength. Translation into performance metrics, however, remains difficult. Acceptability for an outside stucco, for example, differs according to factors such as its nature (water repellent or not), the substrate (capillary or not) and the colour. (Figure 4.35).

Predicting stress and strain and evaluating cracking risk is demanding. The analysis combines transient heat and moisture modelling with stress/strain calculations due to hindered deformation, taking into account cracking conditions, fatigue, creep and relaxation. Nonetheless, the smaller the differences in temperature and relative humidity, the lower stress and strain and the less probable cracking will be. That apparently simple requirement involves two principles: keep temperature and relative humidity fluctuations in loadbearing and massive parts as small as possible and, choose solutions that absorb related loading without cracking for layers subjected to large hygrothermal fluctuations.

Applying the first principle when insulating massive envelopes shows that outside insulation is more effective than inside insulation, because the latter induces much larger temperature and relative humidity fluctuations in the massive parts. Of course, initial shrinkage due to drying of construction moisture and chemical binding cannot be avoided. The second principle implies that extra attention should be given to all layers at the outside of the insulation. Stucco must be deformable, and so needs a reinforcing net. A veneer wall wins in quality when it is frost resisting and somewhat deformable. Scaled and plated elements are fixed in such a way that deformations don't add, and so on. Things become more complicated with constructions where the outside finish has a massive wall as its substrate. The wall then determines how the finish deforms. Coupling

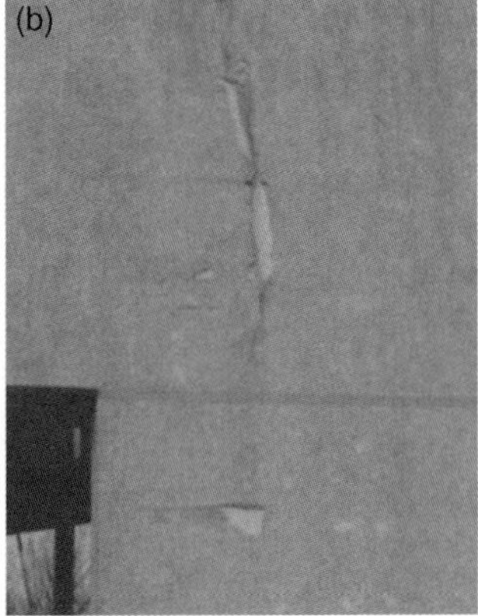

Figure 4.35 (a) Cracking disturbing aesthetics of black-painted stucco, (b) cracking impacting rain-tightness.

therefore should be as weak as possible, which is why the best stucco consists of a base layer with low modulus of elasticity and high deformability and a top layer that is stiffer and harder.

4.9 Transparent parts: solar transmittance

4.9.1 Definition

Transparent parts such as glass act as an opening that transmits part of the incident solar radiation, while what's absorbed warms the panes included the inside surface and thus also causes indirect gains by convection and longwave radiation. Both fix the solar transmittance (g) of the transparent part, see 'Building Physics: Heat, Air and Moisture'. For single glass:

$$g = \tau_S + \alpha_S h_i/(h_i + h_e) \approx \tau_S + 0.205\alpha_S$$

where τ_S is the glass transmissivity and α_S its absorptivity, both short wave, while h_i and h_e are the surface film coefficients, respectively indoors and outdoors. For double glazing, triple glazing and glass plus solar shade combinations, calculating their solar transmittance demands more complex modelling.

4.9.2 Performance requirements

In temperate and cold climates, the solar transmittance must accommodate two conflicting requirements: being as close as possible to 1 for energy efficiency during the heating season, and being as far as possible below 1, though without hindering daylighting, to avoid overheating during the warm half-year and to ensure energy efficiency when active cooling is needed. The best choice may follow from a combined end energy and summer comfort analysis, using transient building energy software. Judging overheating is, for example, done by calculating the number of weighted temperature excess hours (WTE), the sum of the hourly excess factors (EF) during the time a year a space is occupied:

$$\begin{array}{ll} |PMV| \le 0.5 & EF = 0 \\ |PMV| > 0.5 & EF = 0.47 + 0.22|PMV| + 1.3|PMV|^2 + 0.97|PMV|^3 - 0.39|PMV|^4 \end{array} \tag{4.32}$$

PMV in these formulas is the predicted mean vote (see the section on thermal comfort in Chapter 3). All other parameters fixed, the glass area and solar shading are chosen so that the EF-sum does not exceed 100. The following table with f_{gl} the south-facing glass-to-facade ratio (A_{gl}/A_{fac}) helps in rating the solar transmittance (g) that is recommended for the warm half-year:

Inside partitions	gf_{gl}	
	Low ach	High ach
Light	0.12	0.17
Heavy	0.14	0.25

4.9.3 Consequences for the envelope

Solar shading is the way to go, either by using sun absorbing or reflecting glass or applying movable outside screens or fixed outside shading elements. If reflecting glass is used, care should be taken not to design curved facades that focus the reflected solar beams because temperatures at the focal point may then reach unacceptably high values.

Of importance also is to search for an optimal equilibrium between the solar and light transmittance. Neglecting the second in office and school buildings not only increases the electrical energy used for artificial lighting but also lifts the internal gains, which in turn may require more active cooling. Of course, too much solar gain can also end up requiring active cooling.

Further reading

A.J. Metric Handbook, (1969) The Architectural Press, London.

Abuku Masaru (2009) Moisture stress of wind-driven rain on building enclosures, PhD thesis, KU Leuven.

American Hotel & Motel Association (1989) Mold & Mildew in Hotels and Motels.

Andersen, N.E., Christensen, G. and Nielsen, F. (1984) *Bygningers Fugtisolering, SBI-Anvisning 139*, Statens Byggeforskningsinstitut, Horsholm (in Danish).

Anon (1991) Condensation and Energy, Guidelines and Practice, Final Report IEA, EXCO ECBCS Annex 24, ACCO, Leuven.

Anon (1996) Heat, Air and Moisture Transport Through New and Retrofitted Insulated Envelope Parts, Common Exercises, Final Report Task 1 IEA, EXCO ECBCS Annex 24, ACCO, Leuven.

ASHRAE (2001) Handbook of Fundamentals, Chapter 23, Atlanta.

ASHRAE (2005) Handbook of Fundamentals, Chapter 23, Atlanta.

ASHRAE (2009) Handbook of Fundamentals, Chapter 23, Atlanta.

ASHRAE (2013) Handbook of Fundamentals, Chapter 23, Atlanta.

Blocken, B. (2004) Wind-driven rain on buildings, measurements, numerical modelling and applications, PhD thesis, KU Leuven, 323 p.+add.

Blocken, B., Hens, H. and Carmeliet, J. (2002) Methods for the quantification of driving rain on buildings. *ASHRAE Transactions*, **108** (Part 2), 338–350.

Bouwbesluit, Staatsblad 680 (1991) Staatsdrukkerij, Den Haag, Dec. (in Dutch).

Carmeliet, J. (1992) *Duurzaamheid van weefselgewapende pleisters voor buitenisolatie, een probabilistische benadering op basis van de niet locale continue schademechanica*, doctoraal proefschrift, KU Leuven (in Dutch).

Carmeliet, J., Hens, H. and Vermeir, G. (eds.) (2003) *Research in Building Physics*, A.A. Balkema Publishers, Lisse/Abingdon/Exton/Tokyo.

Derome, D., Carmeliet, J. and Karagiosis, A. (2007) Cyclic temperature gradient driven moisture transport in walls with wetted masonry cladding, Proceedings Buildings X Conference, Clearwater Beach (CD-ROM).

Fazio, P., Hua, G., Rao, J. and Desmarais, G. (eds.) (2006) *Research in Building Physics and Building Engineering*, Taylor&Francis, London/Leiden/New York/Philadelphia/ Singapore.

Fazio, P. (2008) Developing a standardized test to evaluate the relative hygrothermal performance of different building envelope systems, Proceedings Building Physics Symposium, Leuven, 29–31 October, pp. 63–66.

Glaser, H. (1958) Wärmeleitung und Feuchtigkeitsdurchgang durch Kühlraumisolierungen, Kältetechnik, 3/1958, pp. 86–91 (in German).

Hagentoft, C.E. (1998) Heat, Air and Moisture Transport Through New and Retrofitted Insulated Envelope Parts, Performances and Practice, Final Report Task 5 IEA, EXCO ECBCS Annex 24, ACCO, Leuven.

Hendriks, L. and Hens, H. (2000) Building Envelopes in a Holistic Perspective, Final Report IEA, EXCO ECBCS Annex 32, ACCO, Leuven.

Hens, H. (1984) Buitenwandoplossingen voor de residentiële bouw, De spouwmuur, Rapport R-D Energie (in Dutch).

Hens, H. (1989) Heat, Air, Moisture Transport in Building Components, Seminar U.Lg.

Hens, H. (1991) Analysis of causes of dampness, influence of salt attack, Seminar 'L'umidita ascendente nelle murature: fenomenologia e sperimentazione', Bari.

Hens, H. (1996) Heat, Air and Moisture Transport Through New and Retrofitted Insulated Envelope Parts, Modeling, Final Report Task 1 IEA, EXCO ECBCS Annex 24, ACCO, Leuven.

Hens, H. (1999) Fungal defacement in buildings, a performance related approach. *International Journal for Heating, Ventilation, Air Conditioning and Refrigeration Research*, **5** (3), 256–280.

Hens, H., Roels, S. and Desadeleer, W. (2005) Glued concrete block veneers with open head joints: rain leakage and hygrothermal performance. Proceedings of the Nordic Building Physics Conference, Rejkjavik, June.

Hens, H. (2007) Does heat, air, moisture modelling really helps in solving hygrothermal problems? Keynote, Rakennusfysiikka, Tampere, 18 October.

Hens, H. (2010) Wind-driven rain: from theory to reality, Proceedings Buildings XI Conference, Clear Water Beach, Florida, 5–9 December (CD-ROM).

Hens, H. (2012) Actual limits of HAM-modelling looking at problems encountered in practice, part 1. *IBPSA-news*, **22** (2), 24–40.

Hens, H. (2013) Heat, Actual limits of HAM-modelling looking at problems encountered in practice, part 2. *IBPSA-news*, **23** (1), 22–33.

Hens, H., Roels, S. and Desadeleer, W. (2004) Rain Leakage through veneer walls, built with concrete blocks, Proceedings CIB-W40 meeting, Glasgow.

Holm, A. (2001) Ermittlung der Genauigkeit von instationären hygrothermischen Bauteilberechnungen mittels eines stochatischen Konzeptes, Doktor-Ingenieurs Abhandlung, Universität Stuttgart, 129 pp. (in German).

Houvenaghel, G., Hens., H. and Horta, A. (2004) The impact of airflow on the hygrothermal behavior of highly insulated pitched roof systems, Proceedings of the Performance of the Exterior Envelopes of Whole Buildings IX Conference, Clear Water Beach, Florida.

Interindustriële studiegroep voor de Bouwnijverheid IC-IB (1979) Prestatiegids voor gebouwen, delen 1 tot 9, WTCB (in Dutch).

Janssens, A. (1997) Reliable Control of Interstitial Condensation in Lightweight Roof Systems. PhD thesis, KU Leuven.

Janssens, A. and Hens, H. (2007) Effects of wind on the transmission heat loss in duo-pitch insulated roofs: a field study. *Energy and Buildings*, **39**, 1047–1054.

Karagiosis, A., Desjarlais, A. and Lstiburek, J. (2007) Scientific Analysis of Vapour Retarder Recommendations for Wall Systems Constructed in North America, Proceedings Buildings X Conference, Clearwater Beach (CD-ROM).

Koci, V., Madera, M., Keppert, M. and Cerny, R. (2010) Mathematical models and computer codes for modelling heat and moisture transport in building materials: a comparison, Proceedings of CESBP 2010, pp. 93–99.

Künzel, H. (1985) *Regenschutz von Außenwänden durch mineralische Putze*, Der Stukkateur, Heft 5/85 (in German).

Künzel, H. (1994) Verfahren zur ein- und zweidimensionalen Berechnung des gekoppelten Wärme- und Feuchtetransport in Bauteilen mit einfachen Kennwerten, Doktor Abhandlung, Universität Stuttgart, 104 pp.+fig. (in German).

Langmans, J. (2013) Feasibility of exterior air barriers in timber frame construction, PhD thesis KU Leuven.

Langmans, J., Nicolai, A., Klein, R. and Roels, S. (2012) A quasi-steady state implementation of air convection in a transient heat and moisture building component model. *Building and Environment*, **58** (12), 208–218.

Lecompte, J. (1989) De invloed van natuurlijke convective op de thermische kwaliteit van geïsoleerde spouwmuurconstructies, doctoraal proefschrift KU Leuven (in Dutch).

Majoor, G.J.M. and de Jong, A. (2002) Herziene normen gepubliceerd voor thermische isolatieberekening en EPN. *Bouwfysica*, **15** (2), (in Dutch).

Mamillan, M. and Boineau, A. (1974) Etude de l'assèchement des murs soumis à des remomtées capillaires, paper 2.2.2, CIB-RILEM 2d International Conference on Moisture Problems in Buildings, Rotterdam (in French).

Neufert (1964) *Bau Entwurfslehre*, Verlag Ullstein, Berlin (in German).

Nevander, L.E. and Elmarsson, B. (1981) *Fukthandbok*, Svensk Byggtjänst, (in Swedish).

Nevander, L.E. and Elmarsson, B. (1991) Fuktdimensionering av träkonstruktioner, Riskanalys, Byggforskningsrådet, rapport R38 (in Swedish).

NRC-Canada (1982) Exterior Walls, Understanding the Problems, Proceedings of the Building Science Forum '82, 63 pp.

NRC-Canada (1983) Humidity, Condensation and Ventilation in Houses, Proceedings of the Building Science Insight '83, 66 pp.

NRC-Canada (1989) An Air Barrier for the Building Envelope, Proceedings of the Building Insight '86, 24 pp.

PATO (1986) Syllabus 'Leidt energiebesparing tot Vochtproblemen', Delft, 30 Sept.–1 Oct. (in Dutch).

Peer, L. (2007) Use of Thermal Breaks in Cladding Support Systems, Proceedings Buildings X Conference, Clearwater Beach (CD-ROM).

Poupeleer, A.S. (2007) Transport and crystallization of dissolved salts in cracked porous building materials, PhD thesis, KU Leuven.

Sagelsdorff, R. (1989) Wasserdampfdiffusion, Grundlagen, Berechnungsverfahren, Diffusionsnachweis, SIA-Dokumentation D 018, Zürich (in German).

Sanders, C. (1996) Heat, Air and Moisture Transport Through New and Retrofitted Insulated Envelope Parts, Environmental Conditions, Final Report Task 2 IEA, EXCO ECBCS Annex 24, ACCO, Leuven.

Sedlbauer, C. (2001) Vorhersage von Schimmelpilzbildung auf und in Bauteilen, Doktor-Ingenieurs Abhandlung, Universität Stuttgart (in German).

Standaert, P. (1984) *Twee- en driedimensionale warmteoverdracht: numerieke methoden, experimentele studie en bouwfysische toepassingen*, doctoraal proefschrift, KU Leuven (in Dutch).

Straube, J.F. (1998) Moisture Control and Enclosure Wall Systems, PhD thesis, University of Waterloo.

Tariku, F., Cornick, S. and Lacasse, M. (2007) Simulation of Wind-driven Rain Penetration Effects on the Performance of a Stucco-clad wall, Proceedings Buildings X Conference, Clearwater Beach (CD-ROM).

Uitvoeringsbesluiten Isolatiedecreet (1991) (in Dutch).

Van Den Bossche, N. (2013) Watertightness of building components: principles, testing and design guidelines, PhD thesis UGent.

Van Mook, F. (2003) Driving rain on building envelopes, Doctoraal proefschrift TU/e (Bouwstenen 69).

Verbruggen, A. (1988) *Investeringsanalyse*, Rationeel energiegebruik in kantoorgebouwen, Wilrijk (in Dutch).

Vereecken, E. (2013) Hygrothermal analysis of interior insulation for renovation projects, PhD thesis KU Leuven.

Vinha, J. (2007) Hygrothermal Performance of Timber-Framed External Walls in Finnish Climatic Conditions, PhD thesis, Tampereen Teknillinen Yliopisto.

Williams, M. (2007) Developing Innovative Drainage and Drying Solutions for the Building Enclosure, Proceedings Buildings X Conference, Clearwater Beach (CD-ROM).

Winnepenninckx, E. (1992) Het Vlaamse isolatie- en ventilatiedecreet, onuitgegeven eindwerk, IHAM (in Dutch).

WTCB-tijdschrift (1982) De rekenmetode van Glaser, 1/82, pp. 13–24 (in Dutch).

Zheng, R., Carmeliet, J., Hens, H., Janssens, A. and Bogaerts, W. (2004) A hot box-cold box investigation of the corrosion behavior of highly insulated zinc roofing systems. *Journal of Thermal Envelope and Building Science*, **28** (1), 27–44.

Zheng, R., Janssens, A., Carmeliet, J., Bogaerts, W. and Hens, H. (2010) Performances of highly insulated compact zinc roofs under a humid-moderate climate: Part I: hygrothermal behavior. *Journal of Building Physics*, **34** (2), 178–191.

Zheng, R., Janssens, A., Carmeliet, J., Bogaerts, W. and Hens, H. (2011) Performances of highly insulated compact zinc roofs under a humid moderate climate – part II: corrosion behaviour. *Journal of Building Physics*, **34** (3), 277–293.

Zillig, W. (2009) Moisture transport in wood using a multiscale approach, PhD thesis KU Leuven.

5 Timber-framed outer wall as an exemplary case

5.1 In general

This chapter illustrates how the heat, air, moisture metrics at the envelope level, discussed in the previous chapter, help in adjusting the design of timber-framed outer walls. Timber-framed construction has gained some popularity in north-west Europe, mainly because it allows the inclusion of thick insulation layers without extending the walls to uneconomic thicknesses. The price paid compared to massive walls insulated outside and to filled brick cavity walls is much poorer thermal storage capacity and higher moisture sensitivity.

5.2 Assembly

Inside to outside, timber-framed outer walls are composed as follows (see also Figure 5.1):

Layer	d	λ	μ	Air permeance $K_a = a\Delta P_a^{b-1}$ kg/(m^2.s.Pa)	
	m	W/(m.K)	–	a	$b-1$
Gypsum board					
No leaks	0.012	0.2	7	3.1×10^{-5}	−0.19
1 leak ϕ 20 mm per m^2				3.8×10^{-4}	−0.39
Thermal insulation (glass wool)	?	0.04	1.2	2.3×10^{-3}	−0.11
Sheathing (plywood)	0.022	0.14	20–30	5.4×10^{-4}	−0.46
Building paper	←		?		→
Cavity	0.04		0		
Timber siding[a)]	0.012	0.14	0[a]	4.1×10^{-4}	−0.32

a) Timber siding, $\mu = 0$? The timber has a diffusion resistance factor far greater than 1, but the many leaks between planks make the sheathing so air permeable that its diffusion resistance can be neglected.

5.3 Heat, air, moisture performances

5.3.1 Airtightness

Assume a 10 cm thick thermal insulation layer. With the gypsum board leak-free and perfectly mounted, air permeance of the whole wall is close to $3\times10^{-5}\,\Delta P_a^{-0.2}$ kg/(m^2.Pa.s), hardly different from the gypsum board. One leak at ϕ 20 mm per square metre lifts that permeance to $2.1\times10^{-4}\Delta P_a^{-0.38}$ kg/(m^2.Pa.s), seven times higher than without but, thanks to the other layers, nonetheless 80% lower than the leaky gypsum board.

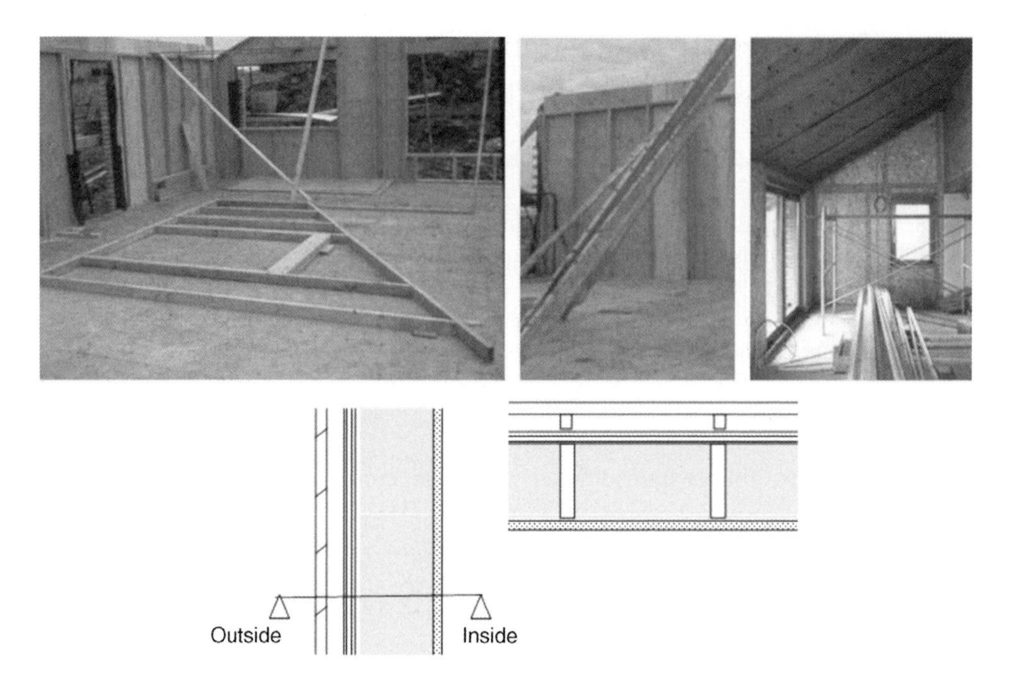

Figure 5.1 Timber-framed construction.

Figure 5.2 and its associated table show related air fluxes in kg/(m^2.h) as a function of the air pressure difference over the wall. At differentials of 5 Pa. the leaky wall sees some 2 m^3 of air an hour passing per square metre, while leak-free, that quantity drops to 0.35 m^3 per hour per square metre.

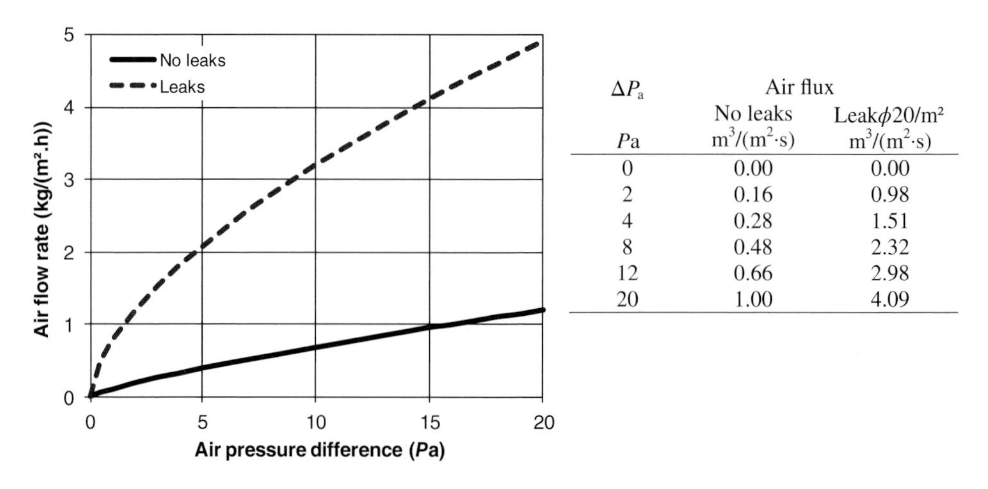

ΔP_a	Air flux	
	No leaks	Leakϕ20/m^2
Pa	m^3/(m^2·s)	m^3/(m^2·s)
0	0.00	0.00
2	0.16	0.98
4	0.28	1.51
8	0.48	2.32
12	0.66	2.98
20	1.00	4.09

Figure 5.2 Air flux across the wall (leaks assumed smeared out).

5.3.2 Thermal transmittance

Was the assembly airtight and thermal bridge free, the clear wall thermal transmittance should equal:

$$U_o = \frac{1}{0.043 + 0.012/0.14 + 0.17 + 0.022/0.14 + X/0.04 + 0.06 + 0.125} = \frac{0.04}{0.026 + X}$$

with X insulation thickness. Also see Figure 5.3 and its associated table, second column.

Timber-framed walls allow low U-values while keeping wall thickness limited (20 cm insulation in a 26 cm thick wall). No thermal bridging of course is a fiction. Studs, bottom, top and connecting plates, all increase heat flow. For an insulation thickness beyond 8 cm, related linear thermal transmittances equal:

Studs	$\psi \approx 0.017\ \text{W/(m.K)}$
Bottom plate	$\psi \approx 0.010\ \text{W/(m.K)}$
Top and connecting plates	$\psi \approx 0.023\ \text{W/(m.K)}$

For a typical timber-framed wall with the studs centre to centre 40 cm apart and a height of 2.6 m, each field contains 2.48 m of studs and 0.4 m of bottom, top and connecting plates. The third column of the table in Figure 5.2 lists the related whole wall thermal transmittances. The thicker the insulation, the more influential is thermal bridging, which was the main motivation for replacing massive timber studs by engineered I-shape studs, see Figure 5.4.

Perfect airtightness is also an illusion. Take a room-high wall ($H = 2.6$ m) with, when leaky, an air permeance $2.1 \times 10^{-4} \Delta P_a^{-0.38}$ kg/(m^2.Pa.s) and when tight $3 \times 10^{-5} \Delta P_a^{-0.2}$ kg/(m^2.Pa.s), in both cases assumed uniformly distributed over each square metre of wall. The room door is closed and the weather windless. Then thermal stack remains the

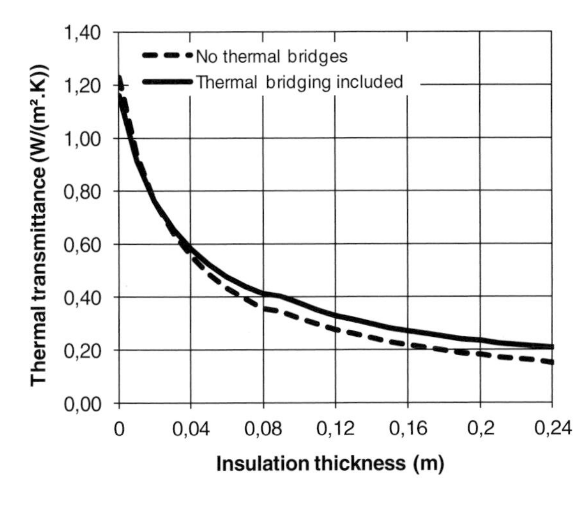

X cm	U_o W/(m².K)	Whole U W/(m².K)
0	1.23	1.16
4	0.55	0.58
8	0.36	0.41
12	0.27	0.33
16	0.22	0.27
20	0.18	0.23
24	0.15	0.20
28	0.13	0.18
32	0.12	0.17

Figure 5.3 Clear and whole wall thermal transmittance.

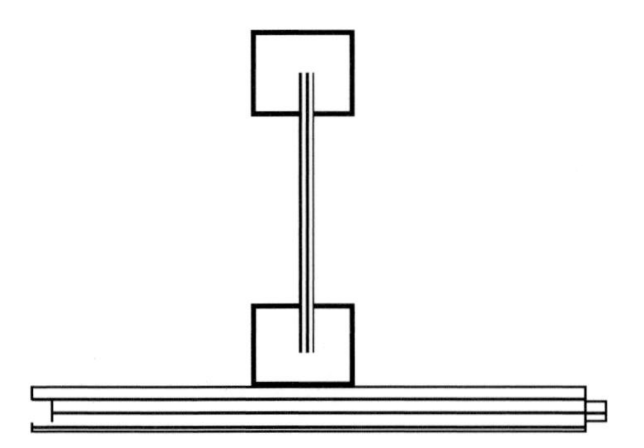

Figure 5.4 I-shape engineered stud.

only drive, giving as pressure over the wall:

$$\Delta P_{\mathrm{a}} = \frac{1}{2}\left[0.043(z - H/2)(\theta_{\mathrm{i}} - \theta_{\mathrm{e}})\right]$$

with z the height ordinate. At 17 °C indoor to outdoor temperature difference, the absolute pressure difference at a height of 0.65 and 1.95 m equals −0.48 and 0.48 Pa, resulting in 0.42 m^3/h inflow and outflow for the leaky wall. Assuming ideal workmanship has kept the lining tight, flows drop to 0.05 m^3/h in and out. Table 5.1 lists the apparent thermal transmittance measured at the inside surface at both heights.

The formula behind the table is:

$$U_{\mathrm{app,si}} = \frac{c_{\mathrm{a}} g_{\mathrm{a}}}{1 - \exp(c_{\mathrm{a}} g_{\mathrm{a}} R_{\mathrm{T}})}$$

With leaks, the apparent thermal transmittance becomes seemingly worse at 0.65 m but better at 1.95 m, while the difference with the clear wall U-value increases with leakage!

Table 5.1 Apparent thermal transmittance on the inside surface, air inflow and outflow.

Insulation thickness cm	U_o W/(m^2.K)	Ideal workmanship, gypsum board tight		Gypsum board leaky	
		$U_{0.65}$ W/(m^2.K)	$U_{1.95}$ W/(m^2.K)	$U_{0.65}$ W/(m^2.K)	$U_{1.95}$ W/(m^2.K)
0	1.23	*1.24*	*1.22*	1.30	1.16
4	0.55	*0.56*	*0.54*	0.62	0.49
8	0.36	*0.36*	*0.35*	0.43	0.29
12	0.27	*0.28*	*0.27*	0.35	0.21
16	0.22	*0.22*	*0.21*	0.29	0.15
20	0.18	*0.19*	*0.17*	0.25	0.12

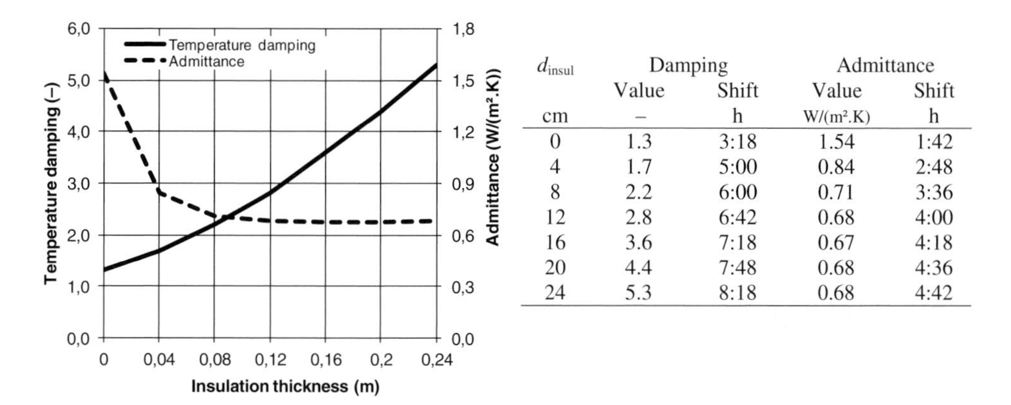

d_{insul}	Damping		Admittance	
	Value	Shift	Value	Shift
cm	–	h	W/(m².K)	h
0	1.3	3:18	1.54	1:42
4	1.7	5:00	0.84	2:48
8	2.2	6:00	0.71	3:36
12	2.8	6:42	0.68	4:00
16	3.6	7:18	0.67	4:18
20	4.4	7:48	0.68	4:36
24	5.3	8:18	0.68	4:42

Figure 5.5 Timber-framed wall: temperature damping and thermal admittance.

So the air inflow and outflow deprives the transmission based thermal transmittance of its value as 'energy' label.

5.3.3 Transient response

Timber-framed walls are too lightweight to act as heat storage volumes. Temperature damping and the admittance are both low. Even in temperate climates, timber-framed dwellings require additional measures to avoid summer overheating, among them restricted glazed surfaces at the sunny side shaded by overhangs or movable exterior screens where needed, massive floors and partition walls, and, that mass combined with night ventilation. How temperature damping and admittances change with insulation thickness is illustrated in Figure 5.5 and the table beside it. Although temperature damping of the airtight wall increases more than linearly with insulation thickness, even with a 24 cm thick insulation package a value of 15 is completely out of reach. Starting from not insulated, the admittance of an insulated wall first decreases and then stabilizes at a very low value once beyond 12 cm insulation. If the wall is leaky, infiltration further decreases, while exfiltration upgrades temperature damping.

5.3.4 Moisture tolerance

5.3.4.1 Construction moisture

This should not be a problem provided the timber is delivered air-dry and the sheathing and inside lining have a moderate vapour resistance, in which case drying, if needed, can go unhindered.

5.3.4.2 Rain control

Drainage planes are the cladding's outside and backside and, the cavity side of the building paper that protects the sheathing and from which the strips are draped as shown in Figure 5.6. Seeping rain blown across the cavity runs down that paper to the bottom,

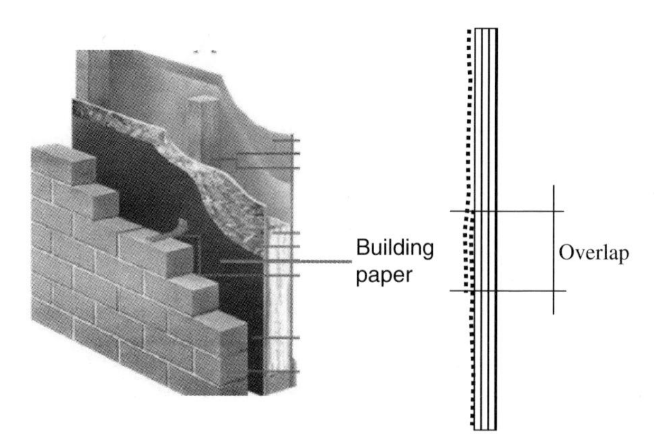

Figure 5.6 Building paper.

where a tray must direct it back outdoors. The paper also helps to increase the wall's air and wind-tightness.

Care must be taken when finishing timber-framed walls with a brick veneer as in Figure 5.6. Solar radiation will drive the rainwater that the veneer buffers as vapour to the inside where it may increase the relative humidity or even give condensation in the insulation and against the vapour retarder, if any. Avoidance demands a correctly balanced diffusion resistance between building paper and vapour retarder.

5.3.4.3 Rising damp

Rising damp is easily avoided by inserting a watertight membrane just above grade, below the bottom plate in all outer and loadbearing partition walls. If termite attack is likely, the membrane has to be replaced by a metal barrier.

5.3.4.4 Hygroscopic moisture and surface condensation

As long as solar driven vapour flow is excluded, too high a hygroscopic moisture content will rarely be a problem in properly insulated timber-framed outer walls. Of course, more insulation enlarges the sorption swings that outside timber claddings experience annually in temperate climates. As a consequence, painted claddings could peel. A proper insulation thickness also makes condensation on the inside surface highly unlikely.

5.3.4.5 Interstitial condensation

Assume that the wall is airtight, so only diffusion matters. If the sheathing is neither hygroscopic nor capillary, vapour pressure reaching saturation at its backside will give interstitial condensation with droplet formation. In reality, plywood and oriented strand board (OSB), – two timber-based materials often used for sheathings – are highly hygroscopic and slightly capillary, turning 'interstitial condensation' into changes in sorption moisture with droplet formation only once both sheathing materials are capillary wet.

Although too simplistic, let's assume no hygroscopic or capillary action. The thermal boundary condition outdoors is best represented by a fictitious equivalent temperature for condensation and drying. Given in Chapter 1 were the values for Uccle, Belgium. Short wave absorptivity of the outside surface is set at 0.8, while the wall looks north and so hardly receives any direct solar radiation. Humid plywood sheathing sees its diffusion resistance factor dropping to a value of 10. The vapour-permeable building paper has a diffusion thickness of 1 cm, while the air in the vented cavity in front shows more or less the same vapour pressure as outdoors. In north-west Europe, the coldest month is January. For Uccle, this gives the following boundary conditions for a residential building (ICC = indoor climate class):

Outside	Mean		Amplitude	January
Temperature	11.4		8.7	2.9
Vapour pressure	1042		430	621
Inside	**Mean**		**Amplitude**	**January**
Temperature	20		3	17.1
Vapour pressure	ICC1	1107	365	749
	ICC2	1312	220	1096
	ICC3	1442	220	1226

Table 5.2 lists the temperatures, saturation pressures and ICC1 vapour pressures in the wall. Nowhere, not even at the backside of the plywood sheathing, is saturation reached. So, no droplets deposit there, but the average relative humidity in the plywood

Table 5.2 ICC1, interstitial condensation.

Layer	R m^2.K/W	ΣR m^2.K/W	Temp. °C	p_{sat} Pa	μ–	$\Sigma\mu d$m	p Pa	$>p_{sat}$?
Indoors			17.1	1949			749	No
h_i	0.13	0.13	16.7	1909		0.007	749	No
Gypsum board	0.06	0.19	16.6	1890		0.091	744	No
Insulation (20 cm)	5.00	5.19	4.0	815	1.2	0.331	731	No
Plywood	0.16	5.35	3.6	792	10	0.551	719	No
Building paper	0.00	5.35	3.6	792	$\mu d = 1$ cm	0.561	621	No
Cavity	0.17	5.52	3.2	769		0.567	621	No
Clad	0.09	5.60	3.0	757				
h_e	0.04	5.64	2.9	752				

Table 5.3 ICC2 and 3, interstitial condensation?

Layer	Temp. °C	p_{sat} Pa	ICC2		ICC3	
			p Pa	p_{corr} Pa	p Pa	p_{corr} Pa
Indoors	17.1	1949	1096	1096	1226	1226
h_i	16.7	1909	1095	1090	1225	1218
Gypsum board	16.6	1890	1078	1019	1203	1113
Insulation (20 cm)	4.0	815	**1030**	**815**	**1141**	**815**
Plywood	3.6	792	985	794	1085	634
Building paper	3.6	792	622	621	622	626
Cavity	3.2	769	621	621	621	621
Clad	3.0	757				
h_e	2.9	752				

nevertheless touches some $100 \times (698 + 626)/(815 + 792) = 82\%$, giving as mean moisture ratio 14–18% kg/kg, which is a little below the mould threshold that in terms of moisture ratio nears 20% kg/kg, or, more correctly, in terms of surface relative humidity a value of around 85%

Table 5.3 summarizes the results for ICC2 and 3. Vapour pressure at the backside of the plywood now reaches saturation so that condensate deposits there. Figure 5.7 shows the January mean saturation and vapour pressures in the wall. The figure also compares the quantities deposited with what's allowable for moisture-proof plywood.

ICC2, which reflects dwellings with a large volume per household member, only gives a small amount of condensate, far below what's allowable. No vapour retarder between insulation and gypsum board is needed. Practitioners who nevertheless believe the opposite mostly mount the synthetic retarder foil so carelessly that vapour tightness remains a fiction. This happily does not matter when airtightness is guaranteed. Instead, in ICC3 buildings, by the end of February the amounts deposited go beyond what's allowed. However, 0.5 m extra vapour resistance on the inside of the thermal insulation suffices to keep the deposit within acceptable limits. A true vapour retarder is therefore not necessary. Thus timber-framed envelopes seem quite tolerant to interstitial condensation.

Unfortunately, this is a premature conclusion. Suppose the plywood or OSB has a vapour resistance factor 30 instead of 10, which is not unlikely. In ICC2, the maximum deposit at the sheathing's backside then reaches an unacceptable 0.8 kg/m^2! In ICC3, it even increases to 1.75 kg/m^2. Or, uncertainty about the sheathing's vapour resistance requires an extra diffusion thickness at the inside of the thermal insulation of more than

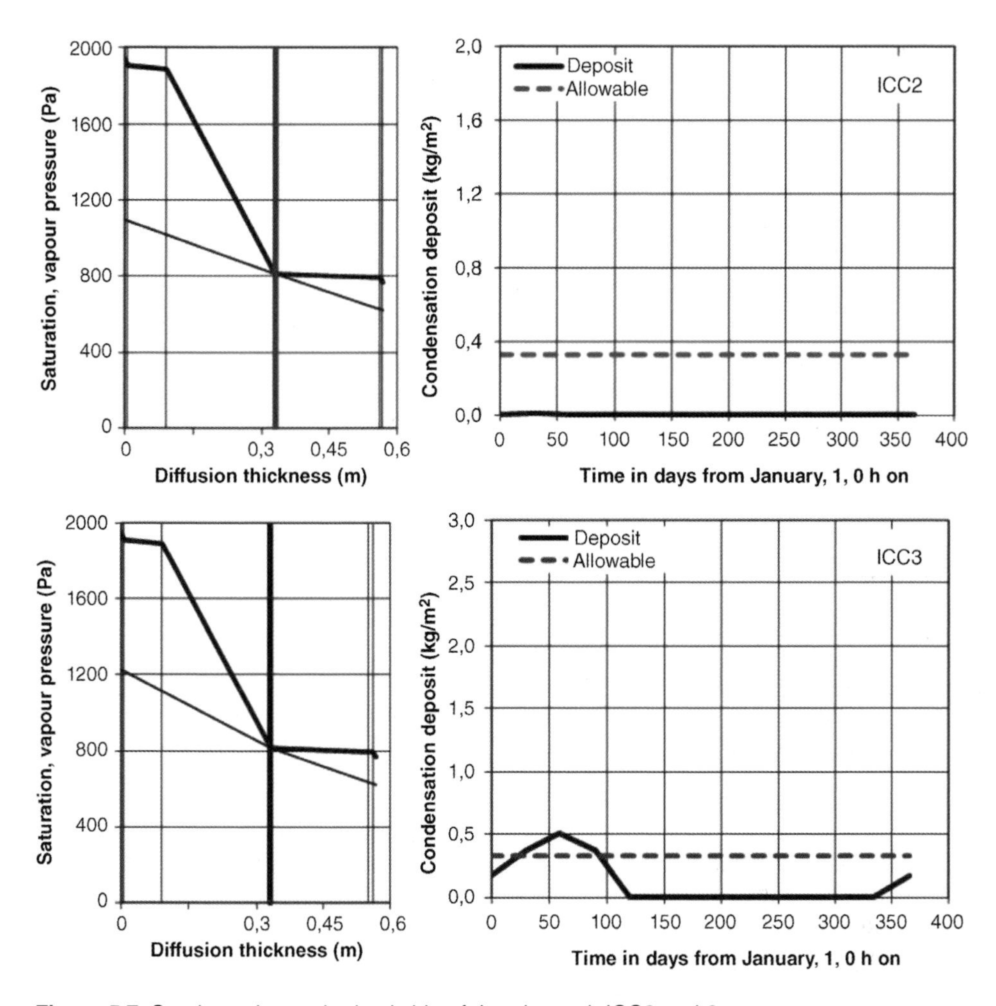

Figure 5.7 Condensation at the backside of the plywood, ICC2 and 3.

1 m in ICC2 and 3 buildings, a value still far below the 40 m that the often applied polyethylene (PE) foil offers when perfectly mounted and not perforated by the electrician when fixing wiring and installing sockets.

A still more realistic picture is that the wall is not airtight. In fact, timber-framed outer walls with perfectly mounted gypsum board inside show some air leakage. When perforated, the permeance increases to $2.1 \times 10^{-4} \Delta P_a^{-0.38}$ kg/(m^2.s.Pa). Diffusion/convection calculations typically consider one cold week. Nonetheless, first, a whole year is scanned, using mean values for wind and stack. In this way the same boundary conditions and allowance criteria as for a diffusion hold. Again, the wall faces north. As the wind blows mainly from the south-west in north-west Europe, north is an outflow orientation. Assume the detached dwelling has a quadratic floor plan. The monthly mean wind and stack pressures (both in Pa) at a height of 1.95 m are then:

	J	F	M	A	M	J	J	A	S	O	N	D
Wind	−1.3	−1.3	−1.6	−1.3	−1.0	−0.9	−0.8	−0.9	−0.9	−1.0	−1.7	−1.5
Stack	−0.4	−0.4	−0.3	−0.3	−0.2	−0.2	−0.2	−0.2	−0.2	−0.3	−0.3	−0.4

Figure 5.8 shows the consequence for an ICC2 situation where leakage is uniformly distributed over each square metre of outer wall and the plywood has a vapour resistance factor of 10. The increase in annual moisture deposit looks dramatic with a peak far beyond acceptable. ICC3 is even worse, with a maximum annual deposit reaching 6.2 kg/m^2.

These results of course do not picture reality. Air leaks are never uniformly distributed. In practice, their exact location remains mostly unknown. Wind also changes continuously in amplitude and direction, making steady state calculations unreliable, which is why the cold week check was put forward. Figure 5.8 also illustrates the deposits at the end of one such week in Uccle (see Chapter 1).

Figure 5.9 shows how, for a fixed air permeance, the deposit varies with the vapour diffusion thickness of the plywood and the building paper. The same figure portrays the impact of the internal lining when a vapour retarding building paper is applied. Air outflow clearly fixes the deposit. Vapour diffusion thicknesses don't matter. Even a combination of an inside lining with high diffusion thickness with a permeable sheathing and building paper doesn't exclude moisture from being deposited at the sheathing's backside, whatever the indoor climate class might be! Obviously, a

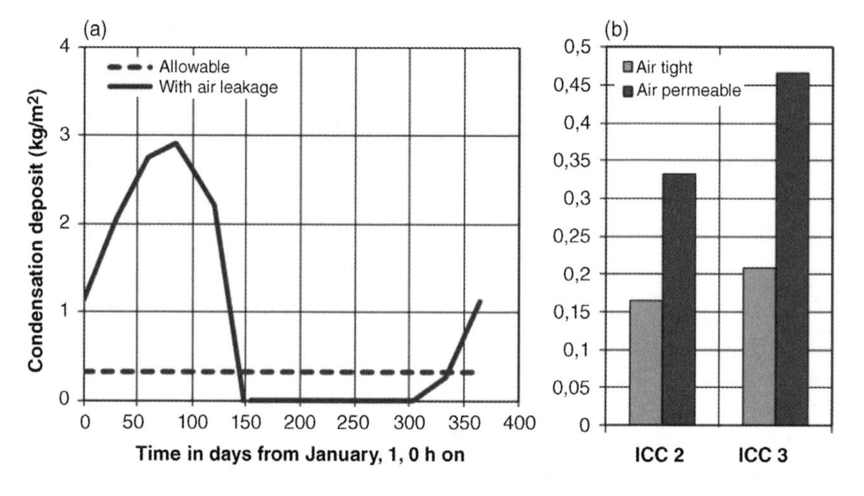

Figure 5.8 (a) A convection/diffusion calculation for ICC2, (b) the cold week deposits.

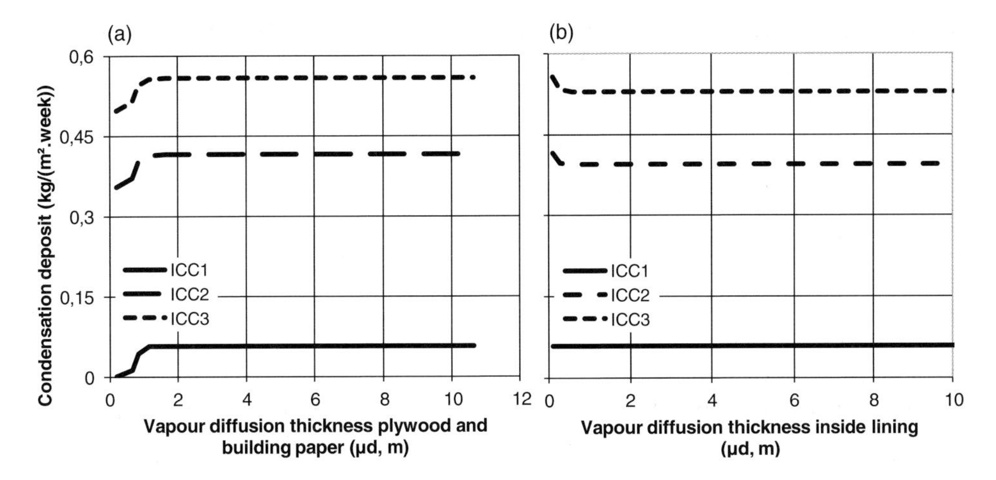

Figure 5.9 Condensation by diffusion/convection: (a) impact of the diffusion thickness of sheathing and building paper, (b) impact of the diffusion thickness of the inside lining for a building paper with diffusion thickness 10 m.

timber-framed outer wall must be as airtight as possible. The safest solution is including a separate air barrier at the backside of the insulation which also serves as a vapour retarder. Typically the PE foil mentioned does the job, though these days, smart air-retarding foils, whose diffusion resistance drops with increasing relative humidity, or carefully taped OSB boards, see Figure 5.1(b), are preferred.

Returning to the supposedly airtight outer wall, assume that it faces south-west, the windy and rainy but sunny orientation in north-west Europe. The wall has a brick veneer but lacks vapour retarding quality on the inside though the building paper used is vapour permeable. Under Uccle conditions, the backside of an unpainted gypsum board lining will not suffer from long-lasting solar-driven condensation, albeit some wetness may deposit during sunny days. But poor drying in winter to the vented cavity behind the veneer will increase the condensation deposit at the backside of the plywood sheathing, compared to no brick veneer as Figure 5.10(a) illustrates. In fact, relative humidity in the cavity approaches 100% for a cavity temperature close to sol-air outdoors.

If the inside gypsum board is painted, solar driven vapour flow will push relative humidity on its backside to values above 80% for a couple of weeks, enough to see mould developing, see Figure 5.10(b). Avoidance requires a building paper with a diffusion resistance in balance with the inside lining and/or the air and vapour retarder inside, that is both either higher or lower.

Note that perforating the air and vapour retarder must be avoided at all costs, a requirement that can only be guaranteed by a service cavity behind, wide enough to contain all wiring and electrical sockets, and post-filled with dense mineral fibre boards before mounting the inside lining.

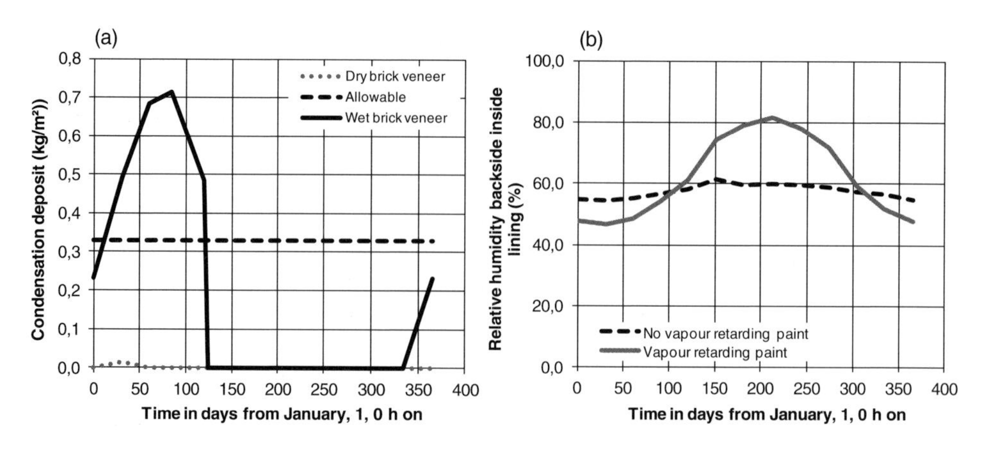

Figure 5.10 (a) ICC2, wet brick veneer, moisture deposit against the plywood, (b) relative humidity at the backside of the inside lining in ICC2.

5.3.4.6 More advanced modelling

As mentioned, in reality so-called interstitial condensation mostly converts into an increase and drying into a decrease of the sorption moisture present. Take an airtight timber-framed outer wall. Inertia is included by transposing the plywood into a chain of hygric capacitances, linked by diffusion resistances with the first and last capacitance coupled to the outdoors and indoors respectively by diffusion resistances representing the building paper and outer cladding on one hand and the insulation, air and vapour retarder, service cavity and inside lining on the other hand. In terms of forward differences the equation per capacitance is:

$$p_{\mathrm{pl}}^{t+\Delta t} = p_{\mathrm{sat},2}^{t+\Delta t}\left[\frac{p_2^t}{p_{\mathrm{sat},2}^t} + \frac{\Delta t}{0.5\rho\xi\left(\Delta d_{1,2} + \Delta d_{2,3}\right)_{\mathrm{pl}}}\left(\frac{p_1^t - p_2^t}{Z_{1,2}} + \frac{p_3^t - p_2^t}{Z_{2,3}}\right)\right]$$

Calculations were done for an ICC3 situation with the timber-framed outer wall facing north. First it misses a vapour retarder at the inside and then it gets one with diffusion thickness 2.5 m. The diffusion thickness of the building paper is 1.8 m, while construction moisture in the plywood corresponds to a relative humidity of 80%. A lathed timber cladding with cavity behind serves as finish. The daily thermal response is assumed to be steady state, which allows the saturation pressure in the plywood to be calculated. The climate outdoors has been measured since 1 December 2004 by a weather station. For the results, see Figure 5.11. Without vapour retarder some condensate is still deposited, forming droplets at the backside of the plywood that first saw its hygroscopic wetness increase to 100% relative humidity, though the amounts are only a fraction of those calculated with the classic steady state diffusion model. Of course, relative humidity in the plywood reaches long-term mean values high enough for mould to germinate. With a vapour retarder, the plywood sees its relative humidity stabilized around 82%, too low to suffer from mould.

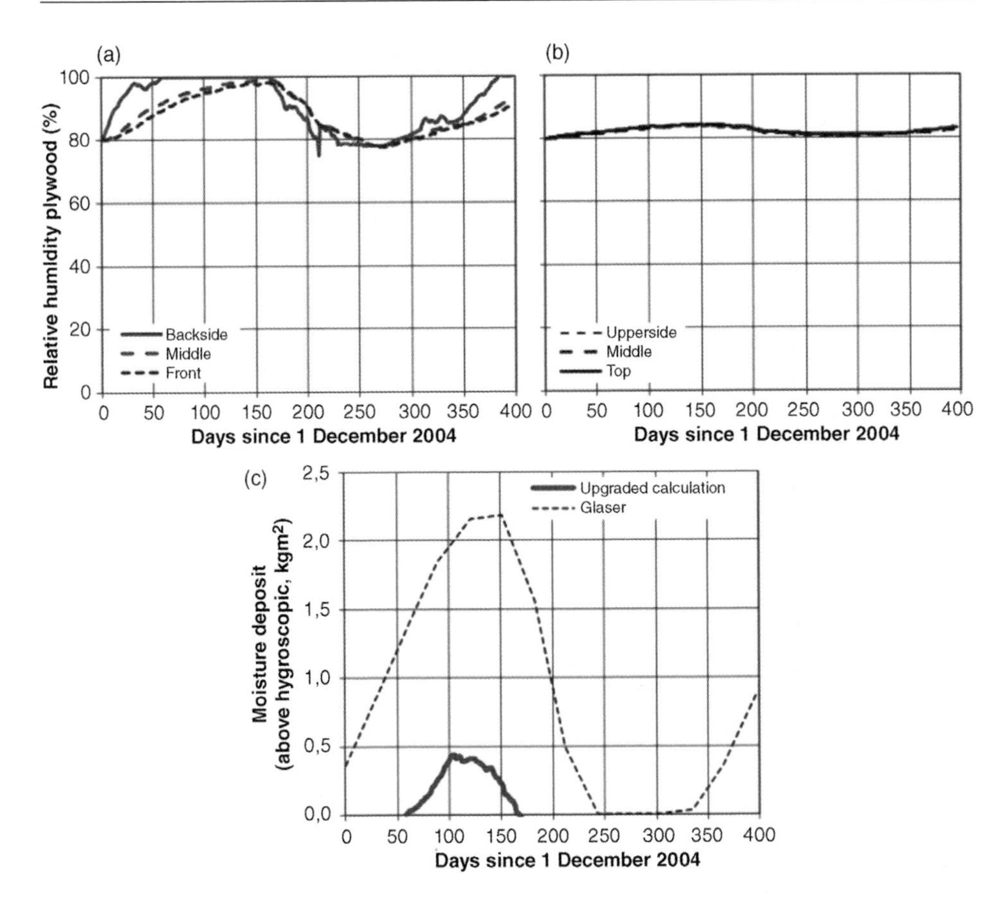

Figure 5.11 Timber-framed outer wall, ICC3, hygric inertia of the plywood accounted for: (a) relative humidity in the plywood without vapour retarder, (b) with vapour retarder. (c) Condensation deposit left without vapour retarder compared to diffusion only (calculated according to the Glaser method).

5.3.4.7 Thermal bridging

There are few problems with too low temperature factors in a well-insulated, correctly ventilated timber-framed building. If in temperate climate regions mould does develop, for example on the inside surfaces of outer walls in bedrooms, the reasons seen in practice were: all inside partition walls and floors insulated, the bedrooms not heated and minimally ventilated. The result was very low temperatures but high relative humidity, high enough to allow mould to germinate on the inside surface of outer walls, certainly on the ones shielded by cupboards.

6 Heat-air-moisture material properties

6.1 Introduction

To describe the heat-air-moisture behaviour of building, insulating or finishing materials, a set of material properties is needed. Density ρ, the mass per unit volume of dry material, and open porosity Ψ_o, the volume per m^3 taken by the pores accessible for water molecules, are two of them. Table 6.1 gives a full list.

This final chapter tabulates design and measured values. While the latter simply refers to listing test results, the design values account for the variance caused by the spread in measured data and the specifics of each application. How it's done is described in standards such as ISO 10456 for the thermal conductivity.

Table 6.1 Array of heat, air, moisture material properties.

	Heat	Air	Moisture
Storage	Specific heat capacity c Volumetric specific heat capacity ρc	Specific moisture ratio ξ Volumetric specific moisture ratio $\rho\xi$	Specific air content
Transport	Thermal conductivity λ Thermal resistance R Radiation: Absorptivity α Emissivity e Reflectivity ρ	Water vapour permeability δ Vapour resistance factor μ Diffusion thickness μd Moisture permeability k_m Thermal moisture diffusion coefficient K_θ	Air permeability k_a Air permeance K_a
Combined	Thermal diffusivity a Contact coefficient b	Moisture diffusivity D_w Water sorption coefficient A	
Consequences	Thermal expansion coefficient α	Hygric expansion ε	

Applied Building Physics: Ambient Conditions, Building Performance and Material Properties, Second Edition. Hugo Hens.

6.2 Dry air and water

Dry air

Gas constant	287.055 J/(kg.K)		
Atmospheric pressure	Normally 101 325 Pa, close to 1 bar. Higher in cyclonic, lower in anticyclonic weather		
Viscosity (η)	**Pressure bar**	**Temperature °C**	**Viscosity Pa.s**
	1	0	0.000 017 4
	1	50	0.000 019 9
	1	100	0.000 022 2
Specific heat capacity (c_p)	1007 J/(kg.K)		
Thermal conductivity (λ)	**Pressure bar**	**Temperature °C**	**λ W/(m.K)**
	1	0	0.024
	1	100	0.031

Water

Density (ρ)	**Pressure bar**	**Temperature °C**	**ρ kg/m^3**
	1	0	999.9
	1	10	999.7
	1	20	998.2
	1	40	992.2
	1	60	983.2
	1	80	971.8
	1	100	958.4
Viscosity (η)	**Pressure bar**	**Temperature °C**	**η Pa.s**
	1	0	1.787×10^{-3}
	1	10	1.307×10^{-3}
	1	20	1.002×10^{-3}
	1	40	0.653×10^{-3}
	1	60	0.467×10^{-3}
	1	80	0.354×10^{-3}
	1	100	0.282×10^{-3}
Surface tension (σ)		$(75.9–0.170\theta)\ 10^{-3}$ N/m with θ temperature in °C	
Specific heat capacity (c)	4187 J/(kg.K)		
Heat of evaporation (l_b)	2 500 000 J/kg at 0 °C		
Heat of solidification (l_s)	33 400 J/kg at °C		
Thermal conductivity	**Pressure bar**	**Temperature °C**	**λ W/(m.K)**
	1	0	0.54
	1	60	0.67
Longwave emissivity	0.95		

6.3 Materials, thermal properties

6.3.1 Definitions

For the thermal conductivity, a difference is made between the declared value, certified versus non-certified, and the design value:

Declared value (λ_D)	The starting point is the thermal conductivity of the dry material, measured in a certified laboratory at an average temperature of 10 °C on a representative number of samples (≥ 20), to allow statistical analysis of the results. The declared value then coincides with the 90% percentile, assuring that that percentage of what's produced has lower thermal conductivity
Certified versus non-certified	A material is certified when the declared thermal conductivity is known. It is non-certified when no certified laboratory did the measurements
Design value (λ_U)	Follows from declared by applying: $\lambda_U = \lambda_D \exp[f_u(X_2 - X_1)]$ or $\lambda_U = \lambda_D \exp[f_\Psi(\Psi_2 - \Psi_1)]$ (6.1) with X_2, Ψ_2 moisture ratio in kg/kg or m^3/m^3 when in use and X_1, Ψ_1 moisture ratio in kg/kg or m^3/m^3 during testing (normally zero). f_u and f_ψ are conversion factors listed in the standards

For the other thermal properties, most data tabulated refers to design values.

6.3.2 Design values

6.3.2.1 Non-certified materials (ISO 10456)

Group	Material	Density (ρ) kg/m^3	Specific heat capacity (c) J/(kg.K)	Thermal conductivity (λ) W/(m.K)
Metals	Aluminium alloys	2800	880	220
	Duralumin	2800	880	160
	Brass	8400	380	120
	Bronze	8700	380	65
	Copper	8900	380	380
	Iron	7900	450	75
	Iron, cast	7500	450	50
	Steel	7800	450	50
	Stainless steel	7900	460	17
	Lead	11300	130	35
	Zinc	7100	380	110

Group	Material	Density (ρ) kg/m^3	Specific heat capacity (c) J/(kg.K)	Thermal conductivity (λ) W/(m.K)
Wood and wood-based materials	Softwood	500	1600	0.13
	Hardwood	700	1600	0.18
	Plywood (from low to high density)	300	1600	0.09
		500	1600	0.13
		700	1600	0.17
		1000	1600	0.24
Wood and wood-based materials	Particle board			
	Soft	300	1700	0.10
	Semi-hard	600	1700	0.14
	Hard	900	1700	0.18
	Particle board, cement bounded	1200	1500	0.23
	OSB	680	1700	0.13
Wood and wood-based materials	Fibre board			
	Soft	400	1700	0.10
	Semi-hard	600	1700	0.14
	Hard	800	1700	0.18
Gypsum	Gypsum blocks	600	1000	0.18
		900	1000	0.30
		1200	1000	0.43
		1500	1000	0.56
	Gypsum board	900	1050	0.25
Mortars	Cement mortar, mixed on site	1800	1100	0.9
		1900	1100	1.0
Plasters	Gypsum plaster			
	Light	600	1000	0.18
	Normal	1000	1000	0.40
	Heavy	1300	1000	0.57
	Lime or gypsum plus sand	1600	1000	0.70
	Cement plus sand	1700	1000	1.0
Concrete	Medium to high density	1800	1000	1.15
		2000	1000	1.35
		2200	1000	1.65
		2400	1000	2.00
	Reinforced 1% steel	2300	1000	2.30
	2% steel	2400	1000	2.50

Group	Material	Density (ρ) kg/m^3	Specific heat capacity (c) J/(kg.K)	Thermal conductivity (λ) W/(m.K)
Stone	Crystalline rock	2800	1000	3.50
	Sedimentary rock	2600	1000	2.30
	Sedimentary, light	1500	1000	0.85
	Lava	1600	1000	0.55
	Basalt	2700–3000	1000	3.50
	Gneiss	2400–2700	1000	3.50
	Granite	2500–3000	1000	2.80
	Marble	2800	1000	3.50
	Slate	2000–2800	1000	2.20
	Limestone			
	Extra soft	1600	1000	0.85
	Soft	1800	1000	1.10
	Semi-hard	2000	1000	1.40
	Hard	2200	1000	1.70
	Extra hard	2600	1000	2.30
	Sandstone	2600	1000	2.30
	Natural pumice	400	1000	0.12
	Artificial stone	1750	1000	1.30
Soils	Clay or silt	1200–1800	1670–2500	1.5
	Sand and gravel	1700–2200	910–1180	2.0
Water, ice, snow	Ice, −10 °C	920	2000	2.30
	Ice, 0 °C	900	2000	2.20
	Water, 10 °C	1000	4187	0.60
	Water, 40 °C	990	4187	0.63
	Snow			
	Fresh (<30 mm)	100	2000	0.05
	Soft (30–70 mm)	200	2000	0.12
	Slightly compacted	300	2000	0.23
	Compacted	500	2000	0.60

Group	Material	Density (ρ) kg/m^3	Specific heat capacity (c) J/(kg.K)	Thermal conductivity (λ) W/(m.K)
Synthetics, solid	Acrylic	1050	1500	0.20
	Polycarbonates	1200	1200	0.20
	PFTE	2200	1000	0.25
Synthetics, solid	PVC	1390	900	0.17
	PMMA	1180	1500	0.18
	Polyacetate	1410	1400	0.30
	Polyamide (nylon)	1150	1600	0.25
	Nylon, 25% glass fibre	1450	1600	0.30
	PE, High density	980	1800	0.50
	PE, low density	920	2100	0.33
	Polystyrene	1050	1300	0.16
	Polypropylene (PP)	910	1800	0.22
	PP, 25% glass fibre	1200	1800	0.25
	Polyurethane	1200	1800	0.25
	Epoxy resin	1200	1200	0.20
	Phenol resin	1300	1700	0.30
	Polyester resin	1400	1200	0.19
Rubbers	Natural	910	1100	0.13
	Neoprene	1240	2140	0.23
	Butyl	1200	1400	0.24
	Foam rubber	60–80	1500	0.06
	Hard rubber	1200	1400	0.17
	EPDM	1150	1000	0.25
	Polyisobutylene	920	1130	0.13
	Polysulphide	1700	1000	0.43
	Butadiene	980	1000	0.25
Glass	Quartz glass	2200	750	1.4
	Normal glass	2500	750	1.0
	Glass mosaic	2000	750	1.2
Gases	Air	1.23	1008	0.025
	Argon	1.70	519	0.017
	Carbon dioxide	1.95	820	0.014
	Sulphur hexafluoride	6.36	614	0.013
	Krypton	3.56	245	0.009
	Xenon	5.68	160	0.0054

Group	Material	Density (ρ) kg/m^3	Specific heat capacity (c) J/(kg.K)	Thermal conductivity (λ) W/(m.K)
Thermal breaks, sealants, weather stripping	Silica gel (desiccant)	720	1000	0.13
	Silicone foam	750	1000	0.12
	Silicone pure	1200	1000	0.35
	Silicone, filled	1450	1000	0.50
	Urethane/PUR (thermal break)	1300	1800	0.21
	PVC, flexible with 40% softener	1200	1000	0.14
	Elastomeric foam, flexible	60–80	1500	0.05
	Polyurethane foam	70	1500	0.05
	Polyethylene foam	70	2300	0.05
Roofing	Asphalt	2100–2300	1000	0.7
	Bitumen, pure	1050	1000	0.17
	Bitumen, felt	1100	1000	0.23
	Clay tiles	2000	800	1.00
	Concrete tiles	2100	1000	1.50
Floor covering	Rubber	1200	1400	0.17
	Plastic	1700	1400	0.25
	Linoleum	1200	1400	0.17
	Carpet/textile	200	1300	0.06
	Underlay, rubber	270	1400	0.10
	Underlay, felt	120	1300	0.05
	Underlay, wool	200	1300	0.06
	Underlay, cork	200	1500	0.05
	Ceramic tiles	2300	840	1.30
	Plastic tiles	1000	1000	0.20
	Cork tiles, light	200	1500	0.050
	Cork tiles, heavy	500	1500	0.065

6.3.2.2 Design values (NBN B62-002 (2001))

The standard differentiates between the thermal conductivity indoors ($_{U\mathrm{i}}$) and outdoors ($_{U\mathrm{e}}$). Indoors prevails between the insulation and inside with, as design value, that for 23 °C and 50% relative humidity. Outdoors applies to all layers outside of the insulation. For capillary materials 10 °C and a moisture ratio 0.75 times the critical value (see Table 6.2) are the conditions considered. For non-capillary materials this is 10 °C and 80% relative humidity.

Table 6.2 Moisture ratio in masonry at indoor and outdoor conditions.

Masonry	Density ρ kg/m^3	Moisture ratio 23 °C, 50% RH		Moisture ratio $0.75(X_{cr}$ or $\Psi_{cr})$		Conversion factors	
		X_i kg/kg	Ψ_I m^3/m^3	X_e kg/kg	Ψ_e m^3/m^3	f_u kg/kg	f_Ψ m^3/m^3
Bricks, perforated	700–2100		0.007		0.075		10
Lime-sand stone	900–2200		0.012		0.090		10
Concrete blocks (massive, perforated)							
Heavy	1600–2400		0.025		0.090		4
Expanded clay	400–1700		0.020		0.090		4
Lightweight	500–1800		0.030		0.090		4
Cellular concrete	300–1000	0.026		0.150		4	

Metals

Material	Density (ρ) kg/m^3	Specific heat capacity (c) J/(kg.K)	Thermal conductivity (λ_{Ui}) W/(m.K)	Thermal conductivity (λ_{Ue}) W/(m.K)
Lead	11 340	130	35	35
Copper	$8300 \leq \rho \leq 8900$	390	384	384
Steel	7800	480–530	45	45
Aluminium, 99%	2700	880	203	203
Iron, cast	7500	530	56	56
Zinc	7000	390	113	113

Stone

Material	Density ρ kg/m^3	Specific heat capacity c J/(kg.K)	Thermal conductivity λ_{Ui} W/(m.K)	Thermal conductivity λ_{Ue} W/(m.K)
Heavy (granite, gneiss, basalt, porphyry)	$2750 \leq \rho \leq 3000$	1000	3.49	3.49
Limestone, hard	2700	1000	2.91	3.49
Marble	2750	1000	2.91	3.49
Sandstone				
Hard	2550	1000	2.21	2.68
Semi	2350	1000	1.74	2.09
Soft	2200	1000	1.40	1.69

Glued masonry, joint width ≤3 mm: Bricks and perforated large format bricks

Density (ρ) kg/m^3	Specific heat capacity (c) J/(kg.K)	Certified Thermal conductivity		Non-certified Thermal conductivity	
		λ_{Ui} W/(m.K)	λ_{Ue} W/(m.K)	λ_{Ui} W/(m.K)	λ_{Ue} W/(m.K)
≤ 700	1000	0.20	0.39	0.22	0.43
$700 < \rho \leq 800$	1000	0.23	0.45	0.25	0.49
$800 < \rho \leq 900$	1000	0.26	0.51	0.28	0.56
$900 < \rho \leq 1000$	1000	0.29	0.57	0.32	0.63
$1000 < \rho \leq 1100$	1000	0.32	0.64	0.35	0.70
$1100 < \rho \leq 1200$	1000	0.35	0.70	0.39	0.77
$1200 < \rho \leq 1300$	1000	0.39	0.76	0.42	0.84
$1300 < \rho \leq 1400$	1000	0.43	0.85	0.47	0.93
$1400 < \rho \leq 1500$	1000	0.46	0.91	0.51	1.00
$1500 < \rho \leq 1600$	1000	0.50	0.99	0.55	1.09
$1600 < \rho \leq 1700$	1000	0.55	1.08	0.60	1.19
$1700 < \rho \leq 1800$	1000	0.59	1.16	0.65	1.28
$1900 < \rho \leq 1900$	1000	0.64	1.27	0.71	1.40
$1900 < \rho \leq 2000$	1000	0.69	1.35	0.76	1.49
$2000 < \rho \leq 2100$	1000	0.74	1.46	0.81	1.61

Lime-sand stone

Density (ρ) kg/m^3	Specific heat capacity (c) J/(kg.K)	Certified λ_{Ui} W/(m.K)	Certified λ_{Ue} W/(m.K)	Non-certified λ_{Ui} W/(m.K)	Non-certified λ_{Ue} W/(m.K)
≤ 900	1000	0.33	0.71	0.36	0.78
$900 < \rho \leq 1000$	1000	0.34	0.74	0.37	0.81
$1000 < \rho \leq 1100$	1000	0.36	0.79	0.40	0.87
$1100 < \rho \leq 1200$	1000	0.41	0.89	0.45	0.97
$1200 < \rho \leq 1300$	1000	0.46	1.01	0.51	1.11
$1300 < \rho \leq 1400$	1000	0.52	1.13	0.57	1.24
$1400 < \rho \leq 1500$	1000	0.60	1.30	0.66	1.43
$1500 < \rho \leq 1600$	1000	0.69	1.50	0.76	1.65
$1600 < \rho \leq 1700$	1000	0.79	1.72	0.87	1.89
$1700 < \rho \leq 1800$	1000	0.91	1.99	1.00	2.19
$1800 < \rho \leq 1900$	1000	1.04	2.26	1.14	2.49
$1900 < \rho \leq 2000$	1000	1.18	2.58	1.30	2.84
$2000 < \rho \leq 2100$	1000	1.35	2.95	1.49	3.25
$2100 < \rho \leq 2200$	1000	1.54	3.37	1.70	3.71

Normal concrete blocks

Density (ρ) kg/m^3	Specific heat capacity (c) J/(kg.K)	Certified Thermal conductivity		Not certified Thermal conductivity	
		λ_{Ui} W/(m.K)	λ_{Ue} W/(m.K)	λ_{Ui} W/(m.K)	λ_{Ue} W/(m.K)
≤ 1600	1000	0.97	1.26	1.07	1.39
$1600 < \rho \leq 1700$	1000	1.03	1.33	1.13	1.47
$1700 < \rho \leq 1800$	1000	1.12	1.45	1.23	1.59
$1800 < \rho \leq 1900$	1000	1.20	1.56	1.33	1.72
$1900 < \rho \leq 2000$	1000	1.32	1.71	1.45	1.88
$2000 < \rho \leq 2100$	1000	1.44	1.86	1.58	2.05
$2100 < \rho \leq 2200$	1000	1.57	2.04	1.73	2.24
$2200 < \rho \leq 2300$	1000	1.72	2.24	1.90	2.46

Expanded clay concrete blocks

Density (ρ) kg/m^3	Specific heat capacity (c) J/(kg.K)	Certified λ_{Ui} W/(m.K)	Certified λ_{Ue} W/(m.K)	Not certified λ_{Ui} W/(m.K)	Not certified λ_{Ue} W/(m.K)
$400 < \rho \leq 500$	1000	0.16		0.18	
$500 < \rho \leq 600$	1000	0.19	0.26	0.21	0.28
$600 < \rho \leq 700$	1000	0.23	0.30	0.25	0.33
$700 < \rho \leq 800$	1000	0.27	0.36	0.30	0.39
$800 < \rho \leq 900$	1000	0.30	0.40	0.33	0.44
$900 < \rho \leq 1000$	1000	0.35	0.46	0.38	0.50
$1000 < \rho \leq 1100$	1000	0.39	0.52	0.43	0.57
$1100 < \rho \leq 1200$	1000	0.44	0.59	0.49	0.65
$1200 < \rho \leq 1300$	1000	0.50	0.66	0.55	0.73
$1300 < \rho \leq 1400$	1000	0.55	0.73	0.61	0.80
$1400 < \rho \leq 1500$	1000	0.61	0.80	0.67	0.88
$1500 < \rho \leq 1600$	1000	0.68	0.90	0.75	0.99
$1600 < \rho \leq 1700$	1000	0.76	1.00	0.83	1.10

Concrete blocks with other lightweight aggregates

≤500	1000	0.27		0.30	
$500 < \rho \leq 600$	1000	0.30	0.39	0.33	0.43
$600 < \rho \leq 700$	1000	0.34	0.43	0.37	0.47
$700 < \rho \leq 800$	1000	0.37	0.47	0.41	0.52
$800 < \rho \leq 900$	1000	0.42	0.53	0.46	0.58
$900 < \rho \leq 1000$	1000	0.46	0.59	0.51	0.65
$1000 < \rho \leq 1100$	1000	0.52	0.66	0.57	0.73
$1100 < \rho \leq 1200$	1000	0.59	0.75	0.64	0.82
$1200 < \rho \leq 1300$	1000	0.65	0.83	0.72	0.91
$1300 < \rho \leq 1400$	1000	0.74	0.95	0.82	1.04
$1400 < \rho \leq 1500$	1000	0.83	1.06	0.92	1.17
$1500 < \rho \leq 1600$	1000	0.94	1.19	1.03	1.31
$1600 < \rho \leq 1800$	1000	1.22	1.55	1.34	1.70

Autoclaved cellular concrete blocks

≤300	1000	0.09		0.10	
$300 < \rho \leq 400$	1000	0.12		0.13	
$400 < \rho \leq 500$	1000	0.14		0.16	
$500 < \rho \leq 600$	1000	0.18	0.29	0.20	0.32
$600 < \rho \leq 700$	1000	0.20	0.33	0.22	0.36
$700 < \rho \leq 800$	1000	0.23	0.38	0.26	0.42
$800 < \rho \leq 900$	1000	0.27	0.44	0.29	0.48

Fresh cellular concrete contains large amounts of production moisture. The values hold once air-dry

Masonry, mortar joints

Calculate the design value λ_U as:

$$\lambda_U = \lambda_{U,block}\, f_{block} + \lambda_{U,mortar}(1 - f_{block}) \tag{6.2}$$

with $\lambda_{U,block}$ the thermal conductivity, f_{block} the surface ratio taken by the blocks and $\lambda_{U,mortar}$ the thermal conductivity of the mortar used. When only block density is known, use the values for glued masonry.

Normal concrete

Material	Density (ρ) kg/m^3	Specific heat capacity (c) J/(kg.K)	Thermal conductivity (λ_{Ui}) W/(m.K)	Thermal conductivity (λ_{Ue}) W/(m.K)
Reinforced	2400	1000	1.7	2.2
Not reinforced	2200	1000	1.3	1.7

These values look low compared to measured data and those given in ISO 10456.

Lightweight concrete used as slab or screed

Concrete with expanded clay, furnace slag, vermiculite, cork, perlite or polystyrene pearls as aggregates Cellular concrete	≤ 350	1000	0.12	
	$300 < \rho \leq 400$	1000	0.14	
	$400 < \rho \leq 450$	1000	0.15	
	$450 < \rho \leq 500$	1000	0.16	
	$500 < \rho \leq 550$	1000	0.17	
	$550 < \rho \leq 600$	1000	0.18	
	$600 < \rho \leq 650$	1000	0.20	0.31
	$650 < \rho \leq 700$	1000	0.21	0.34
	$700 < \rho \leq 750$	1000	0.22	0.36
	$750 < \rho \leq 800$	1000	0.23	0.38
	$800 < \rho \leq 850$	1000	0.24	0.40
	$850 < \rho \leq 900$	1000	0.25	0.43
	$900 < \rho \leq 950$	1000	0.27	0.45
	$950 < \rho \leq 1000$	1000	0.29	0.47
	$1000 < \rho \leq 1100$	1000	0.32	0.52
	$1100 < \rho \leq 1200$	1000	0.37	0.58

Gypsum with and without lightweight aggregates

Density (ρ) kg/m^3	Specific heat capacity (c) J/(kg.K)	Thermal conductivity (λ_{Ui}) W/(m.K)	Thermal conductivity (λ_{Ue}) W/(m.K)
≤ 800	1000	0.22	
$800 < \rho \leq 1100$	1000	0.35	
>1100	1000	0.52	

Plaster

Material	Density (ρ) kg/m^3	Specific heat capacity (c) J/(kg.K)	Thermal conductivity (λ_{Ui}) W/(m.K)	Thermal conductivity (λ_{Ue}) W/(m.K)
Cement plaster	1900	1000	0.93	1.5
Lime plaster	1600	1000	0.70	1.2
Gypsum plaster[a)]	1300	1000	0.52	

a) This is a heavyweight gypsum plaster. Most ready-to-mix gypsum plasters have a density not exceeding 1000 kg/m^3 (see table for gypsum)

Wood and wood-based materials

Material	Density (ρ) kg/m^3	Specific heat capacity (c) J/(kg.K)	Thermal conductivity (λ_{Ui}) W/(m.K)	Thermal conductivity (λ_{Ue}) W/(m.K)
Timber	≤ 600	1880	0.13	0.15
	>600	1880	0.18	0.20
Plywood	≤ 400	1880	0.09	0.11
	$400 < \rho \leq 600$	1880	0.13	0.15
	$600 < \rho \leq 850$	1880	0.17	0.20
	>850	1880	0.24	0.28
Particle board	≤ 450	1880	0.10	
	$450 < \rho \leq 750$	1880	0.14	
	>750	1880	0.18	
Fibre-cement board	1200	1470	0.23	
OSB	650	1880	0.13	
Fibreboard	≤ 375	1880	0.07	
	$375 < \rho \leq 500$	1880	0.10	
	$500 < \rho \leq 700$	1880	0.14	
	>700	1880	0.18	

Insulation materials

Insulation materials may under no circumstances become humid, thus only λ_{Ui} is given

Material	Density (ρ) kg/m^3	Specific heat capacity (c) J/(kg.K)	Thermal conductivity (λ_{Ui}) W/(m.K)
Not certified			
Cork	90–160	1560	0.05
Glass fibre and mineral fibre	10–200	1030	0.045
EPS	10–50	1450	0.045
Polyethylene (PE)	20–65	1450	0.045
Phenol foam	20–50	1400	0.045
PUR, lined	28–55	1400	0.035
XPS	20–65	1450	0.040
Cellular glass	100–140	1000	0.055
Perlite board	200–300	900	0.060

Material	Density (ρ) kg/m^3	Specific heat capacity (c) J/(kg.K)	Thermal conductivity (λ_{Ui}) W/(m.K)
Vermiculite	50–170	1080	0.065
Vermiculite board		900	0.090

Certified

Material	Density (ρ) kg/m^3	Specific heat capacity (c) J/(kg.K)	Thermal conductivity (λ_{Ui}) W/(m.K)
Glass fibre and mineral fibre	10–200	1030	0.040
EPS	10–50	1450	0.040
Phenol foam	20–50	1400	0.025
PUR, lined	28–55	1400	0.028
XPS	20–65	1450	0.034
Cellular glass	100–140	1000	0.048
Perlite board	200–300	900	0.055
VIPs			0.004–0.008

Miscellaneous

Material	Density (ρ) kg/m^3	Specific heat capacity (c) J/(kg.K)	Thermal conductivity (λ_{Ui}) W/(m.K)	Thermal conductivity (λ_{Ue}) W/(m.K)
Glass	2500	750	1.00	1.00
Clay tiles	1700	1000	0.81	1.00
Grès tiles	2000	1000	1.20	1.30
Rubber	1500	1400	0.17	0.17
Linoleum, PVC	1200	1400	0.19	
Fibre cement	$1400 < \rho < 1900$	1000	0.35	0.50
Asphalt	2100	1000	0.70	0.70
Bitumen	1100	1000	0.23	0.23

Perforated blocks, floor elements, gypsum board

Material	Thickness cm	Specific heat capacity (c) J/(kg.K)	Thermal resistance (R_{Ui}) W/(m.K)
Masonry from perforated blocks			
Normal concrete ($\rho > 1200\,kg/m^3$)	14	1000	0.11
	19	1000	0.14
	29	1000	0.20
Lightweight concrete ($\rho < 1200\,kg/m^3$)	14	1000	0.30
	19	1000	0.35
	29	1000	0.45
Clay floor elements			
One cavity along span	8	1000	0.08
	12	1000	-0.11
Two cavities along span	12	1000	-0.13
	16	1000	0.16
	20	1000	0.19
Concrete floor elements	12	1000	0.11
	16	1000	0.13
	20	1000	0.15
Gypsum board	< 1.4	1000	0.05
	≥ 1.4	1000	0.08

6.3.3 Measured data

6.3.3.1 Building materials

Concrete

Density	kg/m^3	2176, $\sigma = 40.5$ Mean for 39 samples		
Specific heat capacity (dry)	J/(kg.K)	840		
Thermal conductivity	W/(m.K)	$2.74 + 0.003\,w$, w: moisture content in kg/m^3		
Absorptivity, reflectivity	—	T(K)	300	6000
		a	0.88	0.6
		ρ	0.12	0.4
Thermal expansion coefficient	$°C^{-1}$	12×10^{-6}		

Lightweight concrete

Density	kg/m^3	$644 \leq \rho \leq 1187$	
Specific heat capacity (dry)	J/(kg.K)	840	
Thermal conductivity	W/(m.K)	$644 \leq \rho \leq 1187\ kg/m^3$ $\theta = 20\ °C$, $w = 0\ kg/m^3$	$0.0414\ \exp(0.00205\rho)$
		$1158 \leq \rho \leq 1187\ kg/m^3$ $\theta = 20\ °C$, $w \leq 74\ kg/m^3$	$0.511 + 0.002\,w$
		$1130 \leq \rho \leq 1138\ kg/m^3$ $\theta = 20\ °C$, $w \leq 144\ kg/m^3$	$0.371 + 0.00\,w$
		$644 \leq \rho \leq 674\ kg/m^3$ $\theta = 20\ °C$, $w \leq 39\ kg/m^3$	$0.161 + 0.001\,w$

Autoclaved aerated concrete

Density	kg/m^3	$455 \le \rho \le 800$	
Specific heat capacity (dry)	J/(kg.K)	840	
Thermal conductivity	W/(m.K)	$598 \le \rho \le 626\,kg/m^3$ $\theta = 10\,°C,\ w \le 425\,kg/m^3$	$0.176 + 0.0008\,w$
		$598 \le \rho \le 626\,kg/m^3$ $\theta = 20\,°C,\ w \le 425\,kg/m^3$	$0.177 + 0.001\,w$
		$455 \le \rho \le 492\,kg/m^3$ $\theta = 20\,°C,\ w \le 298\,kg/m^3$	$0.138 + 0.0009\,w$
As a function of density, temperature and moisture content		$\lambda = 0.172 - 1.67 \times 10^{-3}w - 9.34 \times 10^{-3}\theta - 2.97 \times 10^{-6}\rho + 3.77 \times 10^{-6}\rho w + 1.16 \times 10^{-4}w\theta + 1.6 \times 10^{-5}\rho\theta - 1.62 \times 10^{-7}\rho w\theta$	

Concrete with expanded polystyrene pearls as aggregate

Density	kg/m^3	$259 \le \rho \le 792$
Specific heat capacity (dry)	J/(kg.K)	Depends on the concentration of EPS-pearls: $\rho = 259\,kg/m^3 \quad c = 1370\,J/(kg.K)$ $\rho = 792\,kg/m^3 \quad c = 1018\,J/(kg.K)$
Thermal conductivity	W/(m.K)	see below

Thermal conductivity, W/(m.K):

$259 \le \rho \le 792\,kg/m^3$, $\theta = 20\,°C,\ w \le 425\,kg/m^3$: $0.041\exp(0.00232\rho)$

$\theta = 10\,°C,\ w = 0\,kg/m^3$: $A_1 + A_2 w$

ρ (kg/m^3)	A_1	$A_2 \times 10^{-4}$
259–335	0.074	4.8
357–382	0.111	4.6
407–456	0.126	6.1
641	0.151	7.9
792	0.213	1.0

$\rho = 422\,kg/m^3$, $0 \le \theta \le 30\,°C$: $B_1 + B_2\theta$

w (kg/m^3)	B_1	$B_2 \times 10^{-4}$
0	0.112	1.3
94	0.171	12
262	0.231	15

Lightweight and normal cement mortars

Density	kg/m^3	$1055 \le \rho \le 1822$	
Specific heat capacity (dry)	J/(kg.K)	840	
Thermal conductivity	W/(m.K)	$1055 \le \rho \le 1822$ kg/m^3	$0.088 \exp(0.00125\rho)$
		$\theta = 20$ °C, $w = 0$ kg/m^3	$0.177 + 0.00\,w$
		$\theta = 20$ °C, $w \le 330$ kg/m^3	$0.138 + 0.000\,w$
			$A_1 + A_2 w$
			ρ (kg/m^3) / A_1 / $A_2 \times 10^{-3}$
			1072 / 0.346 / 1.2
			1512 / 0.526 / 3.1
			1800 / 0.854 / 4.5

Lime-sandstone masonry

Density	kg/m^3	$1170 \le \rho \le 1230$				
Specific heat capacity (dry)	J/(kg.K)	840				
Thermal resistance	m^2.K/W	$\theta = 20$ °C	$1/(A_1 + A_2 u)$			
		$u \le u_c$, u in % kg/kg	d (m)	ρ (kg/m^3)	A_1	A_2
			0.14	1140	4.07	0.24

Large format perforated brick masonry

Density	kg/m^3	$860 \leq \rho \leq 1760$				
Specific heat capacity (dry)	J/(kg.K)	840				
Thermal resistance	$m^2.K/W$	$d = 14$ cm $860 \leq \rho \leq 1430$ kg/m^3 $\theta = 20$ °C, $w = 0$ kg/m^3	$1/[0.98 \exp(0.001\rho)]$			
		$d = 19$ cm $830 \leq \rho \leq 1630$ kg/m^3 $\theta = 20$ °C, $w = 0$ kg/m^3	$1/[0.59 \exp(0.0012\rho)]$			
		$\theta = 20$ °C $u \leq u_c$, u in % kg/kg	$1/(A_1 + A_2 u)$			
			d (m)	ρ (kg/m^3)	A_1	A_2
			0.09	1470	7.94	0.40
			0.14	863	2.09	0.09
				1100	3.35	0.18
				1120	2.72	0.28
				1180	3.37	0.31
				1200	3.13	0.32
				1240	4.17	0.38
				1360	2.96	0.34
				1430	4.17	0.42
			0.19	800	1.53	0.10
				830	1.51	0.08
				880	1.83	0.13
				1100	2.41	0.09
				1140	2.26	0.13
				1650	4.00	0.34

Concrete block masonry

Density	kg/m^3	$860 \leq \rho \leq 1650$				
Specific heat capacity (dry)	J/(kg.K)	840				
Thermal resistance	$m^2.K/W$	$d = 14\,cm$, $860 \leq \rho \leq 1650\,kg/m^3$, $\theta = 20\,°C$, $w = 0\,kg/m^3$	$1/[0.73 \exp(0.0014\rho)]$			
		$\theta = 20\,°C$, $u \leq u_c$, u in % kg/kg	$1/(A_1 + A_2 u)$			
			d (m)	ρ (kg/m^3)	A_1	A_2
			0.12	980	2.92	0.12
			0.14	1080	3.12	0.24
			0.14	1115	3.37	0.16
			0.19	860	2.35	0.08

Autoclaved cellular concrete masonry

Density	kg/m^3	$518 \leq \rho \leq 660$				
Specific heat capacity (dry)	J/(kg.K)	840				
Thermal resistance	$m^2.K/W$	$\theta = 20\,°C$, $w \leq w_c$, w in kg/m^3	$1/(A_1 + A_2 u)$			
			d (m)	ρ (kg/m^3)	A_1	A_2
				Mortar		
			0.15	524	1.23	0.007
				660	1.44	0.007
				Glue		
			0.15	518	1.13	0.006
				634	1.20	0.007
				Mortar		
			0.18	550	1.25	0.009

Gypsum plaster

Density	kg/m^3	975	
Specific heat capacity (dry)	J/(kg.K)	840	
Thermal conductivity	W/(m.K)	$569 \leq \rho \leq 981\ kg/m^3$ $\theta = 20\,°C$, w in kg/m^3	$0.263 + 0.001\ w$

Outside rendering

Density	kg/m^3	$878 \leq \rho \leq 1736$	
Specific heat capacity (dry)	J/(kg.K)	840	
Thermal expansion coefficient	K^{-1}	$\rho = 1736\ kg/m^3$	10.9×10^{-6}

Wood

Density	kg/m^3	400 (pine) $\leq \rho \leq$ 690 (beech)	
Specific heat capacity (dry)	J/(kg.K)	1880	
Thermal conductivity	W/(m.K)	Pine $\theta = 20\,°C$, $w = 0\ kg/m^3$	0.11

Particle board

Density	kg/m^3	$570 \leq \rho \leq 800$	
Specific heat capacity (dry)	J/(kg.K)	1880	
Thermal conductivity	W/(m.K)	$500 \leq \rho \leq 702\ kg/m^3$ $\theta = 20\,°C$, $w = 0\ kg/m^3$ $587 \leq \rho \leq 702\ kg/m^3$ $\theta = 20\,°C$, w in kg/m^3	$0.98 + 0.0001(\rho - 590)$ $0.106 + 1.3 \times 10^{-4}\ w$ $+ 3.3 \times 10^{-7}\ w^2$

Plywood

Density	kg/m^3	$445 \le \rho \le 799$	
Specific heat capacity (dry)	J/(kg.K)	1880	
Thermal conductivity	W/(m.K)	$445 \le \rho \le 692$ kg/m^3 $\theta = 20$ °C, $w = 0$ kg/m^3	$0.020 + 1.7 \times 10^{-4} \rho$
		$445 \le \rho \le 799$ kg/m^3 $\theta = 20$ °C, w in kg/m^3	$0.113 + 3.1 \times 10^{-4}\, w$

Fibre cement

Density	kg/m^3	$823 \le \rho \le 2052$	
Specific heat capacity (dry)	J/(kg.K)	840	
Thermal conductivity	W/(m.K)	$823 \le \rho \le 866$ kg/m^3 $\theta = 20$ °C	$0.14 + 5.8 \times 10^{-4}\, w$
		$\rho = 1495$ kg/m^3 $\theta = 20$ °C	$0.42 + 1.2 \times 10^{-3}\, w$

Gypsum board

Weight	kg/m^2	$6.5 \le \rho \le 13$	
Specific heat capacity (dry)	J/(kg.K)	840	
Equivalent thermal conductivity	W/(m.K)	$\theta = 20$ °C, $d = 9.5$ mm w in kg.m^3	$0.07 + 2.1 \times 10^{-4}\, w$

6.3.3.2 Insulation materials

Cork

Density	kg/m^3	111	
Specific heat capacity (dry)	J/(kg.K)	1880	
Thermal conductivity	W/(m.K)	$\theta = 20$ °C, $w = 0$ kg/m^3	0.042

Cellular glass

Density	kg/m^3	$113 \leq \rho \leq 140$	
Specific heat capacity (dry)	J/(kg.K)	840	
Thermal conductivity	W/(m.K)	$113 \leq \rho \leq 139$ kg/m^3 $\theta = 20\,°C$, $\psi = 0\%$ m^3/m^3	$0.037 + 8.8 \times 10^{-5}\rho$
		$126 \leq \rho \leq 134$ kg/m^3 $\theta = 20\,°C$, ψ in % m^3/m^3	$0.047 + 8.2 \times 10^{-4}\psi$
		$\rho = 129$ kg/m^3 $0 \leq \theta \leq 35$	$0.046 + 2.4 \times 10^{-4}\theta$

Glass fibre

Density	kg/m^3	$11.6 \leq \rho \leq 136$	
Specific heat capacity (dry)	J/(kg.K)	840	
Thermal conductivity	W/(m.K)	$\theta = 20\,°C$, $\psi = 0\%$ m^3/m^3	$0.0268 + 4.9 \times 10^{-5}\rho$ $+0.178/\rho$

Mineral fibre

Density	kg/m^3	$32 \leq \rho \leq 191$	
Specific heat capacity (dry)	J/(kg.K)	840	
Thermal conductivity	W/(m.K)	$\theta = 20\,°C$, $\psi = 0\%$ m^3/m^3	$0.0317 + 2.6 \times 10^{-5}\rho$ $+0.206/\rho$
		$\theta = 10\,°C$, $\psi = 0\%$ m^3/m^3	$0.026 + 5.5 \times 10^{-5}\rho$ $+0.331/\rho$

EPS (expanded polystyrene)

Property	Unit	Condition	Value			
Density	kg/m^3	$13 \leq \rho \leq 40$				
Specific heat capacity (dry)	J/(kg.K)	1470				
Thermal conductivity	W/(m.K)	$\psi = 0\%\ m^3/m^3$	$A_1 + A_2\rho + A_3/\rho$			
			θ (°C)	A_1	A_2	$A_3 \times 10^{-4}$
			10	0.017	1.9	0.258
			20	0.021	1.2	0.235
		$\theta = 20\,°C$, ψ in $\%\ m^3/m^3$	$B_1 + B_2\psi$			
			ρ (kg/m^3)	B_1		$B_2 \times 10^{-3}$
			15	0.0390		2.0
			20	0.0348		1.9
			25	0.0326		2.7
			30	0.0331		1.2
		$[1,0]\rho = 15\,kg/m^3$, d in m	$0.029 + 0.017d^{0.25}$			
		θ in °C	$0.0354 + 1.6 \times 10^{-4}\theta$			

XPS (extruded polystyrene)

Property	Unit	Condition	Value
Density	kg/m^3	$25 \leq \rho \leq 55$	
Specific heat capacity (dry)	J/(kg.K)	1470	
Thermal conductivity	W/(m.K)	$\theta = 10\,°C$, $\psi = 0\%\ m^3/m^3$	$0.0174 + 1.6 \times 10^{-5}\rho + 0.263/\rho$
		$\theta = 20\,°C$, $\psi = 0\%\ m^3/m^3$	$0.0404 + 3.9 \times 10^{-4}\rho + 0.029/\rho$
		$\theta = 10\,°C$, $\rho = 35\,kg/m^3$, ψ in $0\%\ m^3/m^3$	$0.0240 + 1.6 \times 10^{-4}\psi + 5.8\ 10^{-5}\psi^2$
		$\theta = 20\,°C$, $\rho = 35\,kg/m^3$, ψ in $0\%\ m^3/m^3$	$0.0251 + 5.2 \times 10^{-5}\psi + 7.0\ 10^{-5}\psi^2$

PUR/PIR (polyurethane foam, polyisocyanurate foam)

Density	kg/m^3	$20 \le \rho \le 40$	
Specific heat capacity (dry)	J/(kg.K)	1470	
Thermal conductivity	W/(m.K)	$\theta = 10\,°C$, $\psi = 0\%\ m^3/m^3$	$-0.112 + 1.9 \times 10^{-3}\rho + 2.36/\rho$
		$\theta = 20\,°C$, $\psi = 0\%\ m^3/m^3$	$-0.008 + 5.1 \times 10^{-4}\rho + 0.436/\rho$

Perlite board

Density	kg/m^3	$135 \le \rho \le 215$				
Specific heat capacity (dry)	J/(kg.K)	1000				
Thermal conductivity	W/(m.K)	$\theta = 20\,°C$, $\psi = 0\%\ m^3/m^3$	$0.046 + 0.00014(\rho - 100)$			
		$\theta = 20\,°C$, ψ in $\%\ m^3/m^3$	$B_1 + B_2\psi$			
			ρ (kg/m^3)	θ (°C)	B_1	$B_2 \times 10^{-3}$
			142	10	0.052	1.5
			142	20	0.053	1.8
			171	10/20	0.059	2.9
			212	20	0.058	5.7
		$\rho = 140\ kg/m^3$	$D_1 + D_2\theta$			
			ψ ($\%\ m^3/m^3$)	D_1	$D_2 \times 10^{-4}$	
			0	0.047	1.2	
			1.6	0.056	5.7	

6.4 Materials, air-related properties

6.4.1 Design values

The 2013 ASHRAE Handbook of Fundamentals, chapter 26, contains a list of air permeability values for several building and insulation materials.

6.4.1.1 Measured values

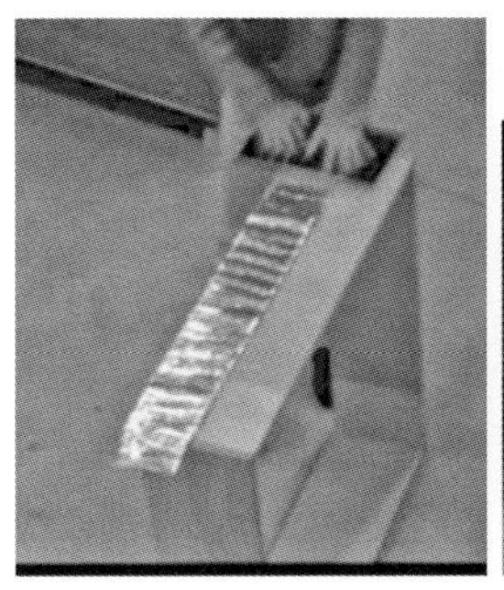

Figure 6.1 Preparing wall samples for measuring the air permeance.

Masonry, air permeance, per m^2 ($K_a = a\Delta P_a^{b-1}$)

Component	a kg/(s.m^2.Pa^b)	b
Facing brick veneer walls		
Bricks of $19 \times 9 \times 4.5$ cm^3, joints not pointed, badly filled	3.24×10^{-4}	0.69
Bricks of $19 \times 9 \times 4.5$ cm^3, joints not pointed, well filled	1.08×10^{-4}	0.75
Bricks of $19 \times 9 \times 4.5$ cm^3, joints pointed, 1	3.48×10^{-5}	0.80
Bricks of $19 \times 9 \times 6.5$ cm^3, joints pointed, 2	3.60×10^{-5}	0.78
Bricks of $19 \times 9 \times 6.5$ cm^3, joints not pointed, badly filled	7.20×10^{-4}	0.68
Bricks of $19 \times 9 \times 6.5$ cm^3, joints not pointed, well filled	1.56×10^{-4}	0.71
Bricks of $19 \times 9 \times 6.5$ cm^3, joints pointed, 1	3.60×10^{-5}	0.81
Bricks of $19 \times 9 \times 6.5$ cm^3, joints pointed, 2	3.48×10^{-5}	0.82
Component	a kg/(s.m^2.Pa^b)	b
Concrete block veneer walls		
Concrete blocks $19 \times 9 \times 9$ cm^3, 1955 kg/m^3, joints pointed	1.44×10^{-4}	0.88
Concrete blocks $19 \times 9 \times 9$ cm^3, 1927 kg/m^3, joints pointed	1.92×10^{-4}	0.86
Concrete blocks $19 \times 9 \times 9$ cm^3, 1881 kg/m^3, joints pointed	2.76×10^{-4}	0.82
Concrete blocks $19 \times 9 \times 9$ cm^3, 2109 kg/m^3. Joints pointed	2.42×10^{-5}	0.85
Concrete blocks $19 \times 9 \times 9$ cm^3, 2153 kg/m^3. Joints pointed	2.29×10^{-5}	0.83
Concrete blocks $19 \times 9 \times 9$ cm^3, 2260 kg/m^3. Joints pointed	4.27×10^{-5}	0.77
Concrete blocks $19 \times 9 \times 9$ cm^3, 2091 kg/m^3. Joints pointed	6.73×10^{-5}	0.86

Component	a kg/(s.m^2.Pa^b)	b
Inside leafs		
Lightweight brickwork 19 × 14 × 14 cm^3, joints not pointed 1	2.64×10^{-3}	0.59
Lightweight brickwork 19 × 14 × 14 cm^3, joints not pointed 2	4.20×10^{-3}	0.57
Lightweight brickwork 19 × 14 × 14 cm^3, joints pointed 1	2.88×10^{-5}	0.72
Lightweight brickwork 19 × 14 × 14 cm^3, joints pointed 2	2.04×10^{-5}	0.81
Lightweight brickwork 19 × 14 × 14 cm^3, joints pointed 3	1.68×10^{-5}	0.82
Lightweight brickwork 19 × 14 × 14 cm^3, plastered at the inside 1	3.72×10^{-7}	0.96
Lightweight brickwork 19 × 14 × 14 cm^3, plastered at the inside 2	2.76×10^{-7}	0.97

Masonry, air permeance, per m^2 *($K_a = a\Delta P_a^{b-1}$), continued*

Component	a kg/(s.m^2.Pa^b)	b
Hollow concrete blocks 39 × 19 × 14 cm^3, 987 kg/m^3, joints not pointed	3.96×10^{-3}	0.58
Hollow concrete blocks, 39 × 19 × 14 cm^3 954 kg/m^3, joints not pointed	5.04×10^{-3}	0.56
Hollow concrete blocks 39 × 19 × 14 cm^3, 910 kg/m^3, joints not pointed	6.00×10^{-3}	0.55
Hollow concrete blocks 39 × 19 × 14, 987 kg/m^3, joints pointed	1.92×10^{-4}	0.91
Hollow concrete blocks 39 × 19 × 14, 954 kg/m^3, joints pointed	3.72×10^{-4}	0.79
Hollow concrete blocks 39 × 19 × 14, 910 kg/m^3, joints pointed	6.00×10^{-4}	0.70
Hollow concrete blocks 39 × 19 × 14, 987 kg/m^3, plastered at the inside	3.48×10^{-7}	0.96
Hollow concrete blocks 39 × 19 × 14, 954 kg/m^3, plastered at the inside	2.64×10^{-7}	0.95
Hollow concrete blocks 39 × 19 × 14, 910 kg/m^3, plastered at the inside	3.12×10^{-7}	0.97
Cellular concrete 60 × 24 × 14 cm^3, 510 kg/m^3, joints glued/not pointed, 1	2.28×10^{-3}	0.61
Cellular concrete 60 × 24 × 14 cm^3, 510 kg/m^3, joints glued/not pointed, 2	1.44×10^{-3}	0.62
Cellular concrete 60 × 24 × 14 cm^3, 510 kg/m^3, joints glued and pointed, 1	9.60×10^{-5}	0.64
Cellular concrete 60 × 24 × 14 cm^3, 510 kg/m^3, joints glued and pointed, 2	1.02×10^{-4}	0.63
Cellular concrete 60 × 24 × 14 cm^3, 510 kg/m^3, joints glued, plastered inside	4.08×10^{-7}	0.97
Sand-limestone, 29 × 14 × 14 cm^3, 1140 kg/m^3, joints not pointed	3.24×10^{-3}	0.61

Component	a kg/(s.m^2.Pa^b)	b
Sand-limestone, 29 × 14 × 14 cm^3, 1140 kg/m^3, joints not pointed	4.20×10^{-3}	0.57
Sand-limestone, 29 × 14 × 14 cm^3, 1140 kg/m^3, joints pointed	2.28×10^{-5}	0.75
Sand-limestone, 29 × 14 × 14 cm^3, 1140 kg/m^3, joints pointed	1.80×10^{-5}	0.80
Sand-limestone, 29 × 14 × 14 cm^3, 1140 kg/m^3, plastered at the inside	3.00×10^{-7}	0.95

Metal Construction, air permeance, per m^2 ($K_a = a\Delta P_a^{b-1}$)

Figure 6.2 The sheet metal component.

Sheet metal component: see Figure 6.2	a kg/(s.m^2.Pa^b)	b
No special care for airtightness	7.9×10^{-5}	0.97
Screw holes at the columns sealed	6.7×10^{-5}	0.90
Screw holes sealed and joints between metal boxes taped	1.6×10^{-5}	0.92

Roof Covers, air permeance, per m² ($K_a = a\Delta P_a^{b-1}$)

Layer	a kg/(s.m².Pa^b)	b
Ceramic tiles, single lock, 9.2 m run of joints per m² (Roman tiles)	0.014	0.55
Ceramic tiles, double lock, 9.2 m run of joints per m² (Roman tiles)	0.019	0.50
Ceramic tiles, double lock, 9.2 m run of joints per m² (Roman tiles), 1 ventilation tile per m²	0.016	0.50
Ceramic tiles, double lock, 8.3 m run of joints per m² (Pan tiles)	0.015	0.50
Concrete tiles, double lock and overlap, 6.2 m run of joints and overlaps per m²	0.011	0.68
Quarry slates, three slates thick	0.014	0.62
Fibre cement slates, three slates thick	0.0081	0.54
Metal tiles, single lock, 3.5 m run of locked joints per me	0.011	0.54
Fibre cement corrugated boards	0.0042	0.79
Metal roof with standing seam, outside the fixing zone	0.0054	0.64
Metal roof with standing seam, in the fixing zone (extends 15 cm at both sides of the fixation)	0.0014	0.70
Roof sandwich element, composed of (top to bottom): – corrugated 0.8 mm thick aluminium sheet as cover – 50 mm EPS with a top surface that follows the corrugations of the aluminium sheet – 0.17 mm thick aluminium vapour barrier Total surface: 4.93 m². Contains two joints, parallel to the slope, and one overlap, orthogonal to the slope. The parallel joints are formed by one corrugation overlapping at the cover side and a profiled insert inside	0.0017	0.79

Underlays, air permeance, per m² ($K_a = a\Delta P_a^{b-1}$)

	a Kg/(s.Pa^b)	b
Fibre/cellulose cement plate, $d = 3.2$ mm, perfectly closed overlap	0.000 42	0.66
Fibre/cellulose cement plate, $d = 3.2$ mm, overlap 1.4 mm open	0.000 42	0.69
Fibre/cellulose cement plate, $d = 3.2$ mm, overlap 3.6 mm open	0.0032	0.60
Fibre/cellulose cement plate, $d = 3.2$ mm, overlap 13 mm open	0.01	0.55
Spun-bonded PE $d = 0.42$ mm, 140 g/m²		
– Sample 1	1.5×10^{-6}	0.97
– Sample 2	2.5×10^{-6}	1.0
– Sample 3	2.8×10^{-6}	1.0
Spun-bonded PE, 11 perforations ϕ1.2 mm per m²	4.6×10^{-6}	1.0
Spun-bonded PE, 11 perforations ϕ1.7 mm per m²	7.6×10^{-6}	0.96
Spun-bonded PE, 11 perforations ϕ2.7 mm per m²	1.7×10^{-5}	0.85
Spun-bonded PE, 11 perforations ϕ3.0 mm per m²	2.2×10^{-5}	0.82
Spun-bonded PE, 11 perforations ϕ5.0 mm per m²	9.2×10^{-5}	0.66
Spun-bonded PE, 11 perforations ϕ8.5 mm per m²	3.2×10^{-4}	0.63
Spun-bonded PE, with overlap	1.3×10^{-5}	0.67
Spun-bonded PE, 11 staples per m²	2.8×10^{-6}	1.0
Spun-bonded PE, 33 staples per m²	3.2×10^{-6}	1.0
Spun-bonded PE, 55 staples per m²	3.3×10^{-6}	1.0
Spun-bonded PE, 111 staples per m²	3.3×10^{-6}	1.0
Spun-bonded PE, overlapping, overlaps continuous sealed		
– Sample 1	0.8×10^{-6}	1.0
– Sample 2	9.5×10^{-6}	0.94
– Sample 3	3.1×10^{-6}	1.0
Bitumen impregnated polypropylene foil, $d = 0.6$ mm, 496 g/m²	7.9×10^{-7}	0.43
Bitumen impregnated polypropylene foil, $d = 0.6$ mm, 496 g/m², with overlap	6.5×10^{-5}	0.37
Bituminous underlay foil		
– Sample 1	5.3×10^{-7}	0.99
– Sample 2	6.1×10^{-7}	0.97
Bituminous underlay foil, 11 perforations ϕ1.2 mm per m²	6.9×10^{-6}	0.74
Bituminous underlay foil, 11 perforations ϕ1.7 mm per m²	1.7×10^{-5}	0.67
Bituminous underlay foil, 11 perforations ϕ2.7 mm per m²	5.1×10^{-5}	0.60
Bituminous underlay foil, 11 perforations ϕ3.0 mm per m²	7.3×10^{-5}	0.59
Bituminous underlay foil, 11 perforations ϕ5.0 mm per m²	2.1×10^{-4}	0.58
Bituminous underlay foil, 11 perforations ϕ8.5 mm per m²	7.0×10^{-4}	0.54
Micro-perforated, glass-fibre reinforced plastic foil	9×10^{-5}	0.77
– with overlap of 20 cm, per m run	0.005	0.2
– with overlap of 30 cm, per m run	0.0025	0.43
– with overlap of 50 cm, per m run	0.0012	0.43

Insulating Materials, air permeability

Material	Air permeability (k_a) kg/(m.s.Pa)
Fiberglas (ρ: density in kg/m^3)	$4.2 \times 10^{-3} \rho^{-1.24}$
Mineral fibre (ρ: density in kg/m^3)	$1.8 \times 10^{-2} \rho^{-1.44}$
Fiberglas 18 kg/m^3, along the thickness	7.2×10^{-5}
Fiberglas, 42 kg/m^3, along the thickness	5.2×10^{-5}
Fiberglas, 42 kg/m^3, orthogonal to the thickness	8.1×10^{-5}
Fiberglas, 78 kg/m^3, along the thickness	3.9×10^{-5}
Fiberglas, 78 kg/m^3, orthogonal to the thickness	5.4×10^{-5}
Fiberglas, 147 kg/m^3, along the thickness	2.2×10^{-5}
Fiberglas, 147 kg/m^3, orthogonal to the thickness	2.9×10^{-5}

Air permeance, per m^2 ($K_a = a\Delta P_a^{b-1}$)

Material	a kg/(s.m^2.Pa^b)	b
EPS, $d = 64.4$ mm, $\rho = 10.2$ kg/m^3	0.0047	0.50
EPS, $d = 63.6$ mm, $\rho = 10.1$ kg/m^3	0.017	0.823
EPS, $d = 39.1$ mm, $\rho = 11.6$ kg/m^3	7.3×10^{-6}	0.94
EPS, $d = 39.1$ mm, $\rho = 11.5$ kg/m^3	8.4×10^{-6}	0.93
EPS, $d = 50.1$ mm, $\rho = 14.1$ kg/m^3	0.000 045	0.94
EPS, $d = 50.2$ mm, $\rho = 14.6$ kg/m^3	0.000 053	0.90
EPS, $d = 40.9$ mm, $\rho = 45.1$ kg/m^3	0.000 03	0.95
EPS, $d = 39.3$ mm, $\rho = 64.2$ kg/m^3	0.000 058	0.90
EPS, $d = 49.0$ mm, $\rho = 64.5$ kg/m^3	0.000 028	0.92
EPS, $d = 49.6$ mm, $\rho = 63.7$ kg/m^3	5.5×10^{-6}	0.93
EPS, $d = 80$ mm, $\rho = 19.8$ kg/m^3	0.000 018	0.93
EPS, $d = 80$ mm, $\rho = 20.2$ kg/m^3	0.000 019	0.86
EPS, $d = 100$ mm, $\rho = 20.6$ kg/m^3	0.000 016	0.90
EPS, $d = 100$ mm, $\rho = 21.2$ kg/m^3	0.000 032	0.88
EPS, $d = 120$ mm, $\rho = 20.5$ kg/m^3	0.000 023	0.97
EPS, $d = 120$ mm, $\rho = 19.7$ kg/m^3	0.000 015	0.93
EPS, $d = 140$ mm, $\rho = 19.9$ kg/m^3	6.8×10^{-6}	0.91
EPS, $d = 140$ mm, $\rho = 20.3$ kg/m^3	6.7×10^{-6}	0.91
EPS, $d = 80$ mm, $\rho = 28.3$ kg/m^3	9.4×10^{-6}	0.89
EPS, $d = 80$ mm, $\rho = 28.5$ kg/m^3	0.000 013	0.83
EPS, $d = 100$ mm, $\rho = 30.5$ kg/m^3	0.000 015	0.91
EPS, $d = 100$ mm, $\rho = 30.9$ kg/m^3	0.000 013	0.87
EPS, $d = 120$ mm, $\rho = 29.7$ kg/m^3	0.000 017	0.89
EPS, $d = 120$ mm, $\rho = 27.7$ kg/m^3	8.9×10^{-6}	0.91
EPS, $d = 140$ mm, $\rho = 29.0$ kg/m^3	4.5×10^{-6}	0.81
EPS with aluminium foil backing, $d = 80$ mm, $\rho = 28.7$ kg/m^3	9.0×10^{-6}	0.74
EPS with aluminium foil backing, $d = 100$ mm, $\rho = 29.7$ kg/m^3	8.2×10^{-7}	0.97
EPS with aluminium foil backing, $d = 120$ mm, $\rho = 30.0$ kg/m^3	7.2×10^{-6}	0.81
EPS with aluminium foil backing, $d = 140$ mm, $\rho = 29.9$ kg/m^3	6.6×10^{-6}	0.80
Glass-fibre bat, $d = 6$ cm, bituminous paper backing, overlaps taped	0.000 065	0.71
	0.000 092	0.70

Air permeance, per m^2 ($K_a = a\Delta P_a^{b-1}$)

Material	a kg/(s.m^2.P$_a$^{b})	b
Glass-fibre bat, $d=6$ cm, bituminous paper backing, overlaps taped, staples alongside the tape	0.000 075	0.74
Glass-fibre bat, $d=6$ cm, bituminous paper backing, overlaps taped, 1 nail ϕ3 mm per m^2 perforating the bat	0.000 071 0.000 097	0.75 0.64
Glass-fibre bat, $d=6$ cm, bituminous paper backing, overlaps taped, bat ripped (L = 40 mm, B-40 mm, 1 per m^2)	0.0002	0.61
Glass-fibre bat, $d=6$ cm, bituminous paper backing, overlaps held together with nailed lath	0.001 0.000 33	0.56 0.76
Glass-fibre bat, $d=6$ cm, bituminous paper backing, overlaps not overlapping	0.0032 0.000 89	1.00 0.85
Glass-fibre bat, $d=6$ cm, aluminium paper backing, overlaps taped	0.000 047	1.00
Glass-fibre bat, $d=6$ cm, aluminium paper backing, overlaps taped, staples alongside the tape	0.000 048	1.00
Glass-fibre bat, $d=6$ cm, aluminium paper backing, overlaps taped, 1 nail ϕ3 mm per m^2 perforating the bat	0.000 05	1.00
Glass-fibre bat, $d=6$ cm, aluminium paper backing, overlaps taped, bat ripped (L = 40 mm, B-40 mm, 1 per m^2)	0.000 085	1.00
Mineral fibre, dense boards, 170 kg/m^3, $d=10$ cm	0.000 23	0.89
XPS, $d=5$ cm, mounted tongue and groove	0.000 40	0.59
XPS, $d=5$ cm, mounted with well closed joints	0.000 47	0.54
XPS, $d=5$ cm, mounted with open joints of 2 mm	0.002	0.55
EPS, $d=12$ m, 9.3 kg/m^3, plain board	0.000 077	0.91
Composite element particle board 4 mm/EPS 88 mm/particle board 4 mm with EPS/EPS joints – 0.5 mm wide – 7.5 mm wide – 15 mm wide	 0.000 71 0.0023 0.010	 0.56 0.60 0.53
Composite element particle board 4 mm/PUR 61 mm/particle board 4 mm with timber/timber joints – 0.3 mm wide – 3.5 mm wide – 8.0 mm wide – 16 mm wide	 0.000 65 0.000 74 0.0011 0.0015	 0.56 0.63 0.61 0.55

Material	a $kg/(s.m^2.Pa^b)$	b
Composite element: particle board 4 mm/PUR 88 mm/particle board 4 mm with PUR/PUR joints		
– 0.2 mm wide	0.000 15	0.66
– 4.3 mm wide	0.000 14	0.65
– 16 mm wide	0.015	0.53
XPS board, $d = 60$ mm, $\rho = 39$ kg/m^3	5.7×10^{-7}	1.00
– Sample 1	4.4×10^{-7}	1.00
– Sample 2	6.7×10^{-7}	1.00
– Sample 3		
XPS-board, $d = 60$ mm, $\rho = 39$ kg/m^3 with gypsum board, glued to the XPS, as inside lining		
– Sample 1	5.6×10^{-7}	1.00
– Sample 2	6.6×10^{-7}	1.00
– Sample 3	4.0×10^{-7}	1.00
– Sample 4	9.0×10^{-7}	1.00

Air permeance, per metre run ($K_a = a\Delta P_a^{b-1}$)

Layer	a kg/(s.m.Pa^b)	b
Composite element: plywood/30 mm air cavity/120 mm mineral fibre insulation/aluminium vapour retarder/plywood. The elements have timber/timber joints, with a wooden insert close to the outside surface and a special joint profile which should guarantee airtightness		
Element loaded with 65 kg/m^2		
– Joint perfectly closed	0.000 018	0.87
– Joint perfectly closed, joint profile removed	0.000 24	0.64
– Joint 1 mm open	0.000 086	0.69
– Joint 2 mm open	0.000 48	0.68
– Joint 3 mm open	0.0013	0.57
Element unloaded	0.000 24	0.66
– Joint 1 mm open	0.000 32	0.75
– Joint 2 mm open	0.0017	0.57
– Joint 5 mm open		
XPS, $d=50$ mm (manufacturer 1), tongue and groove (α)	0.000 17	0.70
	0.000 60	0.60
	0.0014	0.50
– Joint 0 mm open		
– Joint 5 mm open		
– Joint 10 mm open		
XPS, $d=50$ mm (man. 2), tongue and groove (α)		
– Joint width 0 mm	0.000 025	1.0
– Joint width 5 mm	0.000 19	0.70
– Joint width 10 mm	0.000 36	0.60
XPS, $d=50$ mm (manufacturer 2, other batch), tongue and groove (α)		
– Joint width 0 mm	0.000 042	1.0
– Joint width 5 mm	0.000 14	0.70
– Joint width 10 mm	0.000 35	0.60
XPS, $d=50$ mm (manufacturer 3), tongue and groove (α)		
– Joint width 0 mm	8.4×10^{-6}	1.0
– Joint width 5 mm	0.000 12	0.70
– Joint width 10 mm	0.000 96	0.60

Layer	a kg/(s.m.Pa^b)	b
EPs, $d = 50$ mm (manufacturer 1), tongue and groove (α)		
– Joint width 0 mm	0.000 012	1.0
– Joint width 5 mm	0.000 096	1.0
– Joint width 10 mm	0.000 96	0.60
EPs, $d = 40$ mm (manufacturer 2), tongue and groove (α)	4.8×10^{-6}	1.0
– Joint width 0 mm	0.000 026	1.0
– Joint width 5 mm	0.000 24	0.60
– Joint width 10 mm		
Two rebated XPS boards on a rafter, joints between rafter and boards and between boards open		
– Sample 1	6.7×10^{-5}	1.0
– Sample 2	1.8×10^{-3}	0.75

Air permeance, per metre run ($K_a = a \Delta P_a^{b-1}$)

Layer	a kg/(s.m.Pa^b)	b
Two rebated XPS boards on a rafter, joints between rafter and boards sealed with a bituminous tape		
– Sample 1	2.2×10^{-5}	0.98
– Sample 2	1.1×10^{-4}	0.95
– Sample 3	2.7×10^{-5}	0.80
– Sample 4	4.1×10^{-5}	1.0
Two rebated XPS boards on a rafter, joints between rafter and boards and at the top, between the two boards, sealed with a bituminous tape		
– Sample 1	6.6×10^{-6}	0.96
– Sample 2	6.8×10^{-6}	0.95
– Sample 3	11.9×10^{-6}	0.95
Sample 4	4.1×10^{-6}	0.99
Two rebated XPS boards on a rafter, joints between rafter and boards and at the top, between the two boards, sealed with silicone joint filler		
– Sample 1	1.3×10^{-6}	0.97
– Sample 2	0.8×10^{-6}	1.00
– Sample 3	2.1×10^{-6}	0.91
– Sample 4	2.5×10^{-6}	0.92

Vapour Retarders and Air Barriers, air permeance, per m² ($K_a = a\Delta P_a^{b-1}$)

Layers	a Kg/(s.m².Pa^b)	b
PE-foil, $d = 0.2$ mm		
– Sample 1	5.5×10^{-11}	1.0
– Sample 2	3.9×10^{-7}	1.0
– Sample 3	4.8×10^{-7}	1.0
– Sample 4	4.3×10^{-7}	1.0
PE-foil, $d = 0.2$ mm, 11 perforations ϕ1.2 mm per m²	1.1×10^{-6}	0.91
PE-foil, $d = 0.2$ mm, 11 perforations ϕ1.7 mm per m²	2.5×10^{-6}	0.82
PE-foil, $d = 0.2$ mm, 11 perforations ϕ2.7 mm per m²	9.2×10^{-6}	0.71
PE-foil, $d = 0.2$ mm, 11 perforations ϕ3.0 mm per m²	2.4×10^{-5}	0.65
PE-foil, $d = 0.2$ mm, 11 perforations ϕ5.0 mm per m²	1.1×10^{-4}	0.57
PE-foil, $d = 0.2$ mm, 11 perforations ϕ8.5 mm per m²	4.7×10^{-4}	0.53
PE-foil, $d = 0.2$ mm, with overlap, overlap sealed		
– Sample 1 (bad)	6.0×10^{-6}	0.77
– Sample 2	4.9×10^{-7}	1.0
– Sample 3	4.4×10^{-7}	1.0
Plastic foil SD2		
– Sample 1	4.6×10^{-7}	1.0
– Sample 2	5.1×10^{-7}	1.0
– Sample 3	4.7×10^{-7}	1.0
Plastic foil SD2, 11 perforations ϕ1.2 mm per m²	1.9×10^{-6}	0.86
Plastic foil SD2, 11 perforations ϕ1.7 mm per m²	6.6×10^{-6}	0.73
Plastic foil SD2, 11 perforations ϕ2.7 mm per m²	2.4×10^{-5}	0.63
Plastic foil SD2, 11 perforations ϕ3.0 mm per m²	4.6×10^{-5}	0.60
Plastic foil SD2, 11 perforations ϕ5.0 mm per m²	1.2×10^{-4}	0.59
Plastic foil SD2, 11 perforations ϕ8.5 mm per m²	3.5×10^{-4}	0.60
Plastic foil SD2, 22 staples per m²	6.5×10^{-7}	1.0
Plastic foil SD2, 44 staples per m²	7.6×10^{-7}	0.96
Plastic foil SD2, 111 staples per m²	1.1×10^{-6}	0.94
Plastic foil SD2		
– Sample 1	7.3×10^{-7}	0.97
– Sample 2	7.1×10^{-7}	1.0
– Sample 3 (not well done)	5.3×10^{-6}	0.80

Internal Linings, air permeance, per m^2 ($K_a = a\Delta P_a^{b-1}$)

Layers	a kg/(s.m^2.Pa^b)	b
Lathed ceiling, $d = 1$ cm, tongue and groove	0.000 27 0.000 41	0.63 0.68
Lathed ceiling, $d = 1$ cm, tongue and groove, leak ϕ20 mm per m^2	0.000 76	0.63
Gypsum board, no joint	1.5×10^{-7}	0.89
Gypsum board, no joint, painted	1.2×10^{-7}	0.82
Gypsum board, joints reinforced and plastered	0.000 031	0.81
Gypsum board, joints reinforced and plastered, leak ϕ20 mm per m^2	0.00 038	0.61
Gypsum board, joint not plastered	0.000 33	0.73
Gypsum board, joint not plastered, leak ϕ20 mm per m^2	0.000 63	0.73
Gypsum board with alupaper backing, joints reinforced and plastered	0.000 013	1
Gypsum board with alupaper backing, joints reinforced and plastered, leak ϕ20 mm per m^2	0.000 47	0.53
Gypsum board with alupaper backing, joints not plastered, leak ϕ20 mm per m^2	0.000 56	0.59

6.5 Materials, moisture properties

6.5.1 Design values for the vapour resistance factor (ISO 10456)

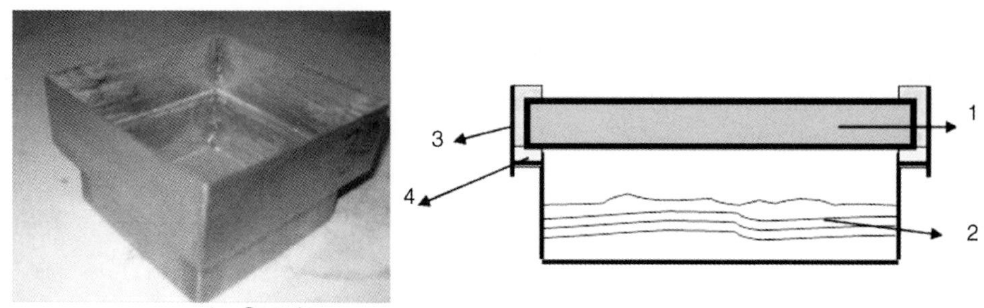

1. Sample
2. Desiccant, saturated salt solution or distilled water
3. Wax-fill
4. Silicon joint

Figure 6.3 Measuring vapour resistance factors.

6.5.1.1 Building and finishing materials

Group	Material	Density (ρ) kg/m^3	Vapour resistance factor μ (–)	
			Wet cup	Dry cup
Metals	Aluminium alloys	2800	∞	∞
	Duralumin	2800		
	Brass	8400		
	Bronze	8700		
	Copper	8900		
	Iron	7900		
	Iron, cast	7500		
	Steel	7800		
	Stainless steel	7900		
	Lead	11 300		
	Zinc	7100		

Group	Material	Density (ρ) kg/m^3	Vapour resistance factor μ (–)	
			Wet cup	Dry cup
Wood and wood-based materials	Softwood	500	50	20
	Hardwood	700	200	50
	Plywood (from low to high density)	300	150	50
		500	200	70
		700	220	90
		1000	250	110
	Particle board			
	– Soft	300	50	10
	– Semi-hard	600	50	15
	– Hard	900	50	20
	OSB	680	50	30
	Fibre board			
	– Soft	400	10	5
	– Semi-hard	600	20	12
	– Hard	800	30	20
Gypsum	Gypsum blocks	600	10	4
		900	10	4
		1200	10	44
		1500	10	
	Gypsum board	900	10	4
Mortars	Normal mortar, mixed at site	1800	10	6
Plasters	Gypsum plaster			
	– Light	600	10	6
	– Normal	1000	10	6
	– Heavy	1300	10	6
	Lime or gypsum, sand	1600	10	6
	Cement, sand	1700	10	6
Concrete	Medium to high density	1800	100	60
		2000	100	60
		2200	120	70
		2400	130	80
	Reinforced 1% steel 2% steel	2300	130	80
		2400	130	80

Group	Material	Density (ρ) kg/m^3	Vapour resistance factor μ (–)	
			Wet cup	Dry cup
Stone	Crystalline rock	2800	10 000	10 000
	Sedimentary rock	2600	250	200
	Sedimentary, light	1500	30	20
	Lava	1600	20	15
	Basalt	2700–3000	10 000	10 000
	Gneiss	2400–2700	10 000	10 000
	Granite	2500–3000	10 000	10 000
	Marble	2800	10 000	10 000
	Slate	2000–2800	1000	800
	Limestone			
	– Extra soft	1600	30	20
	– Soft	1800	40	25
	– Semi-hard	2000	50	40
	– Hard	2200	200	150
	– Extra hard	2600	250	200
	Sandstone	2600	40	302.30
	Natural pumice	400	8	6
	Artificial stone	1750	50	40
Soils	Clay or silt	1200–1800	50	50
	Sand and gravel	1700–2200	50	50
Water, ice, snow	Ice, −10 °C	920	No data	No data
	Ice, 0 °C	900		
	Water, 10 °C	1000		
	Water, 40 °C	990		
	Snow			
	– Fresh (< 30 mm)	100		
	– Soft (30–70 mm)	200		
	– Slightly compacted (70–100 mm)	300		
	– Compacted (> 200 mm)	500		
Plastics, solid	Acrylic	1050	10 000	10 000
	Polycarbonates	1200	5000	5000
	PFTE	2200	10 000	10 000
	PVC	1390	50 000	50 000
	PMMA	1180	50 000	50 000
	Polyacetate	1410	100 000	100 000
	Polyamide (nylon)	1150	50 000	50 000
	Nylon, 25% glass fibre	1450	50 000	50 000
	PE, high density	980	100 000	100 000
	PE, low density	920	100 000	100 000
	Polystyrene	1050	100 000	100 000
	Polypropylene (PP)	910	10 000	10 000

Group	Material	Density (ρ) kg/m^3	Vapour resistance factor μ (−)	
			Wet cup	Dry cup
	PP, 25% glass fibre	1200	10 000	10 000
	Polyurethane	1200	6000	6000
	Epoxy resin	1200	10 000	10 000
	Phenol resin	1300	100 000	100 000
	Polyester resin	1400	10 000	10 000
Rubbers	Natural	910	10 000	10 000
	Neoprene	1240	10 000	10 000
	Butyl	1200	200 000	200 000
	Foam rubber	60–80	7000	7000
	Hard rubber	1200	∞	∞
	EPDM	1150	6000	6000
	Polyisobutylene	920	10 000	10 000
	Polysulphide	1700	10 000	10 000
	Butadiene	980	100 000	100 000
Glass	Quartz glass	2200	∞	∞
	Normal glass	2500	∞	∞
	Glass mosaic	2000	∞	∞
Gases	Air	1.23	1	1
	Argon	1.70	1	1
	Carbon dioxide	1.95	1	1
	Sulphur hexafluoride	6.36	1	1
	Krypton	3.56	1	1
	Xenon	5.68	1	1
Thermal breaks, sealants, weather stripping	Silica gel (desiccant)	720	∞	∞
	Silicone foam	750	10 000	10 000
	Silicone pure	1200	5000	5000
	Silicone, filled	1450	5000	5000
	PUR (thermal break)	1300	60	60
	PVC, flexible with 40% softener	1200	100 000	100 000
	Elastomeric foam, flexible	60–80	10 000	10 000
	Polyurethane foam	70	60	60
	Polyethylene foam	70	100	100
Roofing materials	Asphalt	2100–2300	50 000	50 000
	Bitumen, pure	1050	50 000	50 000
	Bitumen felt	1100	50 000	50 000
	Clay tiles	2000	40	30
	Concrete tiles	2100	100	60

Group	Material	Density (ρ) kg/m^3	Vapour resistance factor μ (−)	
			Wet cup	Dry cup
Floor coverings	Rubber	1200	10 000	10 000
	Plastic	1700	10 000	10 000
	Linoleum	1200	1000	800
	Carpet/textile	200	5	5
	Underlay, rubber	270	10 000	10 000
	Underlay, felt	120	20	15
	Underlay, wool	200	20	15
	Underlay, cork	200	20	15
	Ceramic tiles	2300	∞	∞
	Plastic tiles	1000	10 000	10 000
	Cork tiles, light	200	20	15
	Cork tiles, heavy	500	40	20

6.5.1.2 Insulation materials

Material	Density (ρ) kg/m^3	Vapour resistance factor μ (–)
Cork	90–160	10
Glass– and mineral fibre	10–200	1
EPS	10–50	30–70
XPS	25–65	100–300
PUR	28–55	50–100
VIPs		∞

6.5.2 Measured values

For the test methods used, see the appropriate standards. But not that relative humidity is often maintained using saturated salt solutions. Do not forget that related equilibrium relative humidity depends on temperature, as Table 6.3, which gives test data, underlines.

Table 6.3 Equilibrium relative humidity above saturated salt solutions.

Lithium Chloride	Temperature (°C)	0.2	9.6	19.2	29.6	39.6	46.8
LCl H_2O	RH (%)	14.7	13.4	12.4	11.8	11.8	11.4
Magnesium Chloride	Temperature (°C)	0.4	9.8	19.5	30.2	40.0	48.1
MgCl 6 H_2O	RH (%)	35.2	34.1	33.4	33.2	32.7	31.4
Sodium Dichromate	Temperature (°C)	0.6	10.1	19.8	30.0	37.4	47.3
$Na_2Cr_2O_2$ 2 H_2O	RH (%)	60.4	57.8	55.5	52.4	50.4	48.0
Magnesium Nitrate	Temperature (°C)	0.4	9.9	19.6	30.5	40.2	48.1
$Mg(NO_3)_2$ 6 H_20	RH (%)	60.7	57.5	55.8	51.6	49.7	46.2
Sodium chloride	Temperature (°C)	0.9	10.2	20.3	30.3	39.2	48.3
MaCl	RH (%)	75.0	75.3	75.5	75.6	74.6	74.9
Ammonium sulphate	Temperature (°C)	0.4	10.1	20.0	30.9	40.0	48.0
$(NH_4)_2SO_4$	RH (%)	83.7	81.8	80.6	80.0	80.1	79.2
Potassium nitrate	Temperature (°C)	0.6	10.2	20.0	30.7	40.4	48.1
KNO_3	RH (%)	97.0	95.8	93.1	90.6	88.0	85.6
Potassium sulphate	Temperature (°C)	0.5	10.1	19.8	30.4	39.9	48.1
K_2SO_4	RH (%)	99.0	98.0	97.1	96.8	96.1	96.0

6.5.2.1 Building materials

Stone

Baumberger							
Density	kg/m^3	$\rho = 1980$					
Hygroscopic curve	kg/m^3	$\phi = 0.1$	0.3	0.5	0.65	0.8	0.9
Sorption		8.5	17.6		27.5	35.6	43.1
Capillary moisture content	kg/m^3	210					
Saturation moisture content	kg/m^3	230					
Diffusion resistance factor	–	$\phi = 0.265$	0.535		0.715		0.85
		20	17		14		8.8
Capillary water absorption coeff.	kg/(m^2.s$^{0.5}$)	0.044					

Obernkirchner							
Density	kg/m^3	$\rho = 2150$					
Hygroscopic curve	kg/m^3	$\phi = 0.1$	0.3	0.5	0.65	0.8	0.9
Sorption		0.6	1.3		2.6	3.4	4.3
Capillary moisture content	kg/m^3	110					
Saturation moisture content	kg/m^3	140					
Diffusion resistance factor	—	$\phi = 0.265$	0.535		0.715		0.85
		32	30		28		18
Capillary water absorption coeff.	kg/(m^2.s$^{0.5}$)	0.046					

Stone

Rüthener							
Density	kg/m^3	$\rho = 1950$					
Hygroscopic curve	kg/m^3	$\phi = 0.1$	0.3	0.5	0.65	0.8	0.9
Sorption		1.8	4.5		8.0	12.4	16.9
Capillary moisture content	kg/m^3	200					
Saturation moisture content	kg/m^3	240					
Diffusion resistance factor	—	$\phi = 0.265$	0.535		0.715		0.85
		17	16		13		9.4
Capillary water absorption coeff.	$kg/(m^2.s^{0.5})$	0.30					
Sander							
Density	kg/m^3	$\rho = 2120$					
Hygroscopic curve	kg/m^3	$\phi = 0.1$	0.3	0.5	0.65	0.8	0.9
Sorption		4.4	10.2		15.2		22.6
Capillary moisture content	kg/m^3	130					
Saturation moisture content	kg/m^3	170					
Diffusion resistance factor	—	$\phi = 0.265$	0.535		0.715		0.85
		33	30		22		13
Capillary water absorption coeff.	$kg/(m^2.s^{0.5})$	0.02					
Savonnières							
Density	kg/m^3	$\rho = 1661$ ($\sigma = 34$, 120 samples)					
Capillary moisture content	kg/m^3	160, ($\sigma = 21.5$, 120 samples)					
Saturation moisture content	kg/m^3	382 ($\sigma = 14.2$, 120 samples)					
Capillary water absorption coeff.	$kg/(m^2.s^{0.5})$	0.085 (//, $\sigma = 0.062$), 0.054 ($\perp$, $\sigma = 0.038$)					

Concrete

Density	kg/m^3	$\rho \leq 2176$
Hygroscopic curve	kg/m^3	
– Sorption	$0.2 \leq \phi \leq 0.98$	$147.5\left(1-\frac{\ln\phi}{0.0453}\right)^{-\frac{1}{1.67}}$
– Desorption	$0.2 \leq \phi \leq 0.98$	$147.5\left(1-\frac{\ln\phi}{0.570}\right)^{-\frac{1}{0.64}}$
Critical moisture content	kg/m^3	100–110
Capillary moisture content	kg/m^3	110
Saturation moisture content	kg/m^3	153
Diffusion resistance factor	— $0.2 \leq \phi \leq 0.98$	$\frac{1}{6.8\times10^{-3}+8.21\times10^{-5}\exp(5.66\phi)}$
Capillary water absorption coeff.	$kg/(m^2.s^{0.5})$	0.018
Moisture diffusivity (±)	m^2/s, w in kg/m^3	$1.8\times10^{-11}\exp(0.058w)$

Lightweight concrete

Density	kg/m^3	$644 \leq \rho \leq 1442$
Hygroscopic curve	kg/m^3	
– Sorption	$938 \leq \rho \leq 1442$ $0.2 \leq \phi \leq 0.98$	$110\left(1-\frac{\ln\phi}{0.0277}\right)^{-\frac{1}{2.14}}$
– Desorption	$0.2 \leq \phi \leq 0.98$	$110\left(1-\frac{\ln\phi}{0.0221}\right)^{-\frac{1}{2.91}}$
Critical moisture content	kg/m^3, $\rho = 935$	140
Capillary moisture content	kg/m^3, $872 \leq \rho \leq 980$	97–190
Saturation moisture content	kg/m^3, $\rho = 973$	584
Diffusion resistance factor	— $0.2 \leq \phi \leq 0.98, \rho = 975$	$\frac{1}{6.76\times10^{-2}+1.21\times10^{-3}\exp(3.94\phi)}$
Capillary water absorption coeff.	$kg/(m^2.s^{0.5})$	
	$\rho = 975$	0.08
	$\rho = 1410$	0.029
Moisture diffusivity (±)	m^2/s, w in kg/m^3 $\rho = 975$	$1.3\times10^{-9}\exp(0.035w)$

Cellular concrete 1

Property	Unit / range	Value
Density	kg/m^3	$455 \le \rho \le 800$
Hygroscopic curve	kg/m^3	
– Sorption	$465 \le \rho \le 621$ $0.2 \le \phi \le 0.98$	$300\left(1-\frac{\ln\phi}{0.0011}\right)^{-\frac{1}{1.99}}$
– Desorption	$0.2 \le \phi \le 0.98$	$300\left(1-\frac{\ln\phi}{0.0038}\right)^{-\frac{1}{1.32}}$
Critical moisture content	kg/m^3	180
Capillary moisture content	kg/m^3	$109+0.383\rho$
Saturation moisture content	kg/m^3	$972-0.350\rho$
Diffusion resistance factor	— $0.2 \le \phi \le 0.98$ $458 \le \rho \le 770$	$\frac{1}{1.16\times10^{-1}+6.28\times10^{-3}\exp(4.19\phi)}$
Capillary water absorption coeff.	$kg/(m^2.s^{0.5})$	0.02–0.08
Moisture diffusivity (±)	m^2/s, w in kg/m^3 $\rho=511$	$9.2\times10^{-11}\exp(0.0215w)$

Cellular concrete 2

Property	Unit						
Density	kg/m^3	$\rho=600$					
Hygroscopic curve	kg/m^3	$\phi=0.1$	0.3	0.5	0.65	0.8	0.9
Sorption				7.3	12.5	17	38
Capillary moisture content	kg/m^3	290					
Saturation moisture content	kg/m^3	720					
Diffusion resistance factor	—	$\phi=0.265$	0.535	0.715		0.85	
		7.6		6.7			
Capillary water absorption coeff.	$kg/(m^2.s^{0.5})$	0.09					

Polystyrene concrete

Density	kg/m^3	$259 \leq \rho \leq 792$
Hygroscopic curve	kg/m^3	
– Sorption	$\rho \leq 422$ $0.2 \leq \phi \leq 0.98$	$235\left(1-\dfrac{\ln\phi}{0.0097}\right)^{-\frac{1}{1.55}}$
Saturation moisture content	kg/m^3, $\rho \leq 422$	489
Diffusion resistance factor	— $0.2 \leq \phi \leq 0.98$ $357 \leq \rho \leq 425$	$\dfrac{1}{8.18\times10^{-2}+3.16\times10^{-3}\exp(2.88\phi)}$
Capillary water absorption coeff.	$kg/(m^2.s^{0.5})$ $360 \leq \rho \leq 457$	0.026
Moisture diffusivity (±)	m^2/s, w in kg/m^3 $\rho = 422$	$4.6\times10^{-10}\exp(0.064w)$
Hygric strain	m/m $\rho = 422$	$2.4\times10^{-7}\,\phi^2$

Mortar

Density	kg/m^3	$1050 \leq \rho \leq 1940$		
Hygroscopic curve	kg/m^3			
– Sorption	$\rho = 1940$ $0.2 \leq \phi \leq 0.98$	$283\left(1-\dfrac{\ln\phi}{0.029}\right)^{-\frac{1}{1.39}}$		
– Desorption		$300\left(1-\dfrac{\ln\phi}{0.061}\right)^{-\frac{1}{1.77}}$		
Capillary moisture content	kg/m^3	283		
Diffusion resistance factor	— $0.2 \leq \phi \leq 0.98$	$\dfrac{1}{7.69\times10^{-2}+2.43\times10^{-3}\exp(3.61\phi)}$		
Capillary water absorption coeff.	$kg/(m^2.s^{0.5})$	0.042–0.80		
Moisture diffusivity (±)	m^2/s, w in kg/m^3	$C_1\exp(C_2w)$		
		ρ (kg/m^3)	C_1	C_2
		1072	2.0×10^{-9}	0.022
		1500	2.7×10^{-9}	0.020
		1807	1.4×10^{-9}	0.027

Brick 1

Density	kg/m^3	$\rho = 1700$					
Hygroscopic curve	kg/m^3	$\phi = 0.1$	0.3	0.5	0.65	0.8	0.9
– Sorption				7.5	8.4	18	34
Capillary moisture content	kg/m^3	270					
Saturation moisture content	kg/m^3	380					
Diffusion resistance factor	—	$\phi = 0.265$	0.535		0.715		0.85
		9.5	8.8		8.0		6.9
Capillary water absorption coeff.	kg/(m^2.s$^{0.5}$)	0.25					

Brick 2

Density	kg/m^3	$1505 \le \rho \le 2047$
Hygroscopic curve	kg/m^3	
– Sorption	$0.2 \le \phi \le 0.98$	$200\left(1 - \dfrac{\ln\phi}{1.46\ 10^{-4}}\right)^{-\frac{1}{1.59}}$
Critical moisture content	kg/m^3	100
Capillary moisture content	kg/m^3 $1505 \le \rho \le 2047$	$730.3–0.287\rho$
Saturation moisture content	kg/m^3, ρ idem	$1033–0.404\rho$
Diffusion resistance factor	– $0.2 \le \phi \le 0.98$ $1505 \le \rho \le 1860$	$\dfrac{1}{5.6\ 10^{-2} + 4.67\ 10^{-3}\exp(2.79\phi)}$
Capillary water absorption coeff.	kg/(m^2.s$^{0.5}$)	
– Clay	$1505 \le \rho \le 2000$	$0.653–0.000\,30\rho$
– Loam	$1628 \le \rho \le 1868$	$1.954–0.000\,87\rho$
Moisture diffusivity (±)	m^2/s, w in kg/m^3	$C_1 \exp(C_2 w)$

ρ (kg/m^3)	C_1	C_2
1529	2.1×10^{-9}	0.032
1619	1.9×10^{-8}	0.022
1918	7.4×10^{-9}	0.032

Lime-sandstone 1

Density	kg/m^3	$1685 \leq \rho \leq 1807$
Hygroscopic curve	kg/m^3	
– Sorption	$0.2 \leq \phi \leq 0.98$ $1685 \leq \rho \leq 1726\,kg/m^3$	$210\left(1-\frac{\ln \phi}{3.56 \times 10^{-3}}\right)^{-\frac{1}{2.39}}$
– Desorption		$330\left(1-\frac{\ln \phi}{6.58 \times 10^{-3}}\right)^{-\frac{1}{1.81}}$
Critical moisture content	kg/m^3 $\rho \leq 1807$	120
Capillary moisture content	kg/m^3 $1711 \leq \rho \leq 1777$	233
Diffusion resistance factor	— $0.2 \leq \phi \leq 0.98$ $1505 \leq \rho \leq 1860$	$\frac{1}{4.3 \times 10^{-2} + 4.56 \times 10^{-5} \exp(9.86\phi)}$
Capillary water absorption coeff.	$kg/(m^2.s^{0.5})$ $\rho \leq 1807\,kg/m^3$	0.042
Moisture diffusivity (±)	m^2/s, w in kg/m^3 $\rho \leq 1807\,kg/m^3$	$2.2 \times 10^{-10} \exp(0.027w)$

Lime-sandstone 2

Density	kg/m^3	$\rho = 1900$					
Hygroscopic curve	kg/m^3	$\phi = 0.1$	0.3	0.5	0.65	0.8	0.9
– Sorption				17	18	24.9	40.2
Capillary moisture content	kg/m^3	250					
Saturation moisture content	kg/m^3	290					
Diffusion resistance factor	—	$\phi = 0.265$	0.535		0.715		0.85
		28	24		18		13
Capillary water absorption coeff.	$kg/(m^2.s^{0.5})$	0.045					

Gypsum plaster 1

Density	kg/m^3	975
Hygroscopic curve	kg/m^3	
– Sorption	$0.2 \le \phi \le 0.98$	$310\left(1-\dfrac{\ln\phi}{3.21\times10^{-2}}\right)^{-\frac{1}{1.59}}$
– Desorption		$\dfrac{\phi}{-0.004\phi^2+0.01\phi-0.0004}$
Capillary moisture content	kg/m^3	310
Diffusion resistance factor	— $0.2 \le \phi \le 0.98$	$\dfrac{1}{0.133+7.45\times10^{-4}\exp(5.1\phi)}$
Capillary water absorption coeff.	$kg/(m^2.s^{0.5})$	0.155
Moisture diffusivity (±)	m^2/s, w in kg/m^3	$1.7\times10^{-9}\exp(0.0206w)$

Gypsum plaster 2

Density	kg/m^3	$\rho = 850$					
Hygroscopic curve		$\phi = 0.1$	0.3	0.5	0.65	0.8	0.9
Sorption	kg/m^3			3.6	5.2	6.3	11
Capillary moisture content	kg/m^3	400					
Saturation moisture content	kg/m^3	650					
Diffusion resistance factor	—	$\phi = 0.265$	0.535		0.715		0.85
		8.3			7.3		
Capillary water absorption coeff.	$kg/(m^2.s^{0.5})$	0.29					

EIFS stucco

Property	Unit			
Density	kg/m^3	$878 \leq \rho \leq 1736$		
Capillary moisture content	kg/m^3, $878 \leq \rho \leq 1736$	185		
Diffusion resistance factor	— $880 \leq \rho \leq 1709\,kg/m^3$ $\phi = 86\%$	$1.8\exp(0.0016\rho)$		
		ρ kg/m^3	ϕ_m %	μ —
		1680	86	43
			95	15
Capillary water absorption coeff.	$kg/(m^2.s^{0.5})$	$0.0039 \leq A \leq 0.029$ On average 0.0128		
Moisture diffusivity (±)	$m^2/s^{0.5}$	$C_1 \exp(C_2 w)$		
		ρ kg/m^3	C_1	C_2
		878	4.4×10^{-12}	0.027
		1341	2.1×10^{-10}	0.019

Timber

Property	Unit			
Density	kg/m^3	$400 \leq \rho \leq 690$		
Hygroscopic curve Sorption	kg/m^3 $0.2 \leq \phi \leq 0.98$	$100\left(1 - \frac{\ln \phi}{0.642}\right)^{-\frac{1}{0.64}}$		
Desorption		$120\left(1 - \frac{\ln \phi}{0.248}\right)^{-\frac{1}{1.22}}$		
Diffusion resistance factor	—	ρ kg/m^3	ϕ_m %	μ —
		Oak, 640	30	110
			87	21
		Beech, 435	25	180
			75	20
		Spruce, 390	53	57
			86	13
Capillary water absorption coeff.	$kg/(m^2.s^{0.5})$ $\rho = 390\,kg/m^3$ (spruce)	0.004 ⊥ fibres 0.016//fibres		

Woodwool cement boards

Density	kg/m^3	$314 \leq \rho \leq 767$
Hygroscopic curve – Sorption	% kg/kg $0.2 \leq \phi \leq 0.98$ $\rho = 767$ kg/m^3	$15\left(1 - \frac{\ln \phi}{0.172}\right)^{-\frac{1}{0.84}}$
Capillary moisture content	kg/m^3	180
Saturation moisture content		240
Diffusion resistance factor		≈ 4
Capillary water absorption coeff.	$kg/(m^2.s^{0.5})$	0.007
Moisture diffusivity (±)	m^2/s, w in kg/m^3	$6.2 \times 10^{-12} \exp(0.027w)$

Particle board

Density	kg/m^3	$570 \leq \rho \leq 800$
Hygroscopic curve – Sorption	% kg/kg $0.2 \leq \phi \leq 0.98$	$35\left(1 - \frac{\ln \phi}{0.0328}\right)^{-\frac{1}{1.89}}$
– Desorption		$35\left(1 - \frac{\ln \phi}{0.081}\right)^{-\frac{1}{1.63}}$
Critical moisture content	% kg/kg	85
Capillary moisture content	% kg/kg	90
Saturation moisture content	% kg/kg	99

Diffusion resistance factor	—	$A_1 \exp(A_2\rho)$			
		ϕ %	A_1	A_2	r^2
		25	0.543	0.007 57	0.70
		86	0.508	0.006 76	0.75
Capillary water absorption coeff.	$kg/(m^2.s^{0.5})$ UF, . . . glue used	(UF, UMF) (FF)	0.0035 0.022		
Moisture diffusivity (±)	m^2/s w in kg/m^3	(UF, UNF) (FF)	$2.3 \times 10^{-13} \exp(0.01w)$ $4.5 \times 10^{-12} \exp(0.01w)$		

Plywood

Density	kg/m^3	$445 \le \rho \le 799$		
Hygroscopic curve	% kg/kg			
– Sorption	$0.2 \le \phi \le 0.98$	$75\left(1-\frac{\ln\phi}{6.14\ 10^{-3}}\right)^{-\frac{1}{1.91}}$		
– Desorption		$75\left(1-\frac{\ln\phi}{9.63\ 10^{-3}}\right)^{-\frac{1}{1.93}}$		
Critical moisture content	% kg/kg	75		
Capillary moisture content	% kg/kg	75		
Saturation moisture content	% kg/kg	75		
Diffusion resistance factor	—	15.3 exp [0.0045(ρ – 437)]		
	$437 \le \rho \le 591$ kg/m^3 $\phi = 86\%$	ρ kg/m^3	ϕ_m %	μ —
		548–580	54	97
		437–591	86	24
Capillary water absorption coeff.	kg/(m^2.s$^{0.5}$)	0.003		
Moisture diffusivity (±)	m^2/s$^{0.5}$	3.2×10^{-13} exp (0.015w)		

Fibre cement boards

Density	kg/m^3	$823 \leq \rho \leq 2052$
Hygroscopic curve	kg/m^3	
Sorption	$0.2 \leq \phi \leq 0.98$	
	$\rho = 990\,kg/m^3$	$300\left(1-\frac{\ln\phi}{0.0077}\right)^{-\frac{1}{1.93}}$
	$\rho = 1495\,kg/m^3$	$358\left(1-\frac{\ln\phi}{0.0415}\right)^{-\frac{1}{1.36}}$
Desorption	$\rho = 990\,kg/m^3$	$350\left(1-\frac{\ln\phi}{0.1076}\right)^{-\frac{1}{1.22}}$
	$\rho = 1495\,kg/m^3$	$358\left(1-\frac{\ln\phi}{0.197}\right)^{-\frac{1}{0.8}}$
Critical moisture content	kg/m^3, $\rho = 840$	350
Capillary moisture content	kg/m^3, $\rho = 1495$	358
Saturation moisture content	kg/m^3, $\rho = 1495$	430
Diffusion resistance factor	—, $0.2 \leq \phi \leq 0.98$,	
	$\rho = 840\,kg/m^3$	$\frac{1}{0.0565 + 5.58 \times 10^{-5}\exp(7.85\phi)}$
	$\rho = 1495\,kg/m^3$	$\frac{1}{0.00642 + 1.4 \times 10^{-4}\exp(4.92\phi)}$
Capillary water absorption coeff.	$kg/(m^2.s^{0.5})$	0.024
Moisture diffusivity (±)	$m^2/s^{0.5}$, w in kg/m^3 $840 \leq \rho \leq 1495\,kg/m^3$	$3.4 \times 10^{-11}\exp(0.018w)$

Gypsum board

Weight per m^2	kg/m^2	$6.5 \leq \rho \leq 13$
Hygroscopic curve		
Sorption	kg/m^3 $0.2 \leq \phi \leq 0.98$	$150\left(1-\frac{\ln\phi}{2.99 \times 10^{-4}}\right)^{-\frac{1}{4.81}}$
Desorption		$150\left(1-\frac{\ln\phi}{0.026}\right)^{-\frac{1}{7.86}}$
Diffusion resistance factor	—	
	$d = 9.5$, $d = 12.5\,mm$	$\frac{1}{0.0712 + 2.81 \times 10^{-3}\exp(4.1\phi)}$

Masonry Walls

Bricks

Density kg/m^3 $830 \leq \rho \leq 1760$

Equivalent diffusion thickness

d m	wall	ϕ_m %	μd m
0.09	bricks	54	1.20
	6.5 × 9 × 19 cm	86	0.51
0.09	idem, painted with acrylic paint	54	2.20
0.09	idem, water repellent treated	86	0.65
0.09	Glazed bricks	86	4.00
0.19	Bricks	59	1.60
	6.5 × 9 × 19 cm	81	0.61
0.14	Hollow blocks	58	1.30
	14 × 19 × 29 cm	84	0.84
0.19	Bricks	88	0.53
	6.5 × 9 × 19 cm		

Lime-sandstone

Density kg/m^3 $860 \leq \rho \leq 1650$

Equivalent diffusion thickness

d m	wall	ϕ_m %	μd m
0.14	$\rho = 1140$ kg/m^3	25	3.70
	14 × 14 × 29 cm		

Concrete blocks

Density kg/m^3 $860 \leq \rho \leq 1650$

Equivalent diffusion thickness

d m	wall	ϕ_m %	μd m
0.14	Blocks, $\rho = 960$	60	1.30
	14 × 14 × 29 cm	86	0.54
0.14	Blocks	61	0.61
	$\rho = 1450$ kg/m^3	64	0.58
	14 × 14 × 29 cm	83	0.61
		90	0.28

6.5.2.2 Insulation materials

Cork

Density	kg/m^3	$\rho \leq 111$
Hygroscopic curve	% kg/kg	
Sorption	$0.2 \leq \phi \leq 0.98$	$60\left(1-\dfrac{\ln\phi}{1.02\times10^{-5}}\right)^{-\frac{1}{3.64}}$
Critical moisture content	kg/m^3	60
Diffusion resistance factor		≈ 22

Cellular glass

Density	kg/m^3	$114 \leq \rho \leq 140$
Diffusion resistance factor	—	5000–70 000

Glass fibre

Density	kg/m^3	$12 \leq \rho \leq 133$
Diffusion resistance factor	—, $19 \leq \rho \leq 102$ kg/m^3	1.2

Mineral fibre

Density	kg/m^3	$32 \leq \rho \leq 191$
Diffusion resistance factor	—, $148 \leq \rho \leq 172$ kg/m^3	1.5

Expanded polystyrene (EPS)

Density	kg/m^3	$13 \leq \rho \leq 40$
Diffusion resistance factor	—, $\phi = 0.86$, $15 \leq \rho \leq 40$ kg/m^3	$4.9 + 1.97\rho$

Extruded polystyrene (XPS)

Density	kg/m^3	$25 \leq \rho \leq 55$
Diffusion resistance factor	—, $\phi = 0.86$, $25 \leq \rho \leq 53$ kg/m^3	$48.6 + 3.35\rho$

PUR/PIR

Density	kg/m^3	$20 \le \rho \le 40$
Diffusion resistance factor	—, $\phi = 0.8$, $20 \le \rho \le 40\,kg/m^3$	$1.67 \exp(0.088\rho)$

Expanded perlite

Density	kg/m^3	$135 \le \rho \le 215$
Hygroscopic curve	kg/m^3	
Sorption	$0.2 \le \phi \le 0.98$ $187 \le \rho \le 213\,kg/m^3$	$150\left(1 - \frac{\ln \phi}{1.15 \times 10^{-4}}\right)^{-\frac{1}{2.63}}$
Critical moisture content	kg/m^3	150–200
Capillary moisture content	kg/m^3	550
Diffusion resistance factor	—, $0 \le \phi \le 1$	$\frac{1}{0.0143 + 6.54\,10^{-8} \exp(16.5\phi)}$

6.5.2.3 Finishes

Tongue and groove timber lathing

Weight per m^2	kg/m^2	4.0 ($d = 10$ mm)
Hygroscopic curve	% kg/kg	
Sorption	$0.2 \le \phi \le 0.98$	$50\left(1 - \frac{\ln \phi}{0.0213}\right)^{-\frac{1}{1.96}}$
Desorption		$50\left(1 - \frac{\ln \phi}{0.0397}\right)^{-\frac{1}{1.86}}$

ϕ (—)	Sorption X (% kg/kg)	Desorption X (% kg/kg)
0.33	5.8	7.1
0.52	9.4	12.3
0.75	13.7	15.4
0.86	17.6	20.8
0.97	31.4	37.3
0.98	34.5	

Diffusion thickness	m, $\phi = 0.55$	0.86 ($\sigma = 0.12$)

Wall paper

Weight per m^2, thickness	kg/m^2, mm	Type	kg/m^2	d (mm)
		textile	0.291	0.425
		vinyl	0.216	0.325
		textile	0.333	0.700
		vinyl	0.212	0.450
		paper	0.168	0.280
		paper	0.151	0.280

Hygroscopic curve	% kg/kg Paper $\phi(-)\downarrow$	1	2	3	4	5	6
Sorption	0.33	3.2	1.4	2.9	1.2	1.6	1.8
	0.52	5.5	2.8	5.5	2.3	2.9	4.0
	0.75	7.9	5.0	6.4	3.3	4.6	5.5
	0.86	11.2	8.2	9.8	4.8	7.2	6.8
	0.97	21.4	40.0	24.9	13.3	15.8	16.9
Desorption	0.33	5.3	4.6	4.0	7.0	2.7	3.9
	0.52	8.5	5.6	6.6	8.3	4.2	6.0
	0.75	11.8	8.7	9.9	9.8	6.9	9.4
	0.86	16.1	10.7	14.0	10.0	9.7	12.3
	0.97	42.8	52.2	38.3	17.9	23.8	36.2
Diffusion thickness	M Paper $\phi(-)\downarrow$						
	0.42	0.280	2.14	0.155	0.09	0.035	0.025
	0.75	0.006	0.18	0.019	0.025	0.012	0.008

Paints

Diffusion thickness		ϕ (–)		
		0.425	0.750	0.86
	On gypsum board			
	primer + two layers latex 1 paint			0.17
	Primer + two layers latex 2 paint	4.5		1.10
	Primer + two layers acrylic paint			0.46
	Primer + two layers synthetic paint	3.20		1.00
	Primer + two layers oil paint			0.76
	On cellular concrete			
	Primer + two layers acrylic paint		0.43	
	Structured paint		1.10	

Carpet

Weight per m²	kg/m²	2.18	
Hygroscopic curve Sorption	% kg/kg $0.2 \leq \phi \leq 0.98$	ϕ(–)	X (% kg/kg)
		0.33	8.0
		0.52	9.9
		0.75	13.4
		0.86	17.2
		0.97	29.7
		0.98	29.8

6.5.2.4 Miscellaneous

Newspaper

Weight per m²	kg/m²	0.041	
Hygroscopic curve	% kg/kg $0.2 \leq \phi \leq 0.98$	ϕ (–)	Sorption X (% kg/kg)
		0.33	5.3
		0.52	9.0
		0.75	12.9
		0.86	16.9
		0.97	32.1
		0.98	40.9

Weekly journal

Weight per m^2	kg/m^2	0.047	
Hygroscopic curve	% kg/kg $0.2 \leq \phi \leq 0.98$	ϕ (–)	Sorption X (% kg/kg)
		0.33	2.9
		0.52	4.7
		0.75	6.5
		0.86	8.4
		0.97	16.6
		0.98	19.0

6.5.2.5 Vapour retarders

		Diffusion resistance factor				
PE-foil	ϕ (–) =	*0.28*	*0.52*	*0.70*	*0.75*	*0.86*
	μ =	321 000			289 000	271 000
		Diffusion thickness				
		ϕ (–) = *d (mm) = ↓*	*0.52*	*0.70*	*0.75*	*0.86*
Bituminous paper		*0.1*				1.80 2.80
		0.2	0.70			
		1.4			1.70 6.90	
		0.4		3.90 8.10		
Aluminium paper		*0.1*				2.00 2.80
		0.2				0.17 0.33
		0.24		17.80 77.30		
		—				6.80 17.80
Glass fibre reinforced alupaper		*0.4*		3.80 4.70		
Glass fibre reinforced PVC-foil		*0.4*				12.0 29.0
Stapled PE-foil		*0.15*			7.70	

6.6 Surfaces, radiant properties

Longwave emissivity (e_L) and shortwave absorptivity (α_S) are as follows:

Material, surface	e_L	α_S
Snow		0.15
White paint	0.85	0.25
Black paint	0.97	
Oil paint	0.94	
White washed surface		0.30
Light colours, polished aluminium		0.3–0.5
Yellow brick	0.93	0.55
Red brick	0.93	0.75
Concrete, light coloured floor cover		0.6–0.7
Grass and leaves		0.75
Dark coloured floor cover		0.8–0.9
Carpet		0.8–0.9
Moist bottom		0.90
Dark grey slates		0.90
Bituminous felt	0.92	0.93
Gold, silver and copper polished	0.02	
Copper, oxidized	0.78	
Aluminium, polished	0.05	
Aluminium, oxidized	0.30	
Steel, hot casted	0.77	
Steel, oxidized	0.61	
Steel with silver finish	0.26	
Steel, polished	0.27	
Lead, oxidized	0.28	
Glass	0.92	
Porcelain	0.92	
Plaster	0.93	
Timber, unpainted	0.90	
Marble, polished	0.55	
Paper	0.93	
Water	0.95	
Ice	0.97	

Further reading

ASHRAE (2013) Handbook of Fundamentals.

Bomberg, M. and Kumaran, K. (1986) A test method to determine air flow resistance of exterior membranes and sheathings. *Journal of Thermal Insulation*, **9**, 224–235.

CEN/TC 89 (1998) Building materials and products – Hygrothermal properties – Tabulated design values, working draft.

Hagentoft, C.E. (2001) Introduction to Building Physics, Studentlitteratur, Lund.

Hens, H. (1975) Theoretische en experimentele studie van het hygrothermisch gedrag van bouw- en isolatiematerialen bij inwendige condensatie en droging met toepassing op de platte daken (Theoretical and experimental study of the hygrothermal response of building and insulating materials during interstitial condensation and drying with application on low-sloped roofs), PhD KU Leuven 311 pp. (in Dutch).

Hens, H. (1984) Cataloog van hygrothermische eigenschappen van bouw- en isolatiematerialen (Catalogue of hygrothermal properties of building and insulating materials), rapport E/VI/2, R-D Energy, 100 pp.+appendices (in Dutch).

Hens, H. (1991) Catalogue of Material Properties, Final Report, IEA-EXCO on Energy Conservation in Buildings and Community Systems, Annex 14 'Condensation and Energy'.

ISO 10456 (2007) Building materials and products-Hygrothermal properties-Tabulated design values and procedures for determining declared and design thermal values, 24 p.

Krus, M. (1995) Feuchtetransport- und Speicherkoeffizienten poröser mineralischer Baustoffe. Theoretische Grundlagen und Neue Messtechniken, Doktor Abhandlung, Universität Stuttgart (in German).

KU Leuven, Department of Civil Engineering, Unit Building Physics, 1978–2008, reports on material property measurements (in Dutch and English).

Kumaran, K. (1996) Material Properties, Final Report, IEA, EXCO on Energy Conservation in Buildings and Community Systems, Annex 24 'Heat, Air and Moisture Transfer in Insulated Envelope Parts, Task 3.

Kumaran, K. (2002) A thermal and Moisture Property Database for Common Building and Insulating Materials, Final report ASHRAE Research Project 1018-RP, 231 pp.

NBN B62-002/A1 (2001) Berekening van de warmtedoorgangscoëfficiënten van wanden van gebouwen, 2[e] uitgave (in Dutch).

prCEN/TR 14613 (2003) Thermal performance of building materials and components – principles for the determination of thermal properties of moist materials and components, Final draft.

Roels, S. (2000) Modelling unsaturated moisture transport in heterogeneous limestone, Doctoral Thesis, KU Leuven.

Roels, S. (2008) Experimental Analysis of moisture buffering, Final Report, IEA, EXCO on Energy Conservation in Buildings and Community Systems, Annex 41 'Whole Building Heat, Air and Moisture Response', Task 2.

Roels, S., Carmeliet, J. and Hens, H. (2003) Moisture transfer properties and material characterisation, Final Report EU HAMSTAD project.

Sagelsdorff, R. (1989) Wasserdampfdiffusion, Grundlagen, Berechnungsverfahren, Diffusionsnachweis, SIA-Dokumentation *D* 018, 90 pp. (in German).

Stichting Bouwresearch (1974) Eigenschappen van bouw- en isolatiematerialen (Properties of building and insulating materials), 2^{e}, gewijzigde druk (in Dutch).

Trechsel/Bomberg. ed. (1989) Water Vapor Transmission Through Building Materials and Systems, ASTM, STP 1039.

Wexler, A. and Hasegawa, S. (1954) Relative humidity/temperature relationship of some saturated salt solutions in the temperature range 0 to 50°C. *Journal of Research of the National Bureau of Standards*, **53** (1), 19–26.

Appendix A: Solar radiation for Uccle, Belgium, 50° 51′ north, 4° 21′ east

Table A.1 gives the beam radiation on a surface facing the solar rays under clear sky conditions and Table A.2 lists the diffuse solar radiation on a horizontal surface under clear sky conditions. Both tables look at the 15th of each month. Table A.3 combines beam plus diffuse solar radiation under clear sky conditions for surfaces with different orientation and slope. Table A.4 in turn gives the totals for each month, while Table A.5 gives the total, included reflected solar radiation.

Applied Building Physics: Ambient Conditions, Building Performance and Material Properties, Second Edition. Hugo Hens.

Table A.1 Beam radiation under clear sky conditions on a surface facing the solar rays for the 15th of each month (W/m²).

Hour ↓	J	F	M	A	M	J	J	A	S	O	N	D
4.0	0	0	0	0	0	103	45	0	0	0	0	0
4.5	0	0	0	0	133	223	165	0	0	0	0	0
5.0	0	0	0	29	248	321	270	119	0	0	0	0
5.5	0	0	0	169	355	412	370	246	48	0	0	0
6.0	0	0	0	293	445	489	454	356	192	0	0	0
6.5	0	0	138	399	518	552	524	446	314	112	0	0
7.0	0	0	274	484	578	603	580	519	414	251	0	0
7.5	0	163	383	551	625	645	626	577	494	359	153	0
8.0	102	279	469	604	664	680	664	623	555	445	268	97
8.5	218	378	536	646	695	708	694	660	603	510	360	219
9.0	315	455	587	678	719	730	718	689	640	559	430	310
9.5	389	513	626	704	739	749	738	712	668	596	482	379
10.0	445	557	655	723	754	763	754	730	689	623	519	430
10.5	484	587	676	736	765	774	765	743	704	640	544	464
11.0	509	608	689	745	772	781	773	751	713	651	558	485
11.5	523	620	697	749	776	785	778	756	717	654	561	494
12.0	525	623	698	749	775	784	778	756	716	650	556	491
12.5	513	612	693	740	767	776	770	748	706	632	532	468
13.0	483	592	678	728	756	768	761	739	692	609	500	435
13.5	450	569	654	710	739	752	748	723	670	581	460	392
14.0	403	536	629	690	722	736	733	705	646	543	405	333
14.5	338	490	595	664	700	717	714	683	613	492	331	254
15.0	252	428	549	629	673	693	690	655	571	425	236	156
15.5	142	347	490	586	639	664	661	619	217	336	118	0
16.0	0	243	413	531	596	628	625	573	448	229	0	0
16.5	0	80	315	461	544	584	580	517	359	102	0	0
17.0	0	0	196	373	478	530	526	448	250	0	0	0
17.5	0	0	55	265	398	464	459	358	120	0	0	0
18.0	0	0	0	142	302	385	378	252	0	0	0	0
18.5	0	0	0	0	191	291	282	133	0	0	0	0
19.0	0	0	0	0	70	187	175	0	0	0	0	0
19.5	0	0	0	0	0	75	60	0	0	0	0	0

Table A.2 Diffuse solar radiation under clear sky conditions on a horizontal surface for the 15th of each month (W/m^2).

Hour↓	J	F	M	A	M	J	J	A	S	O	N	D
4.0	0	0	0	0	0	20	9	0	0	0	0	0
4.5	0	0	0	0	25	44	32	0	0	0	0	0
5.0	0	0	0	5	48	64	53	22	0	0	0	0
5.5	0	0	0	31	70	84	74	47	8	0	0	0
6.0	0	0	0	55	90	102	93	69	34	0	0	0
6.5	0	0	24	76	107	118	110	89	57	19	0	0
7.0	0	0	48	95	123	133	125	106	78	43	0	0
7.5	0	27	69	111	137	146	139	121	95	63	24	0
8.0	16	47	87	125	149	158	151	134	109	79	43	15
8.5	34	61	101	137	160	168	162	145	122	93	59	34
9.0	51	79	114	147	169	177	171	155	132	104	72	49
9.5	64	91	124	156	177	185	180	164	141	113	82	61
10.0	74	100	132	163	183	192	187	171	147	120	90	70
10.5	81	107	138	168	188	197	192	176	153	124	95	76
11.0	86	112	142	172	192	200	196	180	156	127	98	80
11.5	89	115	144	173	193	202	198	182	157	128	98	82
12.0	89	116	145	173	193	202	198	182	157	127	97	81
12.5	87	113	141	170	190	199	195	180	154	122	93	77
13.0	81	109	139	166	185	195	191	176	149	117	86	71
13.5	75	104	132	160	179	188	186	170	142	110	78	64
14.0	66	96	125	153	172	181	179	163	135	101	68	53
14.5	55	87	116	144	163	173	171	155	125	89	54	40
15.0	40	74	105	133	153	164	162	145	114	76	38	24
15.5	22	59	92	121	142	153	151	134	101	59	19	2
16.0	0	40	75	107	129	141	139	120	85	39	0	0
16.5	0	13	56	90	114	128	126	106	66	17	0	0
17.0	0	0	34	71	98	113	111	89	45	0	0	0
17.5	0	0	9	49	80	97	95	70	21	0	0	0
18.0	0	0	0	26	59	78	76	48	0	0	0	0
18.5	0	0	0	0	37	58	58	25	0	0	0	0
19.0	0	0	0	0	13	37	34	0	0	0	0	0
19.5	0	0	0	0	0	14	11	0	0	0	0	0

Table A.3 Solar radiation under clear sky conditions for the 15th of each month (direct and diffuse, W/m^2).

Horizontal surface

Hour↓	J	F	M	A	M	J	J	A	S	O	N	D
4.0	0	0	0	0	0	26	8	0	0	0	0	0
4.5	0	0	0	0	36	68	46	0	0	0	0	0
5.0	0	0	0	4	84	122	95	33	0	0	0	0
5.5	0	0	0	50	146	188	157	83	10	0	0	0
6.0	0	0	0	107	219	262	228	147	58	0	0	0
6.5	0	0	37	178	297	338	305	220	117	28	0	0
7.0	0	0	93	256	375	414	381	299	188	79	0	0
7.5	0	37	159	335	451	487	456	376	263	138	34	0
8.0	16	87	230	410	522	559	527	449	337	202	77	16
8.5	54	143	302	479	591	629	597	517	405	265	124	52
9.0	96	201	367	542	656	694	663	582	465	322	171	90
9.5	138	256	424	599	714	748	722	640	518	371	214	128
10.0	176	304	472	647	758	785	766	691	562	410	250	161
10.5	208	342	510	685	787	799	791	730	596	438	276	187
11.0	232	370	537	711	802	798	799	755	618	454	292	205
11.5	245	386	552	724	806	794	799	767	629	460	296	212
12.0	247	391	555	723	805	794	798	768	626	453	290	210
12.5	239	384	545	708	797	799	799	758	610	435	272	197
13.0	220	365	524	680	777	798	794	734	582	406	244	175
13.5	191	335	491	640	742	781	772	698	543	366	207	145
14.0	155	295	448	590	692	741	732	649	485	316	163	109
14.5	115	245	395	533	631	685	676	591	439	258	115	71
15.0	72	190	333	469	564	618	611	527	375	195	69	34
15.5	32	131	264	399	493	548	542	459	304	131	26	0
16.0	0	76	192	323	420	476	470	387	229	73	0	0
16.5	0	28	122	244	344	402	397	311	155	24	0	0
17.0	0	0	62	167	265	327	321	232	89	0	0	0
17.5	0	0	12	98	188	250	244	157	35	0	0	0
18.0	0	0	0	42	119	177	171	92	0	0	0	0
18.5	0	0	0	0	63	113	107	40	0	0	0	0
19.0	0	0	0	0	19	61	55	0	0	0	0	0
19.5	0	0	0	0	0	21	15	0	0	0	0	0

Vertical surface ($s = 90°$) north ($a_s = 180°$)

Hour↓	J	F	M	A	M	J	J	A	S	O	N	D
4.0	0	0	0	0	0	83	49	0	0	0	0	0
5.0	0	0	0	22	119	169	144	61	0	0	0	0
6.0	0	0	0	53	120	163	152	89	25	0	0	0
7.0	0	0	29	52	66	82	77	59	43	25	0	0
8.0	6	25	43	57	68	73	71	62	51	39	23	7
9.0	26	38	49	60	68	72	71	64	55	46	35	25
10.0	36	44	53	62	67	71	69	65	58	50	41	34
11.0	40	48	56	62	67	70	69	65	59	52	44	38
12.0	41	49	56	62	67	70	69	65	59	52	44	39
13.0	39	48	55	62	67	70	69	65	59	50	41	36
14.0	34	44	53	61	68	71	70	65	56	46	35	28
15.0	22	37	48	59	68	73	71	64	52	39	22	13
16.0	0	24	40	55	67	74	71	61	46	24	0	0
17.0	0	0	23	47	101	134	113	58	32	0	0	0
18.0	0	0	0	48	131	179	161	90	0	0	0	0
19.0	0	0	0	0	60	128	113	24	0	0	0	0

Tilted surface, slope 60° ($s = 60°$) north ($a_s = 180°$)

Hour↓	J	F	M	A	M	J	J	A	S	O	N	D
4.0	0	0	0	0	0	82	45	0	0	0	0	0
5.0	0	0	0	21	138	198	164	65	0	0	0	0
6.0	0	0	0	88	198	256	231	138	37	0	0	0
7.0	0	0	39	90	196	254	235	145	59	34	0	0
8.0	8	34	58	81	165	227	208	116	71	53	31	9
9.0	34	50	67	86	129	194	174	94	77	62	46	33
10.0	46	58	73	88	101	162	142	95	81	67	54	44
11.0	51	63	75	89	101	141	121	95	82	69	57	49
12.0	53	64	76	89	101	137	116	95	83	69	57	49
13.0	50	62	75	89	101	151	128	95	81	67	53	46
14.0	43	58	72	87	118	180	154	95	78	62	45	36
15.0	29	49	65	84	155	213	188	94	73	52	28	16
16.0	0	31	55	78	189	244	221	131	63	32	0	0
17.0	0	0	31	96	203	261	239	149	43	0	0	0
18.0	0	0	0	57	163	231	213	115	0	0	0	0
19.0	0	0	0	0	59	135	120	21	0	0	0	0

Tilted surface, slope 30° ($s = 30°$) north ($a_s = 180°$)

Hour↓	J	F	M	A	M	J	J	A	S	O	N	D
4.0	0	0	0	0	0	61	30	0	0	0	0	0
5.0	0	0	0	14	124	180	145	55	0	0	0	0
6.0	0	0	0	108	233	291	258	158	51	0	0	0
7.0	0	0	47	190	319	375	346	246	117	40	0	0
8.0	9	40	91	253	382	439	411	312	172	64	36	10
9.0	39	61	130	300	435	495	466	362	214	78	55	38
10.0	54	72	159	336	472	527	504	401	244	89	64	51
11.0	61	78	176	356	484	521	509	423	261	100	68	57
12.0	62	80	181	360	485	516	506	427	263	99	68	58
13.0	59	77	173	346	478	528	511	416	250	88	63	53
14.0	51	71	152	317	448	513	493	386	224	77	53	42
15.0	33	59	120	275	399	464	445	342	185	63	33	19
16.0	0	37	80	219	340	404	381	286	133	38	0	0
17.0	0	0	36	144	262	330	314	211	70	0	0	0
18.0	0	0	0	55	158	230	215	115	0	0	0	0
19.0	0	0	0	0	44	110	98	12	0	0	0	0

Vertical surface ($s = 90°$) east ($a_s = 90°$)

Hour↓	J	F	M	A	M	J	J	A	S	O	N	D
4.0	0	0	0	0	0	117	70	0	0	0	0	0
5.0	0	0	0	82	228	353	300	156	0	0	0	0
6.0	0	0	0	366	532	562	529	428	243	0	0	0
7.0	0	0	335	582	652	654	645	608	492	287	0	0
8.0	116	293	517	634	648	646	648	640	582	444	245	111
9.0	247	401	527	566	566	568	577	572	529	441	313	221
10.0	257	361	418	425	417	417	435	437	393	327	256	217
11.0	168	228	248	236	222	225	245	251	211	160	128	127
12.0	54	68	78	83	86	88	89	87	78	66	54	48
13.0	44	55	64	69	72	74	75	72	65	55	44	39
14.0	35	46	54	60	64	66	66	63	56	47	35	29
15.0	22	37	47	54	58	60	60	57	49	38	22	13
16.0	0	23	38	46	52	54	54	50	41	22	0	0
17.0	0	0	21	36	44	47	47	42	27	0	0	0
18.0	0	0	0	16	31	38	37	27	0	0	0	0
19.0	0	0	0	0	8	21	20	0	0	0	0	0

Tilted surface, slope 60° ($s = 60°$) east ($a_s = 90°$)

Hour↓	J	F	M	A	M	J	J	A	S	O	N	D
4.0	0	0	0	0	0	112	64	0	0	0	0	0
5.0	0	0	0	73	285	358	300	149	0	0	0	0
6.0	0	0	0	364	560	606	562	435	234	0	0	0
7.0	0	0	329	621	739	760	736	664	510	281	0	0
8.0	106	291	552	740	807	823	809	764	660	475	244	103
9.0	255	437	626	745	801	820	813	769	676	530	346	229
10.0	300	451	582	672	718	731	738	703	603	472	334	258
11.0	248	366	462	534	565	566	585	569	468	344	240	199
12.0	138	225	296	353	386	397	412	391	292	182	107	93
13.0	60	78	114	157	199	232	241	200	107	77	61	53
14.0	46	63	76	86	93	98	98	92	79	63	47	38
15.0	29	49	62	73	80	84	84	78	66	49	28	16
16.0	0	30	48	61	68	73	73	66	53	29	0	0
17.0	0	0	26	46	56	62	61	54	34	0	0	0
18.0	0	0	0	20	40	48	48	35	0	0	0	0
19.0	0	0	0	0	10	27	25	0	0	0	0	0

Tilted surface, slope 30° ($s = 30°$) east ($a_s = 90°$)

Hour↓	J	F	M	A	M	J	J	A	S	O	N	D
4.0	0	0	0	0	0	78	41	0	0	0	0	0
5.0	0	0	0	44	209	273	224	103	0	0	0	0
6.0	0	0	0	268	444	485	450	331	166	0	0	0
7.0	0	0	240	500	636	670	638	549	397	204	0	0
8.0	69	214	446	657	760	790	764	693	569	385	182	68
9.0	198	363	566	735	832	865	844	771	651	485	294	181
10.0	269	429	600	752	843	865	858	795	663	501	330	236
11.0	270	416	567	708	778	776	788	753	616	451	299	226
12.0	214	345	479	607	673	673	685	656	517	354	219	166
13.0	124	236	352	466	546	577	580	522	380	228	115	82
14.0	56	111	207	306	387	439	437	363	229	95	56	45
15.0	34	59	78	150	227	279	277	205	85	59	32	19
16.0	0	34	57	74	88	137	134	84	63	33	0	0
17.0	0	0	29	53	68	76	75	64	38	0	0	0
18.0	0	0	0	23	46	57	56	39	0	0	0	0
19.0	0	0	0	0	12	31	29	0	0	0	0	0

Vertical surface ($s = 90°$) south ($a_s = 0°$)

Hour↓	J	F	M	A	M	J	J	A	S	O	N	D
4.0	0	0	0	0	0	11	4	0	0	0	0	0
5.0	0	0	0	2	28	34	30	14	0	0	0	0
6.0	0	0	0	37	50	52	49	42	33	0	0	0
7.0	0	0	111	139	102	72	70	107	157	137	0	0
8.0	91	195	289	291	238	191	192	246	322	331	217	100
9.0	290	402	466	436	370	316	320	384	473	510	418	293
10.0	471	572	602	555	475	411	423	500	590	635	569	462
11.0	585	675	686	627	527	448	472	570	658	698	644	557
12.0	615	707	709	640	532	451	479	585	667	696	640	568
13.0	561	668	669	592	496	434	460	547	614	630	558	493
14.0	426	558	571	491	405	363	388	455	509	502	401	339
15.0	233	382	422	355	278	245	270	327	366	320	198	145
16.0	0	174	239	202	141	117	139	186	204	128	0	0
17.0	0	0	73	64	156	160	161	57	59	0	0	0
18.0	0	0	0	18	35	42	42	31	0	0	0	0
19.0	0	0	0	0	8	22	20	0	0	0	0	0

Tilted surface, slope 60° ($s = 60°$) south ($a_s = 0°$)

Hour↓	J	F	M	A	M	J	J	A	S	O	N	D
4.0	0	0	0	0	0	15	5	0	0	0	0	0
5.0	0	0	0	3	37	45	40	19	0	0	0	0
6.0	0	0	0	63	84	75	68	62	50	0	0	0
7.0	0	0	134	232	255	238	223	223	217	151	0	0
8.0	85	205	353	440	445	421	406	418	432	376	220	93
9.0	292	439	573	631	627	598	586	604	626	590	438	293
10.0	487	636	743	786	770	726	728	759	775	741	607	471
11.0	612	757	847	881	836	765	787	851	863	817	692	575
12.0	646	795	876	898	843	765	793	871	874	815	688	587
13.0	586	749	826	835	798	753	773	821	806	735	594	505
14.0	438	619	703	702	675	662	680	698	772	580	419	341
15.0	232	416	518	524	501	498	517	527	489	364	200	139
16.0	0	183	291	320	311	315	333	335	274	140	0	0
17.0	0	0	87	124	129	139	154	146	86	0	0	0
18.0	0	0	0	24	47	57	57	42	0	0	0	0
19.0	0	0	0	0	11	29	27	0	0	0	0	0

Tilted surface, slope 30° ($s = 30°$) south ($a_s = 0°$)

Hour↓	J	F	M	A	M	J	J	A	S	O	N	D
4.0	0	0	0	0	0	17	5	0	0	0	0	0
5.0	0	0	0	3	44	54	47	22	0	0	0	0
6.0	0	0	0	92	165	183	158	113	59	0	0	0
7.0	0	0	126	274	354	365	338	293	227	129	0	0
8.0	57	165	331	482	549	555	529	492	436	328	168	62
9.0	220	364	536	669	731	735	711	675	622	520	347	217
10.0	378	536	694	819	873	862	852	828	764	657	489	360
11.0	481	644	791	910	936	893	906	918	847	727	562	445
12.0	510	678	818	927	942	890	909	937	857	725	558	454
13.0	459	636	771	866	900	886	895	889	793	651	478	387
14.0	338	521	657	737	779	800	805	768	666	511	331	256
15.0	172	344	484	565	605	634	641	600	491	317	152	98
16.0	0	146	273	363	412	446	453	408	284	119	0	0
17.0	0	0	83	161	218	258	263	210	96	0	0	0
18.0	0	0	0	27	65	96	99	58	0	0	0	0
19.0	0	0	0	0	13	34	32	0	0	0	0	0

Vertical surface ($s = 90°$) west ($a_s = -90°$)

Hour↓	J	F	M	A	M	J	J	A	S	O	N	D
4.0	0	0	0	0	0	11	4	0	0	0	0	0
5.0	0	0	0	2	26	32	28	13	0	0	0	0
6.0	0	0	0	29	41	44	42	35	20	0	0	0
7.0	0	0	27	43	50	52	50	46	38	24	0	0
8.0	6	25	40	51	56	58	57	53	47	38	23	7
9.0	26	38	49	57	62	63	62	59	54	47	36	26
10.0	37	47	56	65	69	70	69	67	63	56	45	37
11.0	47	56	67	76	81	82	80	77	74	67	55	46
12.0	75	77	106	148	162	145	130	134	157	169	141	98
13.0	100	244	297	349	362	336	323	334	347	335	265	201
14.0	266	370	455	514	529	512	500	502	499	445	312	230
15.0	222	397	539	615	631	621	613	610	579	440	231	147
16.0	0	274	485	623	661	661	657	641	535	274	0	0
17.0	0	0	262	476	583	616	614	550	321	0	0	0
18.0	0	0	0	198	368	453	447	315	0	0	0	0
19.0	0	0	0	0	109	211	200	57	0	0	0	0

Tilted surface, slope 60° ($s = 60°$) west ($a_s = -90°$)

Hour↓	J	F	M	A	M	J	J	A	S	O	N	D
4.0	0	0	0	0	0	14	5	0	0	0	0	0
5.0	0	0	0	2	33	41	36	17	0	0	0	0
6.0	0	0	0	37	52	57	54	44	26	0	0	0
7.0	0	0	34	55	65	68	66	59	49	31	0	0
8.0	8	32	53	67	76	79	77	71	62	50	30	9
9.0	35	50	66	80	89	92	89	83	75	64	48	34
10.0	50	64	80	96	144	158	133	99	91	78	62	49
11.0	64	82	163	270	331	329	310	281	241	191	120	65
12.0	72	241	343	461	513	492	480	470	423	352	251	175
13.0	270	377	499	621	649	666	652	633	571	477	339	250
14.0	298	455	602	722	785	794	779	740	663	530	342	246
15.0	222	429	620	753	812	830	819	776	676	468	229	141
16.0	0	269	506	689	769	796	790	736	567	268	0	0
17.0	0	0	252	486	626	685	680	582	316	0	0	0
18.0	0	0	0	188	370	471	463	311	0	0	0	0
19.0	0	0	0	0	102	208	196	49	0	0	0	0

Tilted surface, slope 30° ($s = 30°$) west ($a_s = -90°$)

Hour↓	J	F	M	A	M	J	J	A	S	O	N	D
4.0	0	0	0	0	0	16	5	0	0	0	0	0
5.0	0	0	0	3	38	47	41	19	0	0	0	0
6.0	0	0	0	42	62	68	64	51	29	0	0	0
7.0	0	0	38	66	81	87	83	72	57	35	0	0
8.0	9	37	63	91	183	216	188	123	78	60	35	10
9.0	41	61	101	238	338	371	340	270	188	101	58	40
10.0	61	123	246	399	501	524	499	432	339	235	125	61
11.0	151	248	389	551	639	635	625	583	482	360	228	147
12.0	234	355	507	672	749	730	726	702	594	455	304	215
13.0	276	421	581	741	830	835	824	779	656	502	330	239
14.0	257	426	598	749	843	877	863	793	660	482	286	201
15.0	167	351	543	698	786	828	817	744	599	377	169	99
16.0	0	196	397	577	679	726	720	641	453	193	0	0
17.0	0	0	178	371	508	577	571	464	230	0	0	0
18.0	0	0	0	130	278	369	361	229	0	0	0	0
19.0	0	0	0	0	69	152	142	28	0	0	0	0

Table A.4 Beam and diffuse solar radiation under permanent clear sky conditions (MJ/(m^2. month)).

Surface↓	J	F	M	A	M	J	J	A	S	O	N	D
Hor.	136	232	424	610	787	823	825	705	497	324	168	111
North												
s = 30°	42	59	150	352	564	643	627	461	236	82	48	36
s = 60°	35	48	77	120	236	330	301	171	90	63	41	31
s = 90°	28	37	56	81	134	170	161	105	64	47	31	24
E + W												
s = 30°	136	226	406	572	731	761	764	659	471	313	167	113
s = 60°	129	207	359	484	604	623	627	552	408	281	156	109
s = 90°	102	159	268	345	419	428	768	388	297	213	122	88
South												
s = 30°	292	410	618	742	844	829	849	801	660	521	334	252
s = 60°	377	489	662	695	708	656	686	709	663	591	419	333
s = 90°	365	442	538	479	415	352	380	450	500	511	396	327
SE + SW												
s = 30°	242	353	554	695	817	817	832	765	605	457	282	209
s = 60°	292	392	564	641	700	673	695	679	586	487	329	258
s = 90°	268	336	444	456	459	422	442	464	440	401	295	242
NE + NW												
s = 30°	52	106	245	422	611	674	663	521	317	167	68	42
s = 60°	40	69	177	270	405	456	446	340	201	106	49	34
s = 90°	29	49	104	175	259	294	287	220	133	73	36	26

Table A.5 Beam, diffuse and reflected solar radiation under permanent clear sky conditions (MJ/(m^2.month), albedo 0.2).

Surface↓	J	F	M	A	M	J	J	A	S	O	N	D
Hor.	136	232	424	610	787	823	825	705	497	324	168	111
North												
s = 30°	43	62	156	360	574	654	638	470	242	86	50	37
s = 60°	42	60	98	150	276	371	342	206	115	79	49	36
s = 90°	41	60	99	142	213	252	243	176	114	80	48	35
E + W												
s = 30°	138	229	412	580	741	772	775	668	478	317	169	115
s = 60°	135	218	380	515	644	664	669	587	433	297	164	115
s = 90°	116	182	310	406	498	510	515	459	347	245	139	99
South												
s = 30°	294	414	624	750	854	840	860	811	667	526	337	254
s = 60°	384	500	683	725	747	697	728	744	688	608	427	338
s = 90°	379	465	580	540	494	434	463	521	550	544	413	338
SE + SW												
s = 30°	244	356	560	703	828	828	843	774	612	461	284	210
s = 60°	299	404	585	671	740	714	736	714	611	503	338	263
s = 90°	283	360	487	517	537	504	525	534	490	433	312	253
NE + NW												
s = 30°	54	110	251	430	622	685	674	531	323	172	71	43
s = 60°	46	81	176	300	444	497	487	375	226	122	57	39
s = 90°	43	73	146	236	338	376	369	291	182	106	53	37